# THOMAS L. FRIEDMAN

# Thank You For Being Late

## *An Optimist's Guide to Thriving in the Age of Accelerations*

PENGUIN BOOKS

PENGUIN BOOKS

UK | USA | Canada | Ireland | Australia
India | New Zealand | South Africa

Penguin Books is part of the Penguin Random House group of companies whose addresses can be found at global.penguinrandomhouse.com.

First published in the United States of America by Picador 2016
First published in Great Britain by Allen Lane 2016
Published in Penguin Books 2017
001



Grateful acknowledgment is made for permission to reprint the following material:
Excerpts from "They're in the Room Where It Happens" from *The New York Times*, December 29, 2015, © 2015 *The New York Times*. All rights reserved. Used by permission and protected by the copyright laws of the United States. The printing, copying, redistribution, or retransmission of this content without express written permission is prohibited.
Excerpt from "The Eye." Words and Music by Tim Hanseroth and Brandi Carlile. © 2015 Southern Oracle Music LLC (ASCAP). All rights administered by WB Music Corp. (ASCAP). All rights reserved. Used by permission of Alfred Music.
Interview with Debra Stone courtesy of the Nathan and Theresa Berman Upper Midwest Jewish Archives, University of Minnesota Libraries.
Excerpts from "Let's Design Social Media That Drives Real Change," by Wael Ghoni a talk delivered at the TEDGlobal>Geneva conference in 2015, courtesy of TED.

Author's note: All the interviews in this book that are not specifically attributed to different news outlets were conducted by me, either for this book or for my own column in *The New York Times*. Occasionally, I have drawn on my own columns and previous books, and where I have used material at length from either source, I have noted it.



Printed in Great Britain by Clays Ltd, St Ives plc

A CIP catalogue record for this book is available from the British Library

ISBN: 978-0-141-98575-6

www.greenpenguin.co.uk

Penguin Random House is committed to a sustainable future for our business, our readers and our planet. This book is made from Forest Stewardship Council® certified paper.

This is my seventh and, who knows, maybe my last book. Since I published *From Beirut to Jerusalem* in 1989, I have been extremely lucky to have had a special group of teacher-friends who have been with me on this journey, many starting with that first book and others on virtually every one since. They have been incredibly generous in helping me think through ideas—over many years, over many hours, over many books and many columns. So this book is dedicated to them: Nahum Barnea, Stephen P. Cohen, Larry Diamond, John Doerr, Yaron Ezrahi, Jonathan Galassi, Ken Greer, Hal Harvey, Andy Karsner, Amory Lovins, Glenn Prickett, Michael Mandelbaum, Craig Mundie, Michael Sandel, Joseph Sassoon, and Dov Seidman. Their intellectual firepower has been awesome, their generosity has been extraordinary, and their friendship has been a blessing.

# *Contents*

# PART I

# REFLECTING

ONE

# *Thank You for Being Late*

Everyone goes into journalism for different reasons—and they're often idealistic ones. There are investigative journalists, beat reporters, breaking-news reporters, and explanatory journalists. I have always aspired to be the latter. I went into journalism because I love being a translator from English to English.

I enjoy taking a complex subject and trying to break it down so that I can understand it and then can help readers better understand it—be that subject the Middle East, the environment, globalization, or American politics. Our democracy can work only if voters know how the world works, so they are able to make intelligent policy choices and are less apt to fall prey to demagogues, ideological zealots, or conspiracy buffs who may be confusing them at best or deliberately misleading them at worst. As I watched the 2016 presidential campaign unfold, the words of Marie Curie never rang more true to me or felt more relevant: "Nothing in life is to be feared, it is only to be understood. Now is the time to understand more, so that we may fear less."

It's no surprise so many people feel fearful or unmoored these days. In this book, I will argue that we are living through one of the greatest inflection points in history—perhaps unequaled since Johannes Gensfleisch zur Laden zum Gutenberg, a German blacksmith and printer, launched the printing revolution in Europe, paving the way for the Reformation. The three largest forces on the planet—technology, globalization, and climate change—are all accelerating at once. As a result, so many aspects of our societies, workplaces, and geopolitics are being reshaped and need to be reimagined.

When there is a change in the pace of change in so many realms at once, as we're now experiencing, it is easy to get overwhelmed by it all. As John E. Kelly III, IBM's senior vice president for cognitive solutions and IBM Research, once observed to me: "We live as human beings in a linear world—where distance, time, and velocity are linear." But the growth of technology today is on "an exponential curve. The only exponential we ever experience is when something is accelerating, like a car, or decelerating really suddenly with a hard braking. And when that happens you feel very uncertain and uncomfortable for a short period of time." Such an experience can also be exhilarating. You might think, "Wow, I just went from zero to sixty miles per hour in five seconds." But you wouldn't want to take a long trip like that. Yet that is exactly the trip we're on, argued Kelly: "The feeling being engendered now among a lot of people is that of always being in this state of acceleration."

In such a time, opting to pause and reflect, rather than panic or withdraw, is a necessity. It is not a luxury or a distraction—it is a way to increase the odds that you'll better understand, and engage productively with, the world around you.

How so? "When you press the pause button on a machine, it stops. But when you press the pause button on human beings they start," argues my friend and teacher Dov Seidman, author of the book *HOW* and CEO of LRN, which advises global businesses on ethics and leadership. "You start to reflect, you start to rethink your assumptions, you start to reimagine what is possible and, most importantly, you start to reconnect with your most deeply held beliefs. Once you've done that, you can begin to reimagine a better path."

But what matters most "is what you do in the pause," he added. "Ralph Waldo Emerson said it best: 'In each pause I hear the call.'"

Nothing sums up better what I am trying to do with this book—to pause, to get off the merry-go-round on which I've been spinning for so many years as a twice-a-week columnist for *The New York Times*, and to reflect more deeply on what seems to me to be a fundamental turning point in history.

I don't remember the exact date of my own personal declaration of independence from the whirlwind, but it was sometime in early 2015, and it was totally serendipitous. I regularly meet friends and interview officials, analysts, or diplomats over breakfast in downtown Washington, D.C., near the *New York Times* bureau. It's my way of packing more learning into a day and not wasting breakfast by eating alone. Once in a while, though, with the D.C. traffic and subways in the morning always a crapshoot, my breakfast guests would arrive ten, fifteen, or even twenty minutes late. They would invariably arrive flustered, spilling out apologies as they sat down: "The Red Line subway was delayed . . ." "The Beltway was backed up . . ." "My alarm failed . . ." "My kid was sick . . ."

On one of those occasions, I realized I didn't care at all about my guest's tardiness, so I said: "No, no, please—don't apologize. In fact, you know what, thank you for being late!"

Because he was late, I explained, I had minted time for myself. I had "found" a few minutes to just sit and think. I was having fun eavesdropping on the couple at the next table (fascinating!) and people-watching the lobby (outrageous!). And, most important, in the pause, I had connected a couple of ideas I had been struggling with for days. So no apology was necessary. Hence: "Thank you for being late."

The first time, I just blurted out that response, not really thinking about it. But after another such encounter, I noticed that it felt good to have those few moments of unplanned-for, unscheduled time, and it wasn't just me who felt better! And I knew why. Like many others, I was beginning to feel overwhelmed and exhausted by the dizzying pace of change. I needed to give myself (and my guests) permission to just slow down; I needed permission to be alone with my thoughts—without having to tweet about them, take a picture of them, or share them with anyone. Each time I reassured my guests that their lateness was not a problem, they would give me a quizzical look at first, but then a lightbulb would suddenly go on in their heads and they would say something like: "I know what you mean . . . 'Thank you for being late!' Hey, you're welcome."

In his sobering book *Sabbath*, the minister and author Wayne Muller observes how often people say to him, "I am so busy." "We say this to one another with no small degree of pride," Muller writes, "as if our exhaustion were a trophy, our ability to withstand stress a mark of real character . . . To be unavailable to our friends and family, to be unable to find time for the sunset (or even to know when the sun has set at all), to whiz through our obligations without time for a single, mindful breath, this has become a model of a successful life."

I'd rather learn to pause. As the editor and writer Leon Wieseltier said to me once: technologists want us to think that patience became a virtue only because in the past "we had no choice"—we had to wait longer for things because our modems were too slow or our broadband hadn't been installed, or because we hadn't upgraded to the iPhone 7. "And so now that we have made waiting technologically obsolete," added Wieseltier, "their attitude is: 'Who needs patience anymore?' But the ancients believed that there was wisdom in patience and that wisdom comes from patience . . . Patience wasn't just the absence of speed. It was space for reflection and thought." We are generating more information and knowledge than ever today, "but knowledge is only good if you can reflect on it."

And it is not just knowledge that is improved by pausing. So, too, is the ability to build trust, "to form deeper and better connections, not just fast ones, with other human beings," adds Seidman. "Our ability to forge deep relationships—to love, to care, to hope, to trust, and to build voluntary communities based on shared values—is one of the most uniquely human capacities we have. It is the single most important thing that differentiates us from nature and machines. Not everything is better faster or meant to go faster. I am built to think about my grandchildren. I am not a cheetah."

It is probably no accident, therefore, that what sparked this book was a pause—a chance encounter I had in, of all places, a parking garage, and my decision not to rush off as usual but to engage with a stranger who approached me with an unusual request.

## *The Parking Attendant*

It was early October 2014. I had driven my car from my home in Bethesda to the downtown there and parked in the public parking garage beneath the Hyatt Regency hotel, where I was meeting a friend at the Daily Grill for breakfast. As required, I got a time-stamped ticket when I arrived. After breakfast, I located my car in the garage and headed for the exit. I drove up to the cashier's booth and handed the man there my ticket, but before studying it, he studied me.

"I know who you are," said the elderly gentleman with a foreign accent and a warm smile.

"Great," I hurriedly responded.

"I read your column," he said.

"Great," I responded, itching to be on my way home.

"I don't always agree," he said.

"Great," I responded. "It means you always have to check."

We exchanged a few more pleasantries; he gave me my change and I drove off, thinking: "It's nice to know the parking guy reads my column in *The New York Times*."

About a week later, I parked in the same garage, as I do roughly once a week to catch the Red Line subway to downtown D.C. from the Bethesda Metro station. I got the same time-stamped ticket, I took the subway to Washington, I spent the day at my office, and I took the Metro back. Then I went down to the garage, located my car, and headed for the exit—and encountered the same attendant in the booth.

I handed him my time-stamped ticket, but this time, before he handed me my change, he said: "Mr. Friedman, I write, too. I have my own blog. Would you look at it?"

"How can I find it?" I asked. He then wrote down the Web address on a small piece of white paper normally used to print out receipts. It said "odanabi.com," and he handed it to me with my change.

I drove off, curious to check it out. But along the way my mind quickly drifted to other thoughts, like: "Holy mackerel! *The parking guy is now my competitor!* The parking guy has his own blog! He's a columnist, too! What's going on here?"

So I got home and called up his website. It was in English and focused on political and economic issues in Ethiopia, where he was from. It concentrated on relations among different ethnic and religious communities, the Ethiopian government's undemocratic actions, and some of the World Bank's activities in Africa. The blog was well designed and displayed a strong pro-democracy bent. The English was good but not perfect. The subject didn't greatly interest me, though, so I didn't spend a lot of time on the site.

But over the next week I kept thinking about this guy: How did he get into blogging? What did it say about our world that such an obviously educated man works as a parking cashier by day but has his own blog by night, a platform that enables him to participate in a global dialogue and tell the whole world about the issues that animate him, that is, Ethiopian democracy and society?

I decided I needed to pause—and learn more about him. The only problem was that I didn't have his personal e-mail, so the only way for me to contact him was to take the subway to work every day and park in the public garage to see if, by chance, I could bump into him again. And that's what I did.

After several days of coming up empty, I was rewarded when one morning I arrived very early and my blogger-parker was there in the cashier's booth. I stopped at the ticket machine, put my car into park, got out, and waved to him.

"Hey, it's Mr. Friedman again," I said. "Can I have your e-mail address? I want to talk to you."

He found a scrap of paper and wrote it down for me. His full name, I discovered, was Ayele Z. Bojia. That same evening I e-mailed him and asked him to tell me a little bit about his background and when he started blogging. I told him I was thinking of writing a book on writing about the twenty-first century and I was interested in how other people got into the blogging/opinion-writing universe.

He e-mailed me back on November 1, 2014: "I consider the first article I posted on Odanabi.com is also the first day I start blogging . . . Of course, if the question is also about what motivates me doing that, there are quite a good number of issues that bother me back home in my country of origin—Ethiopia—on which I would like to reflect my personal perspectives. I hope you would excuse me if I am not able to instantaneously respond to your message as I am doing that in between work. Ayele."

On November 3, I e-mailed him again: "What were you doing in Ethiopia before you came here and what are the issues that bother you most? No rush. Thanks, Tom."

And the same day he wrote back: "Great. I see a big reciprocity here. You are interested to know what issues bother me most while I am interested to learn from you how I can best communicate those issues of my concern to my target constituency and the larger public."

To which I immediately answered: "Ayele, You have a deal! Tom." I promised to share with him all that I could about how to write a column, if he would tell me his life story. He immediately agreed, and we set a date. Two weeks later I came from my office in downtown D.C., near the White House, and Bojia came up from his parking garage, and we met nearby at Peet's Coffee & Tea in Bethesda. He was sitting at a small table by the window. He had salt-and-pepper hair and a mustache and wore a green wool scarf wrapped around his neck. He began by telling me his story of how he became an opinion writer—and then I told him mine—as we each sipped Peet's finest brew.

Bojia, who was sixty-three when we first met, explained that he'd graduated with a BA in economics from Haile Selassie I University, named after the longtime Ethiopian emperor. He is an Orthodox Christian and an Oromo, the largest ethnic group in Ethiopia, with its own distinct language. Dating back from his time as a campus Oromo activist, he explained, he'd been promoting the culture and aspirations of the Oromo people in the context of a democratic Ethiopia.

"All my effort is geared towards making it possible for all peoples of Ethiopia to be proud of whatever nationality they belong to and be a proud Ethiopian by citizenship," Bojia explained. Those efforts drew the ire of the Ethiopian regime and forced him into political exile in 2004.

Bojia, who bore himself with the dignity of an educated immigrant whose day job was just to earn money so he could seriously blog at night, added: "I am not trying to write for the writing sake. I want to learn the techniques. [But] I have a cause to promote."

He named his blog Odanabi.com after a town in Ethiopia near the capital, Addis Ababa. The town is currently being touted to become the administrative and cultural seat of the Oromia regional government. He explained that he began his writing career on various Ethiopian Web platforms—Nazret.com, Ayyaanntu.net, AddisVoice.com, and Gadaa.com, an Oromo site, but their pace and his eagerness to participate in ongoing debates did not match: "I am appreciative of those websites, which gave me an opportunity to express my views, but the process was just too slow." So, he explained, as "a person working at the parking garage with certain financial constraints, I had to open this website [of my own] to have this regular outlet for myself." His site is hosted by Bluehost.com for a small fee.

The political field in Ethiopia is dominated by extremes, Bojia added: "There is no middle ground open to reason." One of the things that impressed him about America and that he wanted to bring to Ethiopia was the way "people stand for their rights, but also see the other guy's points of view." (Maybe you have to be a foreigner from a divided land working in an underground parking garage to see today's America as a country where arguments are bringing people closer, but I loved his optimism!)

He may just be in the cashier's booth making change, he told me, but he's always trying to observe people and how they express themselves and convey their opinions. "Before I came here I never heard of Tim Russert," Bojia said of the late, great *Meet the Press* host. "I don't know him, but when I started following [his program] it was kind of infectious for me. When he engages he doesn't push people in an extreme way. He is merciless in presentation of his facts and very respectful to others' feeling." As a result, Bojia concluded, "by the time he is finished every discussion you feel that he gave us some information"—and triggered something in the mind of the person he interviewed. Tim would have liked that.

Does he know how many people read his blog? I asked.

"From month to month it fluctuates with the issue, but there is a steady audience out there," he informed me, adding that the Web metrics he uses suggest that he is being read in around thirty different countries. But then he added: "If there is any way you can help me manage my website, I will be extremely happy." The thirty-five hours a week he'd spent over the last eight years working in the parking garage were just for "subsistence—my website is where my energy is."

I promised to do what I could to help. Who could resist a parking attendant who knows his Web metrics! But I had to ask: "What's it like for you—parking attendant by day, Web activist by night—to have your own global blog, while sitting in Washington and reaching people in thirty countries"—even if the numbers are small?

"I feel like I am a little bit empowered at this time," Bojia answered without hesitation. "These days I kind of regret that I wasted my time. I would have started some three or four years ago, and not sent stuff here and there. Had I concentrated on developing my own blog by now I would have a bigger audience . . . I have a deep satisfaction from what I am doing. I am doing something positive that helps my country."

## *Heating and Lighting*

So over the next few weeks I e-mailed Bojia two memos on how I went about constructing a column, and I followed up with another meeting at Peet's coffee shop to make certain that he understood what I was trying to say. I can't say how much it helped him, but I learned an enormous amount from our encounters—more than I ever anticipated.

For starters, just entering Bojia's world a tiny bit was an eye-opener. A decade ago the two of us would have had little in common, and now we were colleagues of sorts. Each of us was on a journey to bring our priorities to a wider audience, to participate in the global discussion and to tilt the world our way. We were both also part of a bigger trend. "We have never seen a time when more people could make history, record history, publicize history, and amplify history all at the same time," remarked Dov Seidman. In previous epochs, "to make history you needed an army, to record it you needed a film studio or a newspaper, to publicize it you needed a publicist. Now anyone can start a wave. Now anyone can make history with a keystroke."

And Bojia was doing just that. Artists and writers have moonlighted from time immemorial. What is new today is how many can now moonlight, how many others they can now touch from the moonlight if what they write is compelling, how fast they can go global if they prove they have something to say, and how little money it now costs to do so.

To live up to my side of the bargain with Bojia, I had to think more deeply about the craft of opinion writing than I had ever done before. I had been a columnist for nearly twenty years when we met, after being a reporter for seventeen years, and our encounter forced me to pause and put into words the difference between reporting and opinion writing and what actually makes a column "work."

In my two memos to Bojia I explained that there is no set formula for writing a column, no class you attend, and that everyone does it differently to some degree. But there were some general guidelines I could offer. When you are a reporter, your focus is on digging up facts to explain the visible and the complex and to unearth and expose the impenetrable and the hidden—wherever that takes you. You are there to inform, without fear or favor. Straight news often has enormous influence, but it's always in direct proportion to how much it informs, exposes, and explains.

Opinion writing is different. When you are a columnist, or a blogger in Bojia's case, your purpose is to influence or provoke a reaction and not just to inform—to argue for a certain perspective so compellingly that you persuade your readers to think or feel differently or more strongly or afresh about an issue.

That is why, I explained to Bojia, as a columnist, "I am either in the heating business or the lighting business." Every column or blog has to either turn on a lightbulb in your reader's head—illuminate an issue in a way that will inspire them to look at it anew—or stoke an emotion in your reader's heart that prompts them to feel or act more intensely or differently about an issue. The ideal column does both.

And you can immediately tell when it does—by how readers react. They might say "I didn't know that." That's a good reaction. It means you created some light. "I never looked at the issue that way." You created more light. "I never connected those things." More light. Then there's the columnist's favorite. It happens four times a year: "You said exactly what I felt but didn't know how to say—God bless you." And then there's also: "I want you fired. You're a moron. Who gave you this job? I will dance on your pink slip. I have canceled my subscription." You created heat . . .

But how do you go about generating heat or light? Where do opinions come from? I am sure every opinion writer would offer a different answer. My short one is that a column idea can spring from anywhere: a newspaper headline that strikes you as odd, a simple gesture by a stranger, the moving speech of a leader, the naïve question of a child, the cruelty of a school shooter, the wrenching tale of a refugee. Everything and anything is raw fodder for creating heat or light. It all depends on the connections you make and insights you surface to buttress your opinion.

More broadly speaking, though, I told Bojia, column writing is an act of chemistry—precisely because you must conjure it up yourself. A column doesn't write itself the way a breaking news story does. A column has to be created.

This act of chemistry usually involves mixing three basic ingredients: your own values, priorities, and aspirations; how you think the biggest forces, the world's biggest gears and pulleys, are shaping events; and what you've learned about people and culture—how they react or don't—when the big forces impact them.

When I say your own values, priorities, and aspirations, I mean the things that you care about most and aspire to see implemented most intensely. That value set helps you determine what is important and worth opining about, as well as what you will say. It is okay to change your mind as an opinion writer; what is not okay is to have no mind—to stand for nothing, or for everything, or only for easy and safe things. An opinion writer has to emerge from some framework of values that shapes his or her thinking about what should be supported or opposed. Are you a capitalist, a communist, a libertarian, a Keynesian, a conservative, a liberal, a neocon, or a Marxist?

When I refer to the world's big gears and pulleys, I am talking about what I call "the Machine." (Hat tip to Ray Dalio, the renowned hedge fund investor, who describes the economy as "a machine.") To be an opinion writer, you also always need to be carrying around a working hypothesis of how you think the Machine works—because your basic goal is to take your values and push the Machine in their direction. If you don't have a theory about how the Machine works, you'll either push it in a direction that doesn't accord with your beliefs or you won't move it at all.

And when I say people and culture, I mean how different peoples and cultures are affected by the Machine when it moves and how they, in turn, affect the Machine when they react. Ultimately columns are about people—the crazy things they say, do, hate, and hope for. I like to collect data to inform columns—but never forget: talking to another human being is also data. The columns that get the most response are almost always the ones about people, not numbers. Also, never forget that the best-selling book of all time is a collection of stories about people. It's called the Bible.

I argued to Bojia that the most effective columns emerge from mixing and rubbing these three ingredients together: you can't be an effective opinion writer without a set of values that informs what you're advocating. Dov Seidman likes to remind me of the Talmudic saying "What comes from the heart enters the heart." What doesn't come from your heart will never enter someone else's heart. It takes caring to ignite caring; it takes empathy to ignite empathy. You also can't have an effective column without some "take" on the biggest forces shaping the world in which we live and how to influence them. Your view of the Machine can never be perfect or immutable. It always has to be a work in progress that you are building and rebuilding as you get new information and the world changes. But it is very difficult to persuade people to do something if you can't connect the dots for them in a convincing way—why this action will produce this result, because this is how the gears and pulleys of the Machine work. And, finally, I told Bojia, you'll never have an opinion column that works unless it is inspired and informed by real people. It can't just be the advocacy of abstract principles.

When you put your value set together with your analysis of how the Machine works and your understanding of how it is affecting people and culture in different contexts, you have a worldview that you can then apply to all kinds of situations to produce your opinions. Just as a data scientist needs an algorithm to cut through all the unstructured data and all the noise to see the relevant patterns, an opinion writer needs a worldview to create heat and light.

But to keep that worldview fresh and relevant, I suggested to Bojia, you have to be constantly reporting and learning—more so today than ever. Anyone who falls back on tried-and-true formulae or dogmatisms in a world changing this fast is asking for trouble. Indeed, as the world becomes more interdependent and complex, it becomes more vital than ever to widen your aperture and to synthesize more perspectives.

My own thinking on this subject has been deeply influenced by Lin Wells, who teaches strategy at the National Defense University. According to Wells, it is fanciful to suppose that you can opine about or explain this world by clinging to the inside or outside of any one rigid explanatory box or any single disciplinary silo. Wells describes three ways of thinking about a problem: "inside the box," "outside the box," and "where there is no box." The only sustainable approach to thinking today about problems, he argues, "is thinking without a box."

Of course, that doesn't mean having no opinion. Rather, it means having no limits on your curiosity or the different disciplines you might draw on to appreciate how the Machine works. Wells calls this approach—which I will employ in this book—being "radically inclusive." It involves bringing into your analysis as many relevant people, processes, disciplines, organizations, and technologies as possible—factors that are often kept separate or excluded altogether. For instance, the only way you will understand the changing nature of geopolitics today is if you meld what is happening in computing with what is happening in telecommunications with what is happening in the environment with what is happening in globalization with what is happening in demographics. There is no other way today to develop a fully rounded picture.

These are the main lessons I shared with Bojia in my memos and our coffees. But here is a confession, which I also happily shared with him at our last meeting, which happened as I was completing this book: I had never thought this deeply about my own craft and what makes a column work until our chance encounter prompted me to do so. Had I not paused to engage him, I never would have taken apart, examined, and then reassembled my own framework for making sense of the world in a period of rapid change.

Not surprisingly, the experience set my mind whirring. And not surprisingly, my meetings with Bojia soon led me to start asking myself the same questions I was asking him to explore: What is my value set and where did it come from? How do I think the Machine works today? And what have I learned about how different peoples and cultures are being impacted by the Machine and responding to it?

That's what I started doing—in the pause—and the rest of this book is my answer.

Part II is about how I think the Machine works now—what I think are the biggest forces reshaping more things in more places in more ways on more days. Hint: the Machine is being driven by simultaneous accelerations in technology, globalization, and climate change, all interacting with one another.

And Part III is about how these accelerating forces are affecting people and cultures. That is, how they are reshaping the workplace, geopolitics, politics, ethical choices, and communities—including the small town in Minnesota where I grew up and where my own values were shaped.

Part IV offers the conclusions I draw from it all.

In short, this book is one giant column about the world today. It aims to define the key forces that are driving change around the world, to explain how they are affecting different people and cultures, and to identify what I believe to be the values and responses most appropriate to managing these forces, in order to get the most out of them for the most people in the most places and to cushion their harshest impacts.

So you never know what can result from pausing to talk to another person. To make a short story long—Bojia got a framework for his blog and I got a framework for this book. Think of it as an optimist's guide to thriving and building resilience in this age of accelerations, surely one of the great transformative moments in history.

As a reporter, I am continually amazed that often, when you go back and re-report a story or a period of history, you discover things you never saw the first time. As I began to write this book, it immediately became clear to me that the technological inflection point that is driving the Machine today occurred in a rather innocuous-sounding year: 2007.

What the hell happened in 2007?

# PART II
# ACCELERATING

TWO

# *What the Hell Happened in 2007?*

John Doerr, the legendary venture capitalist who backed Netscape, Google, and Amazon, doesn't remember the exact day anymore; all he remembers is that it was shortly before Steve Jobs took the stage at the Moscone Center in San Francisco on January 9, 2007, to announce that Apple had reinvented the mobile phone. Doerr will never forget, though, the moment he first laid eyes on that phone. He and Jobs, his friend and neighbor, were watching a soccer match that Jobs's daughter was playing in at a school near their homes in Palo Alto. As play dragged on, Jobs told Doerr that he wanted to show him something.

"Steve reached into the top pocket of his jeans and pulled out the first iPhone," Doerr recalled for me, "and he said, 'John, this device nearly broke the company. It is the hardest thing we've ever done.' So I asked for the specs. Steve said that it had five radios in different bands, it had so much processing power, so much RAM [random access memory], and so many gigabits of flash memory. I had never heard of so much flash memory in such a small device. He also said it had no buttons—it would use software to do everything—and that in one device 'we will have the world's best media player, world's best telephone, and world's best way to get to the Web—all three in one.'"

Doerr immediately volunteered to start a fund that would support creation of applications for this device by third-party developers, but Jobs wasn't interested at the time. He didn't want outsiders messing with his elegant phone. Apple would do the apps. A year later, though, he changed his mind; that fund was launched, and the mobile phone app industry exploded. The moment that Steve Jobs introduced the iPhone turns out to have been a pivotal junction in the history of technology—and the world.

Vint Cerf, who was one of the founding fathers of the Internet, remarked to me: "The spread of the smartphone made the Internet more valuable, because it became all the more accessible." But the Internet "also made the smartphone more valuable," because you could use that phone to tap into all the content and computing power of the Internet. The synthesis of these two technologies in one year would have been historically noteworthy all on its own. But it wasn't alone.

There are vintage years in wine and vintage years in history, and 2007 was definitely one of the latter.

Because not just the iPhone emerged in 2007—a whole group of companies emerged in and around that year. Together, these new companies and innovations have reshaped how people and machines communicate, create, collaborate, and think.

Just go down the list. In 2007, a company called VMware went public. VMware's translation software—a kind of digital Rosetta stone—made it possible for the same computer to run multiple operating systems (and applications) all at the same time, rather than requiring a dedicated computer for each different operating system. It was a crucial technology enabling the growth of cloud computing. Meanwhile, storage capacity for computing exploded thanks to the emergence that year of a framework for software called Hadoop, making "big data" possible for all. In 2007, development began on an open-source platform for writing and collaborating on software, called GitHub, that would vastly expand the ability of software to start, as Netscape founder Marc Andreessen once put it, "eating the world." On September 26, 2006, Facebook, a social networking site that had been confined to users on college campuses and at high schools, was opened to everyone at least thirteen years old with a valid e-mail address, and started to scale globally. In 2007, a microblogging company called Twitter, which had been part of a broader start-up, was spun off as its own separate platform and also started to scale globally. Change.org, the most popular social mobilization website, emerged in 2007.

In late 2006, Google bought YouTube, and in 2007 it launched Android, an open-standards platform for devices that would help smartphones scale globally with an alternative operating system to Apple's iOS. In 2007, AT&T, the iPhone's exclusive connectivity provider, invested in something called "software-enabled networks"—thus rapidly expanding its capacity to handle all the cellular traffic created by this smartphone revolution. According to AT&T, mobile data traffic on its national wireless network increased by *more than 100,000 percent* from January 2007 through December 2014.

In 2007, "Satoshi Nakamoto"—the name used by an unknown person or persons—began working on a digital currency and payment system called "Bitcoin." Nakamoto released the concept on October, 31, 2008, in a research paper entitled "Bitcoin: A Peer-to-Peer Electronic Cash System." The paper proposed that "a purely peer-to-peer version of electronic cash would allow online payments to be sent directly from one party to another without going through a financial institution." A decade later, it appears that Bitcoin's digital currency could well become the backbone of the global banking system in the twenty-first century. According to Wikipedia, "Nakamoto claimed that work on the Bitcoin writing of the code began in 2007."

Also in 2007, Amazon released something called the Kindle, onto which, thanks to Qualcomm's 3G technology, you could download thousands of books anywhere in the blink of an eye, launching the e-book revolution. In 2007, Airbnb was conceived in an apartment in San Francisco. In late 2006, the Internet crossed one billion users worldwide, which seems to have been a tipping point. In 2007, Palantir Technologies, the leading company using big data analytics and augmented intelligence to, among other things, help the intelligence community find needles in haystacks, launched its first platform. "Computing power and storage reached a level that made it possible for us to create an algorithm that could make a lot of sense out of things we could not make sense of before," explained Palantir's cofounder Alexander Karp. In 2005, Michael Dell decided to relinquish his job as CEO of Dell and step back from the hectic pace and just be its chairman. Two years later he realized that was bad timing. "I could see that the pace of change had really accelerated. I realized we could do all this different stuff. So I came back to run the company in . . . 2007."

It was also in 2007 that David Ferrucci, who led the Semantic Analysis and Integration Department at IBM's Watson Research Center in Yorktown Heights, New York, and his team began building a cognitive computer called Watson—"a special-purpose computer system designed to push the envelope on deep question and answering, deep analytics, and the computer's understanding of natural language," noted the website HistoryofInformation.com. "'Watson' became the first cognitive computer, combining machine learning and artificial intelligence."

In 2007, Intel introduced non-silicon materials—known as high-k/metal gates (the term refers to the transistor gate electrode and transistor gate dielectric)—into microchips for the first time. This very technical fix was hugely important. Although non-silicon materials were already used in other parts of the microprocessor, their introduction into the transistor helped Moore's law—the expectation that the power of microchips would double roughly every two years—continue on its path of delivering exponential growth in computing power. At that time there was real concern that Moore's law was hitting a wall with traditional silicon transistors.

"By opening the way to non-silicon materials it gave Moore's law another shot in the arm at a time when many people were thinking it was coming to an end," said Sadasivan Shankar, who worked on Intel's material design team at the time and now teaches materials and computational sciences at the Harvard School of Engineering and Applied Sciences. Commenting on the breakthrough, the *New York Times* Silicon Valley reporter John Markoff wrote on January 27, 2007: "Intel, the world's largest chip maker, has overhauled the basic building block of the information age, paving the way for a new generation of faster and more energy-efficient processors. Company researchers said the advance represented the most significant change in the materials used to manufacture silicon chips since Intel pioneered the modern integrated-circuit transistor more than four decades ago."

For all of the above reasons, 2007 was also "the beginning of the clean power revolution," said Andy Karsner, the U.S. assistant secretary of energy for efficiency and renewable energy from 2006 to 2008. "If anyone in 2005 or 2006 told you their predictive models captured where clean tech and renewable energy went in 2007 they are lying. Because what happened in 2007 was the beginning of an exponential rise in solar energy, wind, biofuels, LED lighting, energy efficient buildings, and the electrification of vehicles. It was a hockey stick moment."

And it wasn't just for renewables. It was in 2007 that the shale revolution took off, propelled by big data, advances in GPS and better software that made horizontal drilling to extract natural gas from shale deposits so much more efficient. "Fracking had been used for decades," noted

John Bringardner, in a May 2, 2014, essay in *The New Yorker,* "but in the late nineties, after years of experimenting with variants of the process, a Texas wildcatter named George Mitchell discovered that a combination of sand and water (known as 'high-volume slick water hydrofracturing') was cheaper and more effective than previous methods, which relied on other, heavier fluids." Around 2007, though, he added, "Devon Energy, an Oklahoma oil-and-gas producer that had bought Mitchell's company in 2002, had found that it could extract even more gas from each well by pairing [Mitchell's] method with horizontal drilling. Other companies began copying the combined technique, setting off a natural-gas boom in 2008."

Estimated reserves of natural gas in the United States in 2008 were 35 percent higher than in 2006—a phenomenal boost in reserves in basically two years, according to the Potential Gas Committee, the authority on U.S. gas supplies. "The jump is the largest increase in the 44-year history of reports from the committee," *The New York Times* reported on June 17, 2009.

Last but certainly not least, in 2007 the cost of DNA sequencing began to fall dramatically as the biotech industry shifted to new sequencing techniques and platforms, leveraging all the computing and storage power that was just exploding. This change in instruments was a turning point for genetic engineering and led to the "rapid evolution of DNA sequencing technologies that has occurred in recent years," according to Genome .gov. In 2001, it cost $100 million to sequence just one person's genome. On September 30, 2015, *Popular Science* reported: "Yesterday, personal genetics company Veritas Genetics announced that it had reached a milestone: participants in its limited, but steadily expanding Personal Genetics Program can get their entire genome sequenced for just $1,000."

As the graphs on the next two pages display, 2007 was clearly a turning point for many technologies.

Technology has always moved up in step changes. All the elements of computing power—processing chips, software, storage chips, networking, and sensors—tend to move forward roughly as a group. As their improving capacities reach a certain point, they tend to meld together into a platform, and that platform scales a new set of capabilities, which becomes the new normal. As we went from mainframes to desktops to laptops to smartphones with mobile applications, each generation of technology

## Cost of DNA Sequencing, per Genome

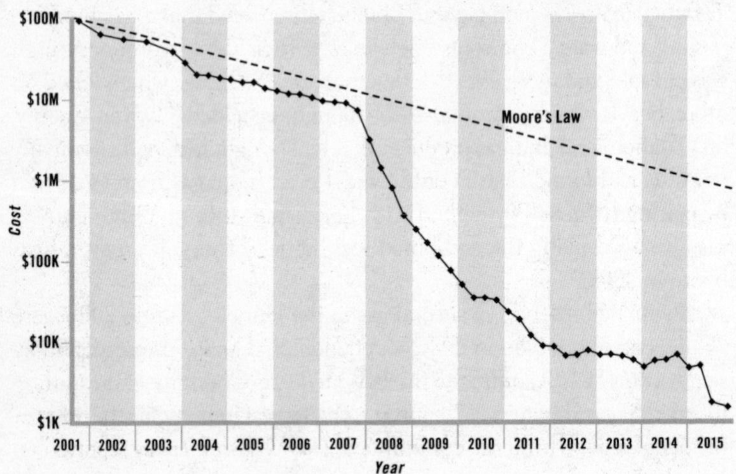


Source: National Human Genome Research Institute

## Utility Patent Grants in Biotech, 1963–2014

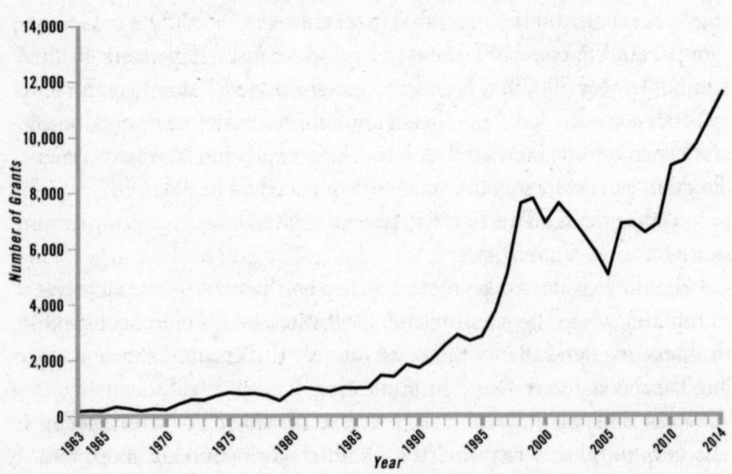


Source: U.S. Patent and Trademark Office

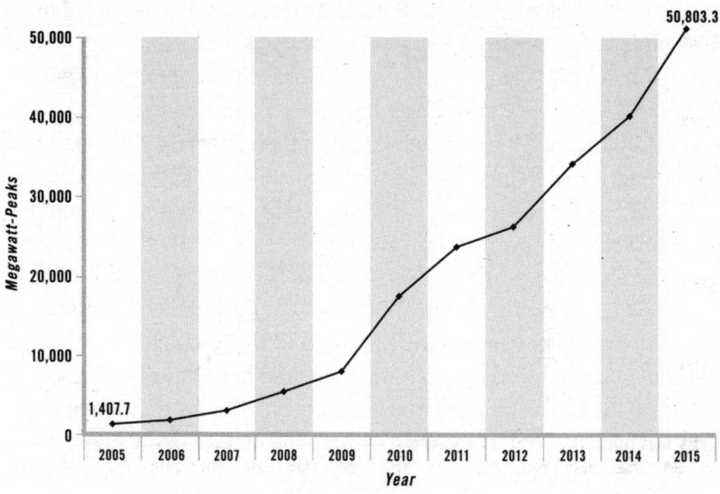

Courtesy of Paula Mints, SVP Market Research

got easier and more natural for people to use than the one before. When the first mainframe computers came out, you needed to have a computer science degree to use them. Today's smartphone can be accessed by young children and the illiterate.

As step changes in technology go, though, the platform birthed around the year 2007 surely constituted one of the greatest leaps forward in history. It suffused a new set of capabilities to connect, collaborate, and create throughout every aspect of life, commerce, and government. Suddenly there were so many more things that could be digitized, so much more storage to hold all that digital data, so many faster computers and so much more innovative software that could process that data for insights, and so many more organizations and people (from the biggest multinationals to the smallest Indian farmers) who could access those insights, or contribute to them, anywhere in the world through their handheld computers—their smartphones.

This is the central technology engine driving the Machine today. It snuck up on us very fast. In 2004, I started writing a book about what I thought then was the biggest force driving the Machine—namely, how the world was getting wired to such a degree that more people in more places had an equal opportunity to compete, connect, and collaborate with more other people for less money with greater ease than ever before. I called that book *The World Is Flat: A Brief History of the Twenty-First Century.* The first edition came out in 2005. I wrote an updated 2.0 edition in 2006 and a 3.0 edition in 2007. And then I stopped, thinking I had built a pretty solid framework that could last me as a columnist for a while.

I was very wrong! Indeed, 2007 was a really bad year to stop thinking.

I first realized just how bad the minute I sat down in 2010 to write my most recent book, *That Used to Be Us: How America Fell Behind in the World It Invented and How We Can Come Back,* which I coauthored with Michael Mandelbaum. As I recalled in that book, the first thing I did when I started working on it was to get the first edition of *The World Is Flat* off my bookshelf—just to remind myself what I was thinking when I started back in 2004. I cracked it open to the index, ran my finger down the page, and immediately discovered that Facebook wasn't in it! That's right—when I was running around in 2004 declaring that the world was flat, Facebook didn't even exist yet, Twitter was still a sound, the cloud was still in the sky, 4G was a parking space, "applications" were what you sent to college, LinkedIn was barely known and most people thought it was a prison, Big Data was a good name for a rap star, and Skype, for most people, was a typographical error. All of those technologies blossomed *after* I wrote *The World Is Flat*—most of them around 2007.

So a few years later, I began updating in earnest my view of how the Machine worked. A crucial impetus was a book I read in 2014 by two MIT business school professors—Erik Brynjolfsson and Andrew McAfee—entitled *The Second Machine Age: Work, Progress, and Prosperity in a Time of Brilliant Technologies.* The first machine age, they argued, was the Industrial Revolution, which accompanied the invention of the steam engine in the 1700s. This period was "all about power systems to augment human muscle," explained McAfee in an interview, "and each successive invention in that age delivered more and more power. But they all required humans to make decisions about them." Therefore, the inventions of that era actually made human control and labor "more valuable and important."

Labor and machines were, broadly speaking, complementary, he added. In the second machine age, though, noted Brynjolfsson, "we are beginning to automate a lot more cognitive tasks, a lot more of the control systems that determine what to use that power for. In many cases today artificially intelligent machines can make better decisions than humans." So humans and software-driven machines may increasingly be substitutes, not complements.

The key, but not the only, driving force making this possible, they argued, was the exponential growth in computing power as represented by Moore's law: the theory first postulated by Intel cofounder Gordon Moore in 1965 that the speed and power of microchips—that is, computational processing power—would double roughly every year, which he later updated to every two years, for only slightly more money with each new generation. Moore's law has held up close to that pattern for fifty years.

To illustrate this kind of exponential growth, Brynjolfsson and McAfee recalled the famous legend of the king who was so impressed with the man who invented the game of chess that he offered him any reward. The inventor of chess said that all he wanted was enough rice to feed his family. The king said, "Of course, it shall be done. How much would you like?" The man asked the king to simply place a single grain of rice on the first square of a chessboard, then two on the next, then four on the next, with each subsequent square receiving twice as many grains as the previous one. The king agreed, noted Brynjolfsson and McAfee—without realizing that sixty-three instances of doubling yields a fantastically big number: something like eighteen quintillion grains of rice. That is the power of exponential change. When you keep doubling something for fifty years you start to get to some very big numbers, and eventually you start to see some very funky things that you have never seen before.

The authors argued that Moore's law just entered the "second half of the chessboard," where the doubling has gotten so big and fast we're starting to see stuff that is fundamentally different in power and capability from anything we have seen before—self-driving cars, computers that can think on their own and beat any human in chess or *Jeopardy!* or even Go, a 2,500-year-old board game considered vastly more complicated than chess. That is what happens "when the rate of change and the acceleration of the rate of change both increase at the same time," said McAfee, and "we haven't seen anything yet!"

So, at one level, my view of the Machine today is built on the shoulders of Brynjolfsson and McAfee's fundamental insight into how the steady acceleration in Moore's law has affected technology—but I think the Machine today is even more complicated. That's because it's not just pure technological change that has hit the second half of the chessboard. It is also two other giant forces: accelerations in the Market and in Mother Nature.

"The Market" is my shorthand for the acceleration of globalization. That is, global flows of commerce, finance, credit, social networks, and connectivity generally are weaving markets, media, central banks, companies, schools, communities, and individuals more tightly together than ever. The resulting flows of information and knowledge are making the world not only interconnected and hyperconnected but interdependent—everyone everywhere is now more vulnerable to the actions of anyone anywhere.

And "Mother Nature" is my shorthand for climate change, population growth, and biodiversity loss—all of which have also been accelerating, as they, too, enter the second halves of their chessboards.

Here again, I am standing on the shoulders of others. I derive the term "the age of accelerations" from a series of graphs first assembled by a team of scientists led by Will Steffen, a climate change expert and researcher at the Australian National University, Canberra. The graphs, which originally appeared in a 2004 book entitled *Global Change and the Earth System: A Planet Under Pressure,* looked at how technological, social, and environmental impacts were accelerating and feeding off one another from 1750 to 2000, and particularly since 1950. The term "Great Acceleration" was coined in 2005 by these same scientists to capture the holistic, comprehensive, and interlinked nature of all these changes simultaneously sweeping across the globe and reshaping the human and biophysical landscapes of the Earth system. An updated version of those graphs was published in the *Anthropocene Review* on March 2, 2015; they appear on pages 166–167 of this book.

"When we started the project it was ten years since the first accelerations had been published, which ran from 1750 to 2000," explained Owen Gaffney, director of strategy for the Stockholm Resilience Centre, and part of the Great Acceleration team. "We wanted to update the graphs to 2010 to see if the trajectory had altered any"—and indeed it had, he said: it had accelerated.

It is the core argument of this book that these simultaneous accelerations in the Market, Mother Nature, and Moore's law together constitute the "age of accelerations," in which we now find ourselves. These are the central gears driving the Machine today. These three accelerations are impacting one another—more Moore's law is driving more globalization and more globalization is driving more climate change, and more Moore's law is also driving more potential solutions to climate change and a host of other challenges—and at the same time transforming almost every aspect of modern life.

Craig Mundie, a supercomputer designer and former chief of strategy and research at Microsoft, defines this moment in simple physics terms: "The mathematical definition of velocity is the first derivative, and acceleration is the second derivative. So velocity grows or shrinks as a function of acceleration. In the world we are in now, acceleration seems to be increasing. [That means] you don't just move to a higher speed of change. The rate of change also gets faster . . . And when the rate of change eventually exceeds the ability to adapt you get 'dislocation.' 'Disruption' is what happens when someone does something clever that makes you or your company look obsolete. 'Dislocation' is when the whole environment is being altered so quickly that everyone starts to feel they can't keep up."

That is what is happening now. "The world is not just rapidly changing," adds Dov Seidman, "it is being dramatically reshaped—it is starting to operate differently" in many realms all at once. "And this reshaping is happening faster than we have yet been able to reshape ourselves, our leadership, our institutions, our societies, and our ethical choices."

Indeed, there is a mismatch between the change in the pace of change and our ability to develop the learning systems, training systems, management systems, social safety nets, and government regulations that would enable citizens to get the most out of these accelerations and cushion their worst impacts. This mismatch, as we will see, is at the center of much of the turmoil roiling politics and society in both developed and developing countries today. It now constitutes probably the most important governance challenge across the globe.

## *Astro Teller's Graph*

The most illuminating illustration of this phenomenon was sketched out for me by Eric "Astro" Teller, the CEO of Google's X research and development lab, which produced Google's self-driving car, among other innovations. Appropriately enough, Teller's formal title at X is "Captain of Moonshots." Imagine someone whose whole mandate is to come to the office every day and, with his colleagues, produce moonshots—turning what others would consider science fiction into products and services that could transform how we live and work. His paternal grandfather was the physicist Edward Teller, designer of the hydrogen bomb, and his maternal grandfather was Gérard Debreu, a Nobel Prize–winning economist. Good genes, as they say. We were in a conference room at X headquarters, which is a converted shopping mall. Teller arrived at our interview on Rollerblades, which is how he keeps up with his daily crush of meetings.

He wasted no time before launching into an explanation of how the accelerations in Moore's law and in the flow of ideas are together causing an increase in the pace of change that is challenging the ability of human beings to adapt.

Teller began by taking out a small yellow 3M notepad and saying: "Imagine two curves on a graph." He then drew a graph with the Y axis labeled "rate of change" and the X axis labeled "time." Then he drew the first curve—a swooping exponential line that started very flat and escalated slowly before soaring to the upper outer corner of the graph, like a hockey stick: "This line represents scientific progress," he said. At first it moves up very gradually, then it starts to slope higher as innovations build on innovations that have come before, and then it starts to soar straight to the sky.

What would be on that line? Think of the introduction of the printing press, the telegraph, the manual typewriter, the Telex, the mainframe computer, the first word processors, the PC, the Internet, the laptop, the mobile phone, search, mobile apps, big data, virtual reality, human-genome sequencing, artificial intelligence, and the self-driving car.

A thousand years ago, Teller explained, that curve representing scientific and technological progress rose so gradually that it could take one hundred years for the world to look and feel dramatically different. For instance, it took centuries for the longbow to go from development into military use in Europe in the late thirteenth century. If you lived in the twelfth century, your basic life was not all that different than if you lived in the eleventh century. And whatever changes were being introduced in major towns in Europe or Asia took forever to reach the countryside, let alone the far reaches of Africa or South America. Nothing scaled globally all at once.

But by 1900, Teller noted, this process of technological and scientific change "started to speed up" and the curve started to accelerate upward. "That's because technology stands on its own shoulders—each generation of invention stands on the inventions that have come before," said Teller. "So by 1900, it was taking twenty to thirty years for technology to take one step big enough that the world became uncomfortably different. Think of the introduction of the car and the airplane."

Then the slope of the curve started to go almost straight up and off the graph with the convergence of mobile devices, broadband connectivity, and cloud computing (which we will discuss shortly). These developments diffused the tools of innovation to many more people on the planet, enabling them to drive change farther, faster, and more cheaply.

"Now, in 2016," he added, "that time window—having continued to shrink as each technology stood on the shoulders of past technologies—has become so short that it's on the order of five to seven years from the time something is introduced to being ubiquitous and the world being uncomfortably changed."

What does this process feel like? In my first book about globalization, *The Lexus and the Olive Tree*, I included a story Lawrence Summers told me that captured the essence of where we'd come from and where we were heading. It was 1988, Summers recalled, and he was working on the Michael Dukakis presidential campaign, which sent him to Chicago to give a speech. A car picked him up at the airport to take him to the event, and when he slipped into the car he discovered a telephone fixed into the backseat. "I thought it was sufficiently neat to have a cell phone in my car in 1988 that I used it to call my wife to tell her that I was in a car with a phone," Summers told me. He also used it to call everyone else he could think of, and they were just as excited.

Just nine years later Summers was deputy treasury secretary. On a trip to the Ivory Coast in West Africa, he had to inaugurate an American-funded health care project in a village upriver from the main city, Abidjan, that was opening its first potable water well. What he remembered most, though, he told me, was that on his way back from the village, as he stepped into a dugout canoe to return downriver, an Ivory Coast official handed him a cell phone and said: "Washington has a question for you." In nine years Summers went from bragging that he was in a car with a mobile phone in Chicago to nonchalantly using one in the backseat of his dugout canoe in Abidjan. The pace of change had not only quickened but was now happening at a global scale.

## *That Other Line*

So that is what is going on with scientific and technological progress—but Teller wasn't done drawing his graph for me. He'd promised two lines, and he now drew the second, a straight line that began many years ago above the scientific progress line but since then had climbed far more incrementally, so incrementally you could barely detect its positive slope.

"The good news is that there is a competing curve," Teller explained. "This is the rate at which humanity—individuals and society—adapts to changes in its environment." These, he added, can be technological changes (mobile connectivity), geophysical changes (such as the Earth warming and cooling), or social changes (there was a time when we weren't okay with mixed-race marriages, at least here in the United States). "Many of those major changes were driven by society, and we have adapted. Some were more or less uncomfortable. But we adapted."

Indeed, the good news is that we've gotten a little bit faster at adapting over the centuries, thanks to greater literacy and knowledge diffusion. "The rate at which we can adapt is increasing," said Teller. "A thousand years ago, it probably would have taken two or three generations to adapt to something new." By 1900, the time it took to adapt got down to one generation. "We might be so adaptable now," said Teller, "that it only takes ten to fifteen years to get used to something new."

Alas, though, that may not good enough. Today, said Teller, the accelerating speed of scientific and technological innovations (and, I would add, new ideas, such as gay marriage) can outpace the capacity of the average human being and our societal structures to adapt and absorb them. With that thought in mind, Teller added one more thing to the graph—a big dot. He drew that dot on the rapidly sloping technology curve just above the place where it intersected with the adaptability line.

He labeled it: "We are here." The graph, as redrawn for this book, can be seen on the next page.

That dot, Teller explained, illustrates an important fact: even though human beings and societies have steadily adapted to change, on average, the rate of technological change is now accelerating so fast that it has risen above the average rate at which most people can absorb all these changes. Many of us cannot keep pace anymore.

"And that is causing us cultural angst," said Teller. "It's also preventing us from fully benefiting from all of the new technology that is coming along every day . . . In the decades following the invention of the internal combustion engine—before the streets were flooded with mass-produced cars—traffic laws and conventions were gradually put into place. Many of those laws and conventions continue to serve us well today, and over the course of a century, we had plenty of time to adapt our laws to new inventions, such as freeways. Today, however, scientific advances are bringing seismic shifts to the ways in which we use our roads; legislatures and municipalities are scrambling to keep up, tech companies are chafing under outdated and sometimes nonsensical rules, and the public is not sure what to think. Smartphone technology gave rise

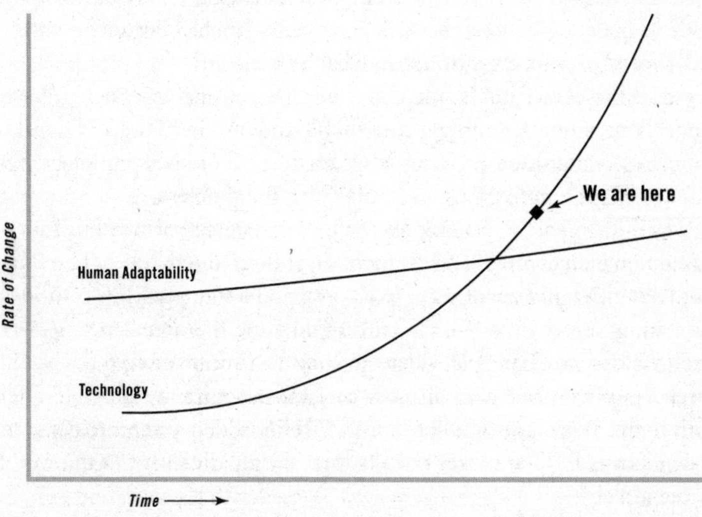

to Uber, but before the world figures out how to regulate ride-sharing, self-driving cars will have made those regulations obsolete."

This is a real problem. When fast gets really fast, being slower to adapt makes you really slow—and disoriented. It is as if we were all on one of those airport moving sidewalks that was going around five miles an hour and suddenly it sped up to twenty-five miles an hour—even as everything else around it stayed roughly the same. That is really disorienting for a lot of people.

If the technology platform for society can now turn over in five to seven years, but it takes ten to fifteen years to adapt to it, Teller explained, "we will all feel out of control, because we can't adapt to the world as fast as it's changing. By the time we get used to the change, that won't even be the prevailing change anymore—we'll be on to some new change."

That is dizzying for many people, because they hear about advances such as robotic surgery, gene editing, cloning, or artificial intelligence, but have no idea where these developments will take us.

"None of us have the capacity to deeply comprehend more than one of these fields—the sum of human knowledge has far outstripped any single individual's capacity to learn—and even the experts in these fields can't predict what will happen in the next decade or century," said Teller. "Without clear knowledge of the future potential or future unintended negative consequences of new technologies, it is nearly impossible to draft regulations that will promote important advances—while still protecting ourselves from every bad side effect."

In other words, if it is true that it now takes us ten to fifteen years to understand a new technology and then build out new laws and regulations to safeguard society, how do we regulate when the technology has come and gone in five to seven years? This is a problem.

Let's take patents as one example of a system that was built for a world in which changes arrived more slowly, explained Teller. The standard patent arrangement was: "We'll give you a monopoly on your idea for twenty years"—usually minus time to issue the actual patent—"in exchange for which people will get to know the information in the patent after it expires." But what if most new technologies are obsolete after four to five years, asked Teller, "and it takes four to five years to get your patents issued? That makes patents increasingly irrelevant in the world of technology."

Another big challenge is the way we educate our population. We go to school for twelve or more years during our childhoods and early adulthoods, and then we're done. But when the pace of change gets this fast, the only way to retain a lifelong working capacity is to engage in lifelong learning. There is a whole group of people—judging from the 2016 U.S. election—who "did not join the labor market at age twenty thinking they were going to have to do lifelong learning," added Teller, and they are not happy about it.

All of these are signs "that our societal structures are failing to keep pace with the rate of change," he said. Everything feels like it's in constant catch-up mode. What to do? We certainly don't want to slow down technological progress or abandon regulation. The only adequate response, said Teller, "is that we try to increase our society's ability to adapt." That is the only way to release us from the society-wide anxiety around tech. "We can either push back against technological advances," argued Teller, "or we can acknowledge that humanity has a new challenge: we must rewire our societal tools and institutions so that they will enable us to keep pace. The first option—trying to slow technology—may seem like the easiest solution to our discomfort with change, but humanity is facing some catastrophic environmental problems of its own making, and burying our heads in the sand won't end well. Most of the solutions to the big problems in the world will come from scientific progress."

If we could "enhance our ability to adapt even slightly," he continued, "it would make a significant difference." He then returned to our graph and drew a dotted line that rose up alongside the adaptability line but faster. This line simulated our learning faster as well as governing smarter, and therefore intersected with the technology/science change line at a higher point.

Enhancing humanity's adaptability, argued Teller, is 90 percent about "optimizing for learning"—applying features that drive technological innovation to our culture and social structures. Every institution, whether it is the patent office, which has improved a lot in recent years, or any other major government regulatory body, has to keep getting more agile—it has to be willing to experiment quickly and learn from mistakes. Rather than expecting new regulations to last for decades, it should continuously reevaluate the ways in which they serve society. Universities are now experimenting with turning over their curriculum much faster and more

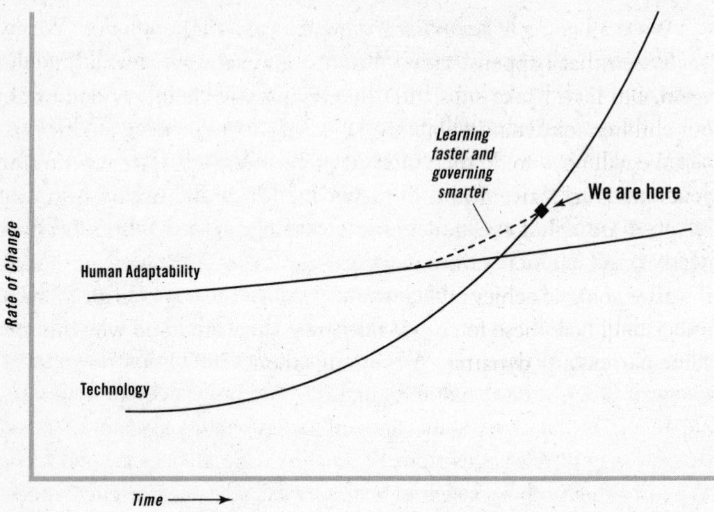


often to keep up with the change in the pace of change—putting a "use-by date" on certain courses. Government regulators need to take a similar approach. They need to be as innovative as the innovators. They need to operate at the speed of Moore's law.

"Innovation," Teller said, "is a cycle of experimenting, learning, applying knowledge, and then assessing success or failure. And when the outcome is failure, that's just a reason to start the cycle over again." One of X's mottos is "Fail fast." Teller tells his teams: "I don't care how much progress you make this month; my job is to cause your rate of improvement to increase—how do we make the same mistake in half the time for half the money?"

In sum, said Teller, what we are experiencing today, with shorter and shorter innovation cycles, and less and less time to learn to adapt, "is the difference between a constant state of destabilization versus occasional destabilization." The time of static stability has passed us by, he added. That does not mean we can't have a new kind of stability, "but the new kind of stability has to be dynamic stability. There are some ways of being, like riding a bicycle, where you cannot stand still, but once you are moving it is actually easier. It is not our natural state. But humanity has to learn to exist in this state."

We're all going to have to learn that bicycle trick.

When that happens, said Teller, "in a weird way we will be calm again, but it will take substantial relearning. We definitely don't train our children for dynamic stability."

We will need to do that, though, more and more, if we want future generations to thrive and find their own equilibrium. The next four chapters are about the underlying accelerations in Moore's law, the Market, and Mother Nature that define how the Machine works today. If we are going to achieve the dynamic stability Teller speaks of, we must understand how these forces are reshaping the world, and why they became particularly dynamic—beginning around 2007.

THREE

# *Moore's Law*

***Lives are changed when people connect. Life is changed when everything is connected.*** ***—Qualcomm motto***

One of the hardest things for the human mind to grasp is the power of exponential growth in anything—what happens when something keeps doubling or tripling or quadrupling over many years and just how big the numbers can get. So whenever Intel's CEO, Brian Krzanich, tries to explain the impact of Moore's law—what happens when you keep doubling the power of microchips every two years for fifty years—he uses this example: if you took Intel's first-generation microchip from 1971, the 4004, and the latest chip Intel has on the market today, the sixth-generation Intel Core processor, you will see that Intel's latest chip offers 3,500 times more performance, is 90,000 times more energy efficient, and is about 60,000 times lower in cost. To put it more vividly, Intel engineers did a rough calculation of what would happen had a 1971 Volkswagen Beetle improved at the same rate as microchips did under Moore's law.

These are the numbers: Today, that Beetle would be able to go about three hundred thousand miles per hour. It would get two million miles per gallon of gas, and it would cost four cents! Intel engineers also estimated that if automobile fuel efficiency improved at the same rate as

Moore's law, you could, roughly speaking, drive a car your whole life on one tank of gasoline.

What makes today's pace of technological change so extraordinary is this: it's not only the computational speed of microchips that's been in steady nonlinear acceleration; it's all the other components of the computer, too. Every computing device today has five basic components: (1) the integrated circuits that do the computing; (2) the memory units that store and retrieve information; (3) the networking systems that enable communications within and across computers; (4) the software applications that enable different computers to perform myriad tasks individually and collectively; and (5) the sensors—cameras and other miniature devices that can detect movement, language, light, heat, moisture, and sound and transform any of them into digitized data that can be mined for insights. Amazingly, Moore's law has many cousins. This chapter will show how the steady acceleration in the power of all five of these components, and their eventual melding into something we now call "the cloud," has taken us somewhere new—to that dot drawn by Astro Teller, the place where the pace of technological and scientific change outstrips the speed with which human beings and societies can usually adapt.

## *Gordon Moore*

Let's begin our story with microchips, also known as integrated circuits, also known as microprocessors. These are the devices that run all of a computer's programs and memory. The dictionary will tell you that a microprocessor is like a mini computational engine built on a single silicon chip, hence its shorthand name, the "microchip," or just the "chip." A microprocessor is built out of transistors, which are tiny switches that can turn a flow of electricity on or off. The computational power of a microprocessor is a function of how fast the transistors actually turn on and off and how many of these you can fit onto a single silicon chip. Before the invention of the transistor, early computer designers relied on bulb-like vacuum tubes, the kind you used to see in the back of an old television, to switch electricity on or off to create computation. This made them very slow and hard to build.

And then suddenly everything changed in the summer of 1958. Jack

Kilby, an engineer at Texas Instruments, "found a solution to this problem," reports NobelPrize.org.

> Kilby's idea was to make all the components and the chip out of the same block (monolith) of semiconductor material . . . In September 1958, he had his first integrated circuit ready . . .
>
> By making all the parts out of the same block of material and adding the metal needed to connect them as a layer on top of it, there was no more need for individual discrete components. No more wires and components had to be assembled manually. The circuits could be made smaller and the manufacturing process could be automated.

A half year later, another engineer, Robert Noyce, came up with his own idea for the integrated circuit—an idea that elegantly solved some of the problems of Kilby's circuit and made it possible to more seamlessly interconnect all the components on a single chip of silicon. And so the digital revolution was born.

Noyce cofounded Fairchild Semiconductor in 1957 (and later Intel) to develop these chips, along with several other engineers, including Gordon E. Moore, who held a doctorate in physical chemistry from the California Institute of Technology and would become director of the research and development laboratories at Fairchild. The company's great innovation was developing a process to chemically print tiny transistors onto a chip of silicon crystal, making them much easier to scale and more suitable for mass production. As Fred Kaplan notes in his book *1959: The Year Everything Changed*, the microchip might not have taken off if it hadn't been for big government programs, notably the race to the moon and the Minuteman ICBM. Both needed sophisticated guidance systems that had to fit inside very small nose cones. The demands of the Defense Department started to create economies of scale for these microchips, and the first person to appreciate that was Gordon Moore.

"Moore was perhaps the first to realize that Fairchild's chemical printing approach to making the microchip meant that they would not only be smaller, more reliable, and use less power than conventional electronic circuits, but also that microchips would be cheaper to produce," noted David Brock in the 2015 special issue of *Core*, the magazine of the Computer History Museum. "In the early 1960s, the entire global semiconductor industry adopted Fairchild's approach to making silicon microchips, and a market emerged for them in military fields, particularly aerospace computing."

I interviewed Moore in May 2015 at the Exploratorium in San Francisco for the fiftieth anniversary of Moore's law. Although eighty-six years old at the time, all of *his* microprocessors were definitely still functioning with tremendous efficiency! In late 1964, Moore explained to me, *Electronics* magazine asked him to submit an article for their thirty-fifth-anniversary edition predicting what was going to happen in the semiconductor component industry in the next ten years. So he took out his notes and surveyed what had happened up to that time: Fairchild had gone from making a single transistor on a chip to a chip with about eight elements—transistors and resistors—while the new chips just about to be released had about twice that number of elements, sixteen, and in their lab they were experimenting with thirty elements and imagining how they would get to sixty! When he plotted it all on a log, it became clear they were doubling every year, so for the article he took a wild guess and predicted this doubling would continue for at least a decade.

As he put it in that now famous *Electronics* article, which appeared on April 19, 1965, entitled "Cramming More Components onto Integrated Circuits": "The complexity for minimum component costs has increased at a rate of roughly a factor of two per year . . . There is no reason to believe it will not remain nearly constant for at least ten years." The Caltech engineering professor Carver Mead, a friend of Moore's, later dubbed this "Moore's law."

Moore explained to me: "I had been looking at integrated circuits—[they] were really new at that time, only a few years old—and they were very expensive. There was a lot of argument as to why they would never be cheap, and I was beginning to see, from my position as head of a laboratory, that the technology was going to go in the direction where we would get more and more stuff on a chip and it would make electronics less expensive . . . I had no idea it was going to turn out to be a relatively precise prediction, but I knew the general trend was in that direction and I had to give some kind of a reason why it was important to lower the cost of electronics." The original prediction looked at ten years, which involved going from about sixty elements on an integrated circuit to sixty thousand—a thousand-fold extrapolation over ten years. But it came

Moore's Law Illustrated by Intel Processors

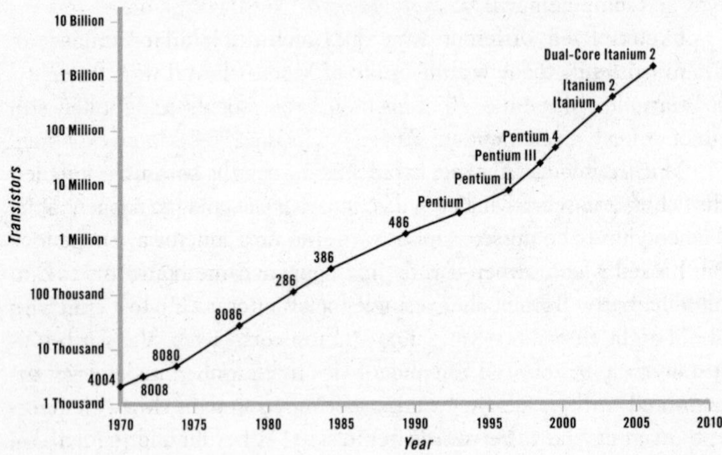


true. Moore realized that pace could not likely be sustained, though, so in 1975 he updated his prediction and said the doubling would happen roughly every two years and the price would stay almost the same.

And it just kept coming true.

"The fact that something similar is going on for fifty years is truly amazing," Moore said to me. "You know, there were all kinds of barriers we could always see that [were] going to prevent taking the next step, and somehow or other, as we got closer, the engineers had figured out ways around these."

What is equally striking in Moore's 1965 article is how many predictions he got right about what these steadily improving microchips would enable:

> Integrated circuits will lead to such wonders as home computers—or at least terminals connected to a central computer—automatic controls for automobiles, and personal portable communications equipment. The electronic wristwatch needs only a display to be feasible today . . .
>
> In telephone communications, integrated circuits in digital

> filters will separate channels on multiplex equipment. [They] will also switch telephone circuits and perform data processing.
>
> Computers will be more powerful, and will be organized in completely different ways . . . Machines similar to those in existence today will be built at lower costs and with faster turn-around.

Moore could fairly be said to have anticipated the personal computer, the cell phone, self-driving cars, the iPad, big data, and the Apple Watch. The only thing he missed, I joked with him, was "microwave popcorn."

I asked Moore, when was the moment he came home and said to his wife, Betty, "Honey, they've named a law after me"?

"For the first twenty years, I couldn't utter the term 'Moore's law'—it was embarrassing," he responded. "It wasn't a law. Finally, I got accustomed to it, where now I could say it with a straight face."

Given that, is there something that he wishes he had predicted—like Moore's law—but didn't? I asked him.

"The importance of the Internet surprised me," said Moore. "It looked like it was going to be just another minor communications network that solved certain problems. I didn't realize it was going to open up a whole universe of new opportunities, and it certainly has. I wish I had predicted that."

There are so many wonderful examples of Moore's law in action that it is hard to pick a favorite. Here's one of the best I have ever come across, offered by the writer John Lanchester in a March 15, 2015, essay in the *London Review of Books* entitled "The Robots Are Coming."

"In 1996," wrote Lanchester, "in response to the 1992 Russo-American moratorium on nuclear testing, the U.S. government started a program called the Accelerated Strategic Computing Initiative [ASCI]. The suspension of testing had created a need to be able to run complex computer simulations of how old weapons were ageing, for safety reasons, and also—it's a dangerous world out there!—to design new weapons without breaching the terms of the moratorium."

In order to accomplish that, Lanchester added:

> ASCI needed more computing power than could be delivered by any existing machine. Its response was to commission a

computer called ASCI Red, designed to be the first supercomputer to process more than one teraflop. A "flop" is a floating point operation, i.e., a calculation involving numbers which include decimal points . . . (computationally much more demanding than calculations involving binary ones and zeroes). A teraflop is a trillion such calculations per second. Once Red was up and running at full speed, by 1997, it really was a specimen. Its power was such that it could process 1.8 teraflops. That's 18 followed by 11 zeros. Red continued to be the most powerful supercomputer in the world until about the end of 2000.

I was playing on Red only yesterday—I wasn't really, but I did have a go on a machine that can process 1.8 teraflops. This Red equivalent is called the PS3 [PlayStation 3]: it was launched by Sony in 2005 and went on sale in 2006. Red was only a little smaller than a tennis court, used as much electricity as eight hundred houses, and cost $55 million. The PS3 fits underneath a television, runs off a normal power socket, and you can buy one for under two hundred [pounds]. Within a decade, a computer able to process 1.8 teraflops went from being something that could only be made by the world's richest government for purposes at the furthest reaches of computational possibility, to something a teenager could reasonably expect to find under the Christmas tree.

Now that Moore's law has entered the second half of the chessboard, how much farther can it go? A microchip, or chip, as we said, is made up of transistors, which are tiny switches; these switches are connected by tiny copper wires that act like pipes through which electrons flow. The way a chip operates is that you push electrons as fast as possible through many copper wires on a single chip. When you send electrons from one transistor to another, you are sending a signal to turn a given switch on and off and thus perform some kind of computing function or calculation. With each new generation of microchips, the challenge is to push electrons through thinner and thinner wires to more and smaller switches to shut the electron flow on and off faster and faster to generate more computing power with as little energy and heat as possible for as low a cost as possible in as small a space as possible.

"Someday it has to stop," said Moore. "No exponential like this goes on forever."

We are not there yet, though.

For fifty years the industry has kept finding new ways to either shrink transistor dimensions by roughly 50 percent at roughly the same cost, thus offering twice the transistors for the same price, or fit the same number of transistors for half the cost. It has done so by shrinking the transistors and making the wires thinner and more closely spaced. In some cases, this has involved coming up with new structures and materials, all to keep that exponential growth roughly on track every twenty-four months or so. Just one example: the earliest integrated circuits used one layer of aluminum wire pipes; today they use thirteen layers of copper pipes, each placed on top of the other with nanoscale manufacturing.

"I have probably seen the death of Moore's law predicted a dozen times," Intel's CEO, Brian Krzanich, told me. "When we were working at three microns [one-thousandth of a millimeter: 0.001 millimeters, or about 0.000039 inches], people said, 'How will we get below that—can we make film thickness thin enough to make such devices and could we reduce the wavelength of light to pattern such small features?' But each time we found breakthroughs. It is never obvious beforehand and it is not always the answer that is first prescribed that provides the breakthrough. But every time we have broken through the next barrier."

Truth be told, said Krzanich, the last two iterations of Moore's law were accomplished after closer to two and a half years rather than two, so there has been some slowing down. Even so, whether the exponential is happening every one, two, or three years, the important point is that thanks to this steady nonlinear improvement in microchips, we keep steadily making machines, robots, phones, watches, software, and computers smarter, faster, smaller, cheaper, and more efficient.

"We are at the fourteen-nanometer generation, which is way below anything you can see with the human eye," Krzanich explained, referring to Intel's latest microchip. "The chip might be the size of your fingernail and on that chip will be over one billion transistors. We know how to get to ten nanometers pretty well, and we have most of the answers for seven and even five. Beyond five nanometers there are a bunch of ideas that people are thinking about. But that is how it has always been through time."

Bill Holt, Intel's executive vice president of technology and manufacturing, is the man in charge of keeping Moore's law going. On the tour he gave me of Intel's Portland, Oregon, chip fabrication plant, or fab, I watched through windows into the clean room where twenty-four hours a day robots move the chips from one manufacturing process to the next, while men and women in white lab coats make sure the robots are happy. Holt, too, has little patience for those who are sure Moore's law is running out. So much work is being done now with new materials that can pack more transistors that use less energy and create less heat, says Holt, that he is confident in ten years "something" will come along and lead the next generation of Moore's law.

The sheer amount of brainpower that Intel continues to throw at the challenge is staggering. It now has some two thousand PhDs working on different aspects of the problem at its Portland lab alone. The sheer amount of processing power that Moore's Law is now producing is also staggering.

As I write this in the summer of 2017, Intel's main workhorse microprocessor is that 14-nanometer chip it introduced in 2014. It packs a mind-boggling 37.5 million transistors per square millimeter. By the end of 2017, explained Mark Bohr, Intel Senior Fellow for Technology and Manufacturing, Intel will begin producing and distributing a 10nm chip that will pack "100 million transistors per square millimeter—more than double the previous density with less heat and power usage." When you multiply these vastly more powerful chips over multiple motherboards and multiple racks in multiple servers and multiple server farms, well, if you think the world is fast now just wait a year . . .

For example, this move from 14nm to 10nm chips will enable auto manufacturers to shrink the brain of a self-driving car—a brain that has to constantly take in sensor data from 360 degrees and then instantaneously process whether it's a dog, a human, a biker, or another car and then decide how to proceed—from something trunk-size to a box that can fit under the front seat. That is precisely what will enable self-driving vehicles to scale.

As Moore's law pushes against the limits of physics and material science, it is important to remember that from the very beginning the processing power in microchips was also improved by software advances, not just silicon. "More powerful chips were what enabled more sophisticated software, and some of that more sophisticated software was then used to make the chips themselves get faster through new designs and optimization of all the complexity that was growing on the chip itself," remarked Craig Mundie.

And it is these mutually reinforcing breakthroughs in chip design and software that have laid the foundation for the recent breakthroughs in artificial intelligence, or AI. They are also advancing next-generation quantum computing. Because machines are now able to absorb and process data at previously unimagined rates and amounts, they can now recognize patterns and learn much as our biological brains do.

But it all started with that first microchip and Moore's law. "Plenty of people have predicted the end of Moore's law plenty of times," Holt concluded, "and they predicted it for different reasons. The only thing they all have in common is that they were all wrong."

## *Sensors: Why Guessing Is Officially Over*

There was a time when you might have referred to someone as "dumb as a fire hydrant" or "dumb as a garbage can."

I wouldn't do that anymore.

One of the major and perhaps unexpected consequences of technological acceleration is this: fire hydrants and garbage cans are now getting really smart. For instance, consider the Telog Hydrant Pressure Recorder, which attaches to a fire hydrant and broadcasts its water pressure wirelessly straight to the desktop of the local utility, greatly reducing blowouts and hydrant breakdowns. And now you can pair that with Bigbelly garbage cans, which are loaded with sensors that wirelessly announce when they are full and in need of being emptied—so the garbage collectors can optimize their service routes and the city can become cleaner for less money. Yes, even the garbageman is a tech worker now. The company's website notes that "each Bigbelly receptacle measures 25" W × 26.8" D × 49.8" H and uses built-in solar panels to run motorized compactors, which dramatically reduce waste volumes to help create greener, cleaner streets . . . The receptacles have built-in cloud computing technology to digitally signal to trash collectors that they have reached capacity and need immediate attention."

That garbage can could take an SAT exam!

What is making hydrants and garbage cans so much smarter is

another acceleration, not directly related to computing per se but critical for expanding what computing can now do—and that is sensors. WhatIs.com defines a sensor as "a device that detects and responds to some type of input from the physical environment. The specific input could be light, heat, motion, moisture, pressure, or any one of a great number of other environmental phenomena. The output is generally a signal that is converted to human-readable display at the sensor location or transmitted electronically over a network for reading or further processing."

Thanks to the acceleration of the miniaturization of sensors, we are now able to digitize four senses—sight, taste, touch, and hearing—and are working on the fifth: smell. A wirelessly connected fire hydrant pressure sensor creates a digital measurement that tells the utility when the pressure is too high and too low. A temperature sensor tracks the expansion and contraction of the liquid in a thermometer to create a digital temperature readout. Motion sensors emit regular energy flows—microwaves, ultrasonic waves, or light beams—but send out a digital signal when that flow is interrupted by a person or car or animal entering its path. Police now bounce sensor beams off cars to measure their speed, and bounce sound waves off buildings to locate the source of a gunshot. The light sensor on your computer measures the light in your work area and then adjusts the screen brightness accordingly. Your Fitbit is a combination of sensors measuring the number of steps you take, the distance you've gone, the calories you've burned, and how vigorously you move your limbs. The camera in your phone is a still and video camera capturing and transmitting images from anywhere to anywhere.

This vast expansion in our ability to sense our environment and turn it into digitized data was made possible by breakthroughs in materials science and nanotechnology that created sensors so small, cheap, smart, and resistant to heat and cold that we could readily install them and fasten them to measure stress under extreme conditions and then transmit the data. Now we can even paint them—using a process called 3-D inking—on any parts of any machine, building, or engine.

To better understand the world of sensors I visited General Electric's huge software center in San Ramon, California, to interview Bill Ruh, GE's chief digital officer. That in itself is a story. GE, thanks in large part to its accelerating ability to put sensors all over its industrial equipment, is becoming more of a software company, with a big base now in Silicon Valley. Forget about washing machines—think intelligent machines. GE's ability to install sensors everywhere is helping to make possible the "industrial Internet," also known as the "Internet of Things" (IoT), by enabling every "thing" to carry a sensor that broadcasts how it is feeling at any moment, thus allowing its performance to be immediately adjusted or predicted in response. This Internet of Things, Ruh explained, "is creating a nervous system that will allow humans to keep up with the pace of change, make the information load more usable," and basically "make every thing intelligent."

General Electric itself gathers data from more than 150,000 GE medical devices, 36,000 GE jet engines, 21,500 GE locomotives, 23,000 GE wind turbines, 3,900 gas turbines, and 20,700 pieces of oil and gas equipment, all of which wirelessly report to GE how they are feeling every minute.

This new industrial nervous system, argued Ruh, was originally accelerated by advances in the consumer space—such as camera-enabled smartphones with GPS. They are to the industrial Internet in the twenty-first century, said Ruh, what the moonshot was to industrial progress in the twentieth century—they drove a great leap forward in an array of interlinked technologies and materials, making all of them smaller, smarter, cheaper, and faster. "The smartphone enabled sensors to get so cheap that they could scale, and we could put them everywhere," said Ruh.

And now those sensors are churning out insights at a level of granularity we have never had before. When all of these sensors transmit their data to centralized data banks, and then increasingly powerful software applications look for the patterns in that data, we can suddenly see weak signals before they become strong ones, and we can see patterns before they cause problems. Those insights can then be looped back for preventive action—when we empty the garbage bins at the optimal moment or adjust the pressure in a fire hydrant before a costly blowout, we are saving time, money, energy, and lives and generally making humanity more efficient than we ever imagined we could be.

"The old approach was called 'condition-based maintenance'—if it looks dirty, wash it," explained Ruh. "Preventive maintenance was: change the oil every six thousand miles, whether you drive it hard or not." The new approach is "predictive maintenance" and "prescriptive maintenance." We can now predict nearly the exact moment when a tire, engine, car or truck battery, turbine fan, or widget needs to be changed, and we can prescribe the exact detergent that works best for that particular engine operating under different circumstances.

If you look at the GE of the past, added Ruh, it was based on mechanical engineers' belief that by using physics you could model the whole world and right away get insights into how things worked. "The idea," he explained, "was that if you know exactly how the gas turbine and combustion engine work, you can use the laws of physics and say: 'This is how it is going to work and when it is going to break.' There was not a belief in the traditional engineering community that the data had much to offer. They used the data to verify their physics models and then act upon them. The new breed of data scientists here say: 'You don't need to understand the physics to look for and find the patterns.' There are patterns that a human mind could not find, because the signals are so weak early on that you won't see them. But now that we have all this processing power, those weak signals just pop out at you. And so as you get that weak signal, it now becomes clear that it is an early indication that something is going to break or is becoming inefficient."

In the past, the way we detected weak signals was with intuition, added Ruh. Experienced workers knew how to process weak data. But now, with big data, "with a much finer grain of fidelity, we can make finding the needle in the haystack *the norm*"—not the exception. "And we can then augment the human worker with machines, so they work as colleagues, and enable them to process weak signals together and overnight become like a thirty-year veteran."

Think about that. The intuition about how a machine is operating on a factory floor used to come from working there for thirty years and being able to detect a slightly different sound signature emanating from the machine, telling you something might not be exactly right. That is a weak signal. Now, with sensors, a new employee can detect a weak signal on the first day of work—without any intuition. The sensors will broadcast it.

This ability to generate and apply knowledge so much faster is enabling us to get the most not only out of humans but also out of cows. Guessing is over for dairy farmers, too, explained Joseph Sirosh, corporate vice president of the Data group in Microsoft's Cloud and Enterprise Division. Sounds like a pretty brainy job—managing bits and bytes. But when I sat down with Sirosh to learn about the acceleration in sensing, he chose to explain it to me with a very old example: cows.

Okay, it wasn't that simple. He wanted to talk about "the connected cow."

The story Sirosh tells goes like this: Dairy farmers in Japan approached the Japanese computer giant Fujitsu with a question. Could they improve the odds for successfully breeding cows in large dairy farms? It turns out that cows go into heat, or estrus—their period of sexual receptivity and fertility when they can be successfully artificially inseminated—only for a very short window: twelve to eighteen hours roughly every twenty-one days, and often primarily at night. This can make it enormously difficult for a small farmer with a large herd to monitor all his cows and identify the ideal time to artificially inseminate each one. If this can be done well, dairy farmers can ensure uninterrupted milk production from each cow throughout the year, maximizing the per capita output of the farm.

The solution Fujitsu came up with, explained Sirosh, was to fit the cows with pedometers connected by radio signal to the farm. The data was transmitted to a machine-learning software system called GYUHO SaaS running on Microsoft Azure, the Microsoft cloud. Fujitsu's research had established that a big increase in the number of steps per hour was a 95 percent accurate signal for the onset of estrus in dairy cows. When the GYUHO system detected a cow in heat, it would send a text alert to the farmers on their mobile phones, enabling them to administer artificial insemination at exactly the right times.

"It turns out that there is a simple secret of when the cow is in heat—the number of steps she takes picks up," said Sirosh. "That is when AI [artificial intelligence] meets AI [artificial insemination]." Having this system at their fingertips made the farmers more productive not only in expanding their herds—"you get a huge improvement in conception rates," said Sirosh—but also in saving time: it liberated them from having to rely on their own eyes, instincts, expensive farm labor, or the *Farmers' Almanac* to identify cows in heat. They could use the labor savings for other productive endeavors.

All the data being generated from the cows' sensors revealed another, even more important insight, said Sirosh: Fujitsu researchers found that within the sixteen-hour ideal window for artificial insemination, if you performed that function in the first four hours, there was a "seventy percent probability you got a female calf, and if you did it in the second four hours there was a higher probability that you got a male." So this could enable a farmer "to shape the mix of cows and bulls in his herd according to his needs."

The data just kept spitting out more insights, said Sirosh. By studying the pattern of footsteps, the farmers were able to gain early detection of eight different cow diseases, enabling early treatment and improving the overall health and longevity of the herd. "A little ingenuity can transform even the oldest of industries like farming," concluded Sirosh.

If a cow with a sensor makes a dairy farmer into a genius, a locomotive enabled with sensors is no longer a dumb train but an IT system on wheels. It can suddenly sense and broadcast the quality of the tracks every one hundred feet. It can sense the slope and how much energy it needs to go over each mile of terrain, putting on the gas a little less when it goes downhill, and generally maximizing fuel efficiency or velocity to get from point A to point B. And now all GE locomotives are being equipped with cameras to better monitor how the engineers are operating the engines at every curve. GE now also knows that if you have to run your engine at 120 percent on a hot day, certain parts will need to have their predictive maintenance moved up.

"We are constantly enriching and training our nervous system, and everyone benefits from the data," said Ruh. But it's not only the learning you can do with sensors and software; it's also the transforming you can do with sensors and software together. Today, explained Ruh, "we no longer need to build physical changes into every product to improve their performance, we just do it with software. I take a dumb locomotive and throw sensors and software into it, and suddenly I can do predictive maintenance, I can make it operate up and down the tracks at the optimal speeds to save gasoline, I schedule all the trains more efficiently and even park them more efficiently." Suddenly a dumb locomotive gets faster, cheaper, and smarter—without replacing a screw, a bolt, or an engine. "I can use sensor data and software to make the machine act more efficiently as though we [manufactured] a whole new generation," added Ruh.

In a plant, he added, "you can get tunnel vision into the job you are doing. But what if the machine is watching out for you, thanks to the fact that we will have a camera on everything—everything will have eyes and ears? We talk about the five senses. What people don't realize yet is that I am going to give the five senses to machines to interact with humans in the same way we interact with colleagues today."

And there's money in them thar hills—lots of it, explained GE's CEO, Jeff Immelt, in an interview with McKinsey & Company in October 2015:

> Every CEO of a railroad could tell you their [fleet] velocity. The velocity tends to be, let's say, between twenty and twenty-five miles per hour. This tends to be the average miles per hour that a locomotive travels in a day—twenty-two. Doesn't seem very good. And the difference between twenty-three and twenty-two for, let's say, Norfolk Southern, is worth two hundred fifty million dollars in annual profit. That's huge for a company like that. That's one mile [per hour]. So that's all about scheduling better. It's all about less downtime. It's all about not having broken wheels, being able to get through Chicago faster. That's all analytics.

With every passing day, explained John Donovan, AT&T's chief strategy officer, we are turning more and more "digital exhaust into digital fuel" and generating and applying the insights faster and faster. The American department store owner John Wanamaker was an early twentieth-century pioneer in both retailing and advertising. He once famously observed: "Half the money I spend on advertising is wasted; the trouble is I don't know which half." That needn't be the case today.

Latanya Sweeney, the then chief technology officer for the Federal Trade Commission, explained on National Public Radio on June 16, 2014, how sensing and software are transforming retail: "What a lot of people may not realize is that, in order for your phone to make a connection on the Internet, it's constantly sending out a unique number that's embedded in that phone, called the MAC address, to say, 'Hey, any Wi-Fis out there?' . . . And by using these constant probe requests by the phone looking for Wi-Fis, you could actually track where that phone has been, how often that phone comes there, down to a few feet." Retailers now use this information to see what displays you lingered over in their stores and which ones tempted you to make a purchase, leading them to adjust displays regularly during the day. But that's not the half of it—big data now allows retailers to track who drove by which billboard and then shopped in one of their stores.

As *The Boston Globe* reported on May 19, 2016:

> Now the nation's largest billboard company, Clear Channel Outdoor Inc., is bringing customized pop-up ads to the interstate. Its Radar program, up and running in Boston and 10 other US cities, uses data AT&T Inc. collects on 130 million cellular subscribers, and from two other companies, PlaceIQ Inc. and Placed Inc., which use phone apps to track the comings and goings of millions more.
>
> Clear Channel knows what kinds of people are driving past one of their billboards at 6:30 p.m. on a Friday—how many are Dunkin' Donuts regulars, for example, or have been to three Red Sox games so far this year.

It can then precisely target ads to them.

Sorry, Mr. Wanamaker. You lived in the wrong era. Guessing is so twentieth century. *Guessing is officially over.*

But so might be privacy. When you think of all the data that is being vacuumed up by giant firms—Facebook, Google, Amazon, Apple, Alibaba, Tencent, Microsoft, IBM, Netflix, Salesforce, General Electric, Cisco, and all the telephone companies—and how efficiently they can now mine that data for insights, you have to wonder how anyone will be able to compete with them. No one else will have that much digital exhaust as raw material to analyze and fuel better and better predictions. And digital exhaust is now power. We need to keep a close eye on the monopoly power that big data can create for big companies. It is not just how they can dominate a market with their products now, but how they can reinforce that domination with all the data they can collect.

## *Storage/Memory*

As we've seen, sensors hold great power. But all those sensors gathering all that data would have been useless without parallel breakthroughs in storage. These breakthroughs have given us chips that can store more data and software that can virtually interconnect millions of computers and make them store and process data as if they were a single desktop.

Just how big did that storage have to get and how sophisticated did the software have to become? Consider this May 11, 2014, talk by Randy Stashick, the then president of engineering at UPS, who spoke at the Production and Operations Management Society Conference on the importance of big data. He began by showing a number 199 digits long.

"Any idea what that number represents?" he asked the audience.

"Let me tell you a couple of things it does not represent," Stashick continued.

> It's not the number of hot dogs the famous Varsity restaurant, just up the street from us, has sold since opening in 1928. Nor is it the number of cars on Atlanta's infamous interstates at five o'clock on a Friday afternoon. Actually, that number, 199 digits in all, represents the number of discrete routes a UPS driver could conceivably take while making an average of one hundred twenty daily stops. Now, if you really want to get crazy, take that number and multiply it by fifty-five thousand. That's the number of U.S. routes our drivers are covering each business day. To display that number, we'd probably need that high-definition screen at AT&T Stadium in Dallas, where the Cowboys play. But somehow UPS drivers find their way to more than nine million customers every day, to deliver nearly seventeen million packages filled with everything from a new iPad for a high school graduate in Des Moines, to insulin for a diabetic in Denver, to two giant pandas relocating from Beijing to the Atlanta Zoo. How do they do it? The answer is operations research.
>
> More than two hundred sensors in the vehicle tell us if the driver is wearing a seat belt, how fast the vehicle is traveling, when the brakes are applied, if the bulkhead door is open, if the package car is going forward or backing up, the name of the street it's traveling on, even how much time the vehicle has spent idling versus its time in motion. Unfortunately, we don't know if the dog sitting innocently by the front door is going to bite.

To work through a number of routing options that is 199 digits long and also take into account data fed from two hundred sensors in each UPS truck requires *a lot* of storage, computing, and software capacity—more than anything available, even imaginable, to the average company as recently as fifteen years ago. Now it is available to any company. And therein lies a really important story about how a combination of storage chips hitting the second half of the chessboard and a software breakthrough named after a toy elephant put the "big" into "big data" analytics.

Microchips, as we have noted, are simply collections of more and more transistors. You can program those transistors for computation or for transmission or for memory. Memory chips come in two basic forms—DRAM, or dynamic random access memory, which does the temporary shoving of bits of data around as they are being processed, or "flash" memory, which permanently stores data when you press "save." Moore's law applies also to memory chips—we have been steadily packing more transistors storing more bits of memory on each chip for less money and using less energy. Today's average cell phone camera might have a sixteen-gigabyte memory, meaning it is storing sixteen billion bytes of information (a byte is eight bits) on a flash memory chip. Ten years ago flash memory density was not advanced enough to store a single photo on a phone—that is how fast all of this has accelerated, thereby making so many other things faster.

"Big data would not be here without Moore's law," said Intel's senior fellow Mark Bohr. "It gave us the bigger memory, more intensive computing, and the power, efficiency, and reliability that large server farms require to handle all that processing power. If those servers were made out of vacuum tubes, it would take one Hoover Dam to operate just one server farm."

But it wasn't just hardware that put the "big" in big data. It was also a software innovation—perhaps the most important to emerge in the last decade that you've never heard about. That software allowed millions of computers strung together to act like one computer, and it also made all that data searchable down to the level of finding those needles in the haystack. It was made by a company whose founder named it Hadoop—after his two-year-old son's favorite toy elephant, so that the name would be easy to remember. Remember that name: Hadoop. It has helped to change the world—but with a huge assist from Google.

The father of that little boy and the founder of Hadoop is Doug Cutting, who describes himself as a "catalyst" for software innovation. Cutting grew up in rural Napa County in California—and had not seen a computer until he entered Stanford in 1981, a school he had to borrow money to attend. There, he studied linguistics but also took courses in computer science, learned how to program, "and found it fun." He also found that programming would be the best way to pay off his student loans. So instead of going to graduate school, he got a job at the legendary Xerox PARC research center, where he was directed to join the linguistics team working on artificial intelligence and a relatively new field at the time called "search."

People forget that "search" as a field of inquiry existed before Google. Xerox had missed the personal computer business market, even though it had many great tech ideas, said Cutting, so the company was "trying to figure out how to transition from copy paper and toner to the digital world. It came up with the idea that copiers would replace filing cabinets. You would just scan everything and then search it. Xerox had this paper-oriented view of the world. It was the classic example of a company that could not move away from its cash cow—paper was its lifeblood—and it was trying to figure out how to move paper into the digital world. That was its rationale for looking into search. This is before the Web happened."

When the Web emerged, companies, led by Yahoo, started to organize it for consumers. Yahoo began as a directory of directories. Anytime someone put up a new website, Yahoo would add it to its directory, and then it started breaking websites down into groups—finance, news, sports, business, entertainment, et cetera. "And then search came along," said Cutting, "and Web search engines, like AltaVista, started cropping up. It had cataloged twenty million Web pages. That was a lot—and for a while it leapfrogged everyone. That was happening around 1995 to '96. Google showed up shortly thereafter [in 1997] with a small search engine, but claiming much better methods. And gradually it proved itself."

As Google took off, Cutting explained, he wrote an open-source search program in his spare time to compete with Google's proprietary system. The program was called Lucene. A few years later he and some colleagues started Nutch, which was the first big open-source Web search engine competitor to Google.

Open source is a model for developing software where anyone in the community can contribute to its ongoing improvement and freely use the collective product, usually under license, as long as they share their improvements with the wider community. It takes advantage of the commons and the notion that all of us are smarter than one of us; if everyone works on a program or product and then shares their improvements, that product will get smarter faster and then drive more change even faster.

Cutting's desire to create an open-source search program had to overcome a very basic problem: "When you have one computer—and you can store as much data on that computer as its hard drive can hold and you can process data as far and fast as the processor in that computer can process—that naturally limits the size and rate of the computation you can perform," Cutting explained.

But with the emergence of Yahoo and AOL, billions and billions of bits and bytes of data were piling up on the Web, requiring steadily increasing amounts of storage and computation power to navigate them. So people just started combining computers. If you could combine two computers, you could store twice as much and process twice as fast. With computer memory drives and processors getting cheaper, thanks to Moore's law, businesses started realizing that they could create football-field-sized buildings stocked with processors and drives from floor to ceiling, known as server farms.

But what was missing, said Cutting, was the ability to hook those drives and processors together so they could all work in a coordinated manner to store lots of data and also run computations across the whole body of that data, with all the processors running together in parallel. The really hard part was reliability. If you have one computer, it might crash once a week, but if you had one thousand it would happen one thousand times more often. So, for all of this to work, you needed a software program that could run the computers together seamlessly and another program to make the giant ocean of data that was created searchable for patterns and insights. Engineers in Silicon Valley like to wryly refer to a problem like this as a SMOP—as in, "We had all the hardware we needed—there was just this Small Matter Of Programming [SMOP] we had to overcome."

We can all thank Google for coming up with both of those programs in order to scale its search business. Google's true genius, said Cutting, was "to describe a storage system that made one thousand drives look like one drive, so if any single one failed you didn't notice," along with a software package for processing all that data they were storing in order to make it useful. Google had to develop these itself, because at the time there was no commercial technology capable of addressing its ambitions to store, process, and search all the world's information. In other words, Google had to innovate in order to build the search engine it felt the world wanted. But it used these programs exclusively to operate its own business and did not license them for anyone else.

However, in the time-honored tradition of programming engineers, Google, proud of what it had built, decided to share the basics with the public. And so it published two papers outlining in a general way the two key programs that enabled it to amass and search so much data at once. One paper, published in October 2003, outlined GFS, or Google File System. This was a system for managing and accessing huge amounts of data stored in clusters of cheap commodity computer hard drives. Because of Google's aspiration to organize all the world's information, it required petabytes and eventually exabytes (each of which is approximately one quintillion—1,000,000,000,000,000,000—bytes of data) to be stored and accessed.

And that required Google's second innovation: Google MapReduce, which was released in December 2004. Google described it as "a programming model and an associated implementation for processing and generating large data sets . . . Programs written in this functional style are automatically parallelized and executed on a large cluster of commodity machines. The . . . system takes care of the details of partitioning the input data, scheduling the program's execution across a set of machines, handling machine failures, and managing the required inter-machine communication. This allows programmers without any experience with parallel and distributed systems to easily utilize the resources of a large distributed system." In plain language, Google's two design innovations meant we could suddenly store more data than we ever imagined and could use software applications to explore that mountain of data with an ease we never imagined.

In the computing/search world, Google's decision to share these two basic designs—but not the actual proprietary code of its GFS and MapReduce solutions—with the wider computing community was a very,

very, very big deal. Google was, in effect, inviting the open-source community to build on its insights. Together these two papers formed the killer combination that has enabled big data to change nearly every industry. They also propelled Hadoop.

"Google described a way to easily harness lots of affordable computers," said Cutting. "They did not give us the running source code, but they gave us enough information that a skilled person could reimplement it and maybe improve on it." And that is precisely what Hadoop did. Its algorithms made hundreds of thousands of computers act like one giant computer. So anyone could just go out and buy commodity hardware in bulk and storage in bulk, run it all on Hadoop, and presto, do computation in bulk that produced really fine-grained insights.

Soon enough, Facebook and Twitter and LinkedIn all started building on Hadoop. And that's why they all emerged together in 2007! It made perfect sense. They had big amounts of data streaming through their business, but they knew that they were not making the best use of it. They couldn't. They had the money to buy hard drives for storage, but not the tools to get the most out of those hard drives, explained Cutting. Yahoo and Google wanted to capture Web pages and analyze them so people could search them—a valuable goal—but search became even more effective when companies such as Yahoo or LinkedIn or Facebook could see and store every click made on a Web page, to understand exactly what users were doing. Clicks could already be recorded, but until Hadoop came along no one besides Google could do much with the data.

"With Hadoop they could store all that data in one place and sort it by user and by time and all of a sudden they could see what every user was doing over time," said Cutting. "They could learn what part of a site was leading people to another. Yahoo would log not only when you clicked on a page but also everything on that page that could be clicked on. Then they could see what you did click on and did not click on but skipped, depending on what it said and depending on where it was on the page. This gave us big data analytics: when you can see more, you can understand more, and if you can understand more, you can make better decisions rather than blind guesses. And so data tied to analytics gives us better vision. Hadoop let people outside of Google realize and experience that, and that then inspired them to write more programs around Hadoop and start this virtuous escalation of capabilities."

So now you have Google's system, which is a proprietary closed-source system that runs only in Google's data centers and that people use for everything from basic search to facial identification, spelling correction, translation, and image recognition, and you have Hadoop's system, which is open source and run by everyone else, leveraging millions of cheap servers to do big data analytics. Today tech giants such as IBM and Oracle have standardized on Hadoop and contribute to its open-source community. And since there is so much less friction on an open-source platform, and so many more minds working on it—compared with a proprietary system—it has expanded lightning fast.

Hadoop scaled big data thanks to another critical development as well: the transformation of unstructured data.

Before Hadoop, most big companies paid little attention to unstructured data. Instead, they relied on Oracle SQL—a computer language that came out of IBM in the seventies—to store, manage, and query massive amounts of structured data and spreadsheets. "SQL" stood for "Structured Query Language." In a structured database the software tells you what each piece of data is. In a bank system it tells you "this is a check," "this is a transaction," "this is a balance." They are all in a structure so the software can quickly find your latest check deposit.

Unstructured data was anything you could not query with SQL. Unstructured data was a mess. It meant you just vacuumed up everything out there that you could digitize and store, without any particular structure. But Hadoop enabled data analysts to search all that unstructured data and find the patterns. This ability to sift mountains of unstructured data, without necessarily knowing what you were looking at, and be able to query it and get answers back and identify patterns was a profound breakthrough.

As Cutting put it, Hadoop came along and told users: "Give me your digits structured and unstructured and we will make sense of them. So, for instance, a credit card company like Visa was constantly searching for fraud, and it had software that could query a thirty- or sixty-day window, but it could not afford to go beyond that. Hadoop brought a scale that was not there before. Once Visa installed Hadoop it could query four or five years and it suddenly found the biggest fraud pattern it ever found by having a longer window. Hadoop enabled the same tools that people already knew how to use to be used at a scale and affordability that did not exist before."

That is why Hadoop is now the main operating system for data analytics supporting both structured and unstructured data. We used to throw away data because it was too costly to store, especially unstructured data. Now that we can store it all and find patterns in it, everything is worth vacuuming up and saving. "If you look at the quantity of data that people are creating and connecting to and the new software tools for analyzing it—they're all growing at least exponentially," said Cutting.

Before, small was fast but irrelevant, and big had economies of scale and of efficiency—but was not agile, explained John Donovan of AT&T. "What if we can now take massive scale and turn it into agility?" he asked. In the past, "with large scale you miss out on agility, personalization, and customization, but big data now allows you all three." It allows you to go from a million interactions that were impersonal, massive, and unactionable to a million individual solutions, by taking each pile of data and leveraging it, combing it, and defining it with software.

This is no small matter. As Sebastian Thrun, the founder of Udacity and one of the pioneers of massive open online courses (MOOCs) when he was a professor at Stanford, observed in an interview in the November/December 2013 issue of *Foreign Affairs*:

> With the advent of digital information, the recording, storage, and dissemination of information has become practically free. The previous time there was such a significant change in the cost structure for the dissemination of information was when the book became popular. Printing was invented in the fifteenth century, became popular a few centuries later, and had a huge impact in that we were able to move cultural knowledge from the human brain into a printed form. We have the same sort of revolution happening right now, on steroids, and it is affecting every dimension of human life.

And we're just at the end of the beginning. Hadoop came about because Moore's law made the hardware storage chips cheaper, because Google had the self-confidence to share some of its core insights and to dare the open-source community to see if they could catch up and leapfrog—and because the open-source community, via Hadoop, rose to the challenge. Hadoop's open-source stack was never a pure clone of Google's, and by today it has diverged in many creative ways. As Cutting put it: "Ideas are important, but implementations that bring them to the public are just as important. Xerox PARC largely invented the graphical user interface, with windows and a mouse, the networked workstation, laser printing, et cetera. But it took Apple and Microsoft's much more marketable implementations for these ideas to change the world."

And that is the story of how Hadoop gave us the big data revolution—with help from Google, which, ironically, is looking to offer its big data tools to the public as a business now that Hadoop has leveraged them to forge this whole new industry.

"Google is living a few years in the future," Cutting concluded, "and they send us letters from the future in these papers and we are all following along and they are also now following us and it's all beginning to be two-way."

Indeed, everyone is now in the big data business—or at least every company that wants to survive.

"Data is the new oil," explained Brian Krzanich, the Intel CEO. "Oil used to underlay everything—the automotive industry, plastics, chemicals, electrification, and transportation," and there were huge economic benefits derived from its infrastructure—from the ships, pipelines, refineries, and gas stations that were required to move all the oil around. Oil and gas infused themselves into every aspect of life and commerce.

"You can say the same today about data," added Krzanich. Instead of oil wells, though, it is microchips and servers; instead of refineries, it is data centers and software; and instead of pipelines, it is bandwidth and fiber-optic cables, but the data they pump out are infusing every aspect of life and commerce.

And just as with oil, those who are most adept at drilling for this data, that is digitizing it, amassing it, storing it—and then using algorithms to *analyze, optimize, customize, prophesize, and automatize* to improve every possible service, design, customer experience, or manufacturing process—will be the winners.

And those who don't, concluded Krzanich, "will be dead in five years."

Because the difference between those who use big data to create AI to *analyze, optimize, customize, prophesize, and automatize* and those who don't will be mammoth. Those who can analyze massive amounts of data will be able to spot trends that could never have been seen before; those who can optimize the flight path of an airliner will get more energy savings than ever before; those who can customize their products or services for every individual customer will dominate rivals like never before; and those who can prophesize when an elevator or airplane engine part will break and replace it before it does will save their customers money like never before. Finally, those who can model an idea on computers—that is, create a digital twin for anything from a bridge to a nuclear weapon—and test out digitally how it will work before you build it, will save time, money, and resources like never before.

And all of those things will only get better as the chips get faster, the software smarter, and the networking faster.

"The more data you have, the better your product," explained Kai-Fu Lee, president of Sinovation Ventures' Artificial Intelligence Institute, in a June 24, 2017 essay in *The New York Times*. "The better your product, the more data you can collect; the more data you can collect, the more talent you can attract; the more talent you can attract, the better your product. It's a virtuous circle, and the United States and China have already amassed the talent, market share, and data to set it in motion."

## *Software: Making Complexity Invisible*

It is impossible to talk about the acceleration in the development and diffusion of software without talking about the singular contribution of Bill Gates and his cofounder of Microsoft, Paul Allen. Software had been around for a long time before Bill Gates. It's just that the users of computers never really noticed, because it came loaded into the computer you bought, a kind of necessary evil with all that gleaming hardware. Mssrs. Gates and Allen changed all of that, starting in the 1970s, with their first adventures in writing an interpreter for a programming language called BASIC and then the operating system DOS.

Back in the day, hardware companies mostly contracted out or produced their own software, with each running its own operating system and proprietary applications on its own machines. Allen and Gates concluded that if you had a common software that could run on all kinds of different machines—which would one day be Acer, Dell, IBM, and hundreds of others—the software itself would have value and not just be something that was given away with the hardware. It is hard to remember today what a radical idea this was then. But Microsoft was born on this proposition—that people should not just pay one time for the software to be developed as part of a machine; rather, each individual user should pay to have the capabilities of each software program. What the DOS operating system did, in essence, was abstract away the differences in hardware between every computer. It didn't matter if you bought a Dell, an Acer, or an IBM. They all suddenly had the same operating system. This made desktop and laptop computers into commodities—the last thing their manufacturers wanted. Value then shifted to whatever differentiated software you could write that would work on top of DOS—and that you could charge each individual to use. That was how Microsoft got very rich.

We now take software so much for granted that we forget what it actually does. "What is the business of software?" asks Craig Mundie, who for many years worked alongside Gates as Microsoft's chief of research and strategy and has been my mentor on all things software and hardware. "Software is this magical thing that takes each emerging form of complexity and abstracts it away. That creates the new baseline that the person looking to solve the next problem just starts with, avoiding the need to master the underlying complexity themselves. You just get to start at that new layer and add your value. Every time you move the baseline up, people invent new stuff, and the compounding effect of that has resulted in software now abstracting complexity everywhere."

Think for second about a software application such as Google Photos. Today it can pretty much recognize everything in every photograph that you've ever stored on your computer. Twenty years ago, if your spouse said to you, "Honey, find me some photos of our vacation on the beach in Florida," you would have to manually go through photo album after photo album, and shoe box after shoe box, to find them. Then photography became digital and you were able to upload all your photos online. Today, Google Photos backs up all your digital photos, organizes them, labels them, and, using recognition software, enables you to find any beach scene you're looking for with a few clicks or gestures, or maybe even by just describing it verbally. In other words, the software has abstracted away all the complexity in that sorting and retrieval process and reduced it to a few keystrokes or touches or voice commands.

Think for another second about what it was like to catch a taxi five years ago. "Taxi, taxi," you shouted from the curb, perhaps standing in the rain, as taxi after taxi whizzed by with passengers already inside. So you then called the taxi company from a nearby phone booth, or maybe a cell phone, and, after keeping you on hold for five minutes, they told you that it would be a twenty-minute wait—and you didn't believe what they said and neither did they. Today, we all know how different that is: all the complexity associated with calling, locating, scheduling, dispatching, and paying for and even rating the driver of your taxi has been abstracted away—hidden, layer by layer—and now reduced to a couple of touches of the Uber app on your smartphone.

The history of computers and software, explains Mundie, "is really the history of abstracting away more and more complexity through combinations of hardware and software." What enables application developers to perform that magic are APIs, or application programming interfaces. APIs are the actual programming commands by which computers fulfill your every wish. If you want the application you're writing to have a "save" button so that when you touch it your file is stored in the flash drive, you create that with a set of APIs—the same with "create file," "open file," "send file," and on and on.

Today, APIs from many different developers, websites, and systems have become much more seamlessly interactive; companies share many of their APIs with one another so developers can design applications and services that can interface with and operate on one another's platforms. So I might use Amazon's APIs to enable people to buy books there by clicking items on my own website, ThomasL-Friedman.com.

"APIs make possible a sprawling array of Web-service 'mashups,' in which developers mix and match APIs from the likes of Google or Facebook or Twitter to create entirely new apps and services," explains the developer website ReadWrite.com. "In many ways, the widespread availability of APIs for major services is what's made the modern Web experience possible. When you search for nearby restaurants in the

Yelp app for Android, for instance, it will plot their locations on Google Maps instead of creating its own maps," by interfacing with the Google Maps API.

This type of integration is called "seamless," explains Mundie, "since the user never notices when software functions are handed from one underlying Web service to another . . . APIs, layer by layer, hide the complexity of what is being run inside an individual computer—and the transport protocols and messaging formats hide the complexity of melding all of this together horizontally into a network." And this vertical stack and these horizontal interconnections create the experiences you enjoy every day on your computer, tablet, or phone. Microsoft's cloud, Hewlett Packard Enterprise, not to mention the services of Facebook, Twitter, Google, Uber, Airbnb, Skype, Amazon, TripAdvisor, Yelp, Tinder, or NYTimes.com—they are all the product of thousands of vertical and horizontal APIs and protocols running on millions of machines talking back and forth across the network.

Software production is accelerating even faster now not only because tools for writing software are improving at an exponential rate. These tools are also enabling more and more people within and between companies to collaborate to write ever more complex software and API codes to abstract away ever more complex tasks—so now you don't just have a million smart people writing code, you have a million smart people *working together* to write all those codes.

And that brings us to GitHub, one of today's most cutting-edge software generators. GitHub is the most popular platform for fostering collaborative efforts to create software. These efforts can take any form—individuals with other individuals, closed groups within companies, or wide-open open source. It has exploded in usage since 2007. Again, on the assumption that all of us are smarter than one of us, more and more individuals and companies are now relying on the GitHub platform. It enables them to learn quicker by being able to take advantage of the best collaborative software creations that are already out there for any aspect of commerce, and then to build on them with collaborative teams that draw on brainpower both inside and outside of their companies.

GitHub today is being used by more than twelve million programmers to write, improve, simplify, store, and share software applications and is growing rapidly—it added a million users between my first interview there in early 2015 and my last in early 2016.

Imagine a place that is a cross between Wikipedia and Amazon—just for software: You go online to the GitHub library and pick out the software that you need right off the shelf—for, say, an inventory management system or a credit card processing system or a human resources management system or a video game engine or a drone-controlling system or a robotic management system. You then download it onto your company's computer or your own, you adapt it for your specific needs, you or your software engineers improve it in some respects, and then you upload your improvements back into GitHub's digital library so the next person can use this new, improved version. Now imagine that the best programmers in the world from *everywhere*—either working for companies or just looking for a little recognition—are all doing the same thing. You end up with a virtuous cycle for the rapid learning and improving of software programs that drives innovation faster and faster.

Originally founded by three grade-A geeks—Tom Preston-Werner, Chris Wanstrath, and P. J. Hyett—GitHub is now the world's largest code host. Since I could not visit any major company today without finding programmers using the GitHub platform to collaborate, I decided I had to visit the source of so much source code at its San Francisco headquarters. By coincidence, I had just interviewed President Barack Obama in the Oval Office about Iran a week earlier. I say that only because the visitor lobby at GitHub is *an exact replica* of the Oval Office, right down to the carpet!

They like to make their guests feel special.

My host, GitHub's CEO, Chris Wanstrath, began by telling me how the "Git" got into GitHub. Git, he explained, is a "distributed version control system" that was invented in 2005 by Linus Torvalds, one of the great and somewhat unsung innovators of our time. Torvalds is the open-source evangelist who created Linux, the first open-source operating system that competed head-to-head with Microsoft Windows. Torvalds's Git program allowed a team of coders to work together, all using the same files, by letting each programmer build on top of, or alongside, the work of others, while also allowing each to see who made what changes—and to save them, undo them, improve them, and experiment with them.

"Think of Wikipedia—that's a version control system for writing an open-source encyclopedia," explained Wanstrath. People contribute to each entry, but you can always see, improve, and undo any changes. The only rule is that any improvements have to be shared with the whole community. Proprietary software—such as Windows or Apple's iOS—is also produced by a version control system, but it is a closed-source system, and its source code and changes are not shared with any wider community.

The open-source model hosted by GitHub "is a distributed version controlled system: anyone can contribute, and the community basically decides every day who has the best version," said Wanstrath. "The best rises to the top by the social nature of the collaboration—the same way books get rated by buyers on Amazon.com. On GitHub the community evaluates the different versions and hands out stars or likes, or you can track the downloads to see whose version is being embraced most. Your version of software could be the most popular on Thursday and I could come in and work on it and my version might top the charts on Friday, but meanwhile the whole community will enjoy the benefits. We could merge them together or go off on our different paths, but either way there is more choice for the consumer."

How did he get into this line of work? I asked Wanstrath, age thirty-one. "I started programming when I was twelve or thirteen years old," he said. "I wanted to make video games. I loved video games. My first program was a fake AI program. But video games were way too difficult for me then, so I learned how to make websites." Wanstrath enrolled at the University of Cincinnati as an English major, but he spent most of his time writing code instead of reading Shakespeare, and participating in the rudimentary open-source communities online. "I was desperate for mentorship and looking for programs that needed help, and that led me to a life of building developer tools," he explained.

So Wanstrath fired off his open-source résumé and examples of his work to various software shops in Silicon Valley, looking for a junior-level programming job. Eventually a manager at CNET.com, a media platform that hosts websites, decided to take a chance on him, based not on his grades from college but on the "likes" on his programming from different open-source communities. "I didn't know much about San Francisco," he said. "I thought it was beaches and Rollerbladers." He soon found out it was bits and bytes.

So, in 2007 "I was a software engineer using open-source software to build our products for CNET." Meanwhile, in 2007, Torvalds went to Google and gave a Tech Talk one day about Git—his tool for collaborative coding. "It was on YouTube and so a bunch of my open-source colleagues said, 'We're going to try this Git tool and get away from all these different servers serving different communities.'"

Up to that point the open-source community was very open but also very balkanized. "Back then there was really no open-source *community*," recalled Wanstrath. "It was a collection of open-source communities, and it was based on the project, not the people. That was the culture. And all the tools, all the ideology, were focused on how you run and download this project and not how people work together and talk to each other. It was all project-centric." Wanstrath's emerging view was: Why not be able to work on ten projects at the same time in the same place and have them all share an underlying language, so they could speak to one another and programmers could go from one to the next and back?

So he began talking about a different approach with his CNET colleague P. J. Hyett, who had a computer science degree, and Tom Preston-Werner, with whom Wanstrath had collaborated on open-source projects long before they ever met in person.

"We were saying to ourselves: 'It is just so freaking hard to use this Git thing. What if we made a website to make it easier?" recalled Wanstrath. "And we thought: 'If we can get everyone using Git, we can stop worrying about what tools we are using and start focusing on what we are writing.' I wanted to do it all with one click on the Web, so I could leave comments about a program and follow people and follow code the same way I follow people on Twitter—and with the same ease." That way if you wanted to work on one hundred different software projects, you didn't have to learn one hundred different ways to contribute. You just learned Git and you could easily work on them all.

So in October 2007, the three of them created a hub for Git—hence "GitHub." It officially launched in April 2008. "The core of it was this distributed version control system with a social layer that connected all the people and all the projects," said Wanstrath. The main competitor at that time—SourceForge—took five days to decide whether to host your open-source software. GitHub, by contrast, was just a share-your-code-with-the-world kind of place.

"Say you wanted to post a program called 'How to Write a Column,'" he explained to me. "You just publish it under your name on GitHub. I would view that online and say: 'Hey, I have few points I would like to add.' In the old days, I would probably write up the changes I wanted to make and pitch them in the abstract to the community. Now I actually take your code into my sandbox. That is called a 'fork.' I work on it and now my changes are totally in the open—it's my version. If I want to submit the changes back to you, the original author, I make a pull request. You look at the new way I have laid out 'How to Write a Column'; you can see all the changes. And if you like it, you press the 'merge' button. And then the next viewer sees the aggregate version. If you don't like all of it, we have a way to discuss, comment, and review each line of code. It is curated crowdsourcing. But ultimately you have an expert—the person who wrote the original program—'How to Write a Column'—who gets to decide what to accept and what to reject. GitHub will show that I worked on this, but you get to control what is merged with your original version. Today, this is the way you build software."

A decade and a half ago Microsoft created a technology called .NET—a proprietary closed-source platform for developing serious enterprise software for banks and insurance companies. In September 2014, Microsoft decided to open-source it on GitHub to see what the community could add. Within six months Microsoft had more people working on .NET for free than they had had working on it inside the company since its inception, said Wanstrath.

"Open source is not people doing whatever they wanted," he quickly added. "Microsoft established a set of strategic goals for this program, told the community where they wanted to go with it, and the community made fixes and improvements that Microsoft then accepted. Their platform originally only ran on Windows. So one day Microsoft announced that in the future they would make it work on Mac and Linux. The next day the community said, 'Great, thank you very much. We'll do one of those for you.'" The GitHub community just created the Mac version themselves—overnight. It was a gift back to Microsoft for sharing.

"When I use Uber," concluded Wanstrath, "all I am thinking about now is where I want to go. Not how to get there. It is the same with GitHub. Now you just have to think about what problem do you want to solve, not what tools." You can now go to the GitHub shelf, find just what you need, take it off, improve it, and put it back for the next person. And in the process, he added, "we are getting all the friction out. What you are seeing from GitHub, you are seeing in every industry."

When the world is flat you can put all the tools out there for everyone, but the system is still full of friction. But *the world is fast* when the tools disappear, and all you are thinking about is the project. "In the twentieth century, the constraint was all about the hardware and making the hardware faster—faster processors, more servers," said Wanstrath. "The twenty-first century is all about the software. We cannot make more humans, but we can make more developers, and we want to empower people to build great software by lifting the existing ones up and opening up the world of development to create more coders . . . so they can create the next great start-up or innovation project."

There is something wonderfully human about the open-source community. At heart, it's driven by a deep human desire for collaboration and a deep human desire for recognition and affirmation of work well done—not financial reward. It is amazing how much value you can create with the words "Hey, what you added is really cool. Nice job. Way to go!" Millions of hours of free labor are being unlocked by tapping into people's innate desires to innovate, share, and be recognized for it.

In fact, what is most exciting to see today, said Wanstrath, "is the people behind the projects discovering each other now on GitHub. It's companies finding developers, developers finding each other, students finding mentors, and hobbyists finding co-conspirators—it's everything. It is becoming a library in the holistic sense. It is becoming a community in the deepest sense of the word." He added: "People meet each other on GitHub and discover they are living in the same town and then go out and share pizza and talk all night about programming."

Still, even open source needs money to operate, especially when you have twelve million users, so GitHub devised a business model. It charges companies for using its platform for private business accounts, where companies create private software repositories with their proprietary business codes and decide who they want to let collaborate on them. A great many major companies have both private and public repositories on GitHub now, because it enables them to move faster, making use of the most brainpower.

"We built our cloud architecture on open-source software, called

OpenStack, so we can leverage the community, and we have a hundred thousand developers who don't work for us—but what they can do in a week, we couldn't do in a year," said Meg Whitman, president and CEO of Hewlett Packard Enterprise. "I am convinced that the world is driven by validation and that's what makes these communities so powerful. People are driven by their desire for others in the community to validate their work. You like me? *Really?* Most people don't get tons of validation. I learned this at eBay. People went crazy about their feedback. Where else can you wake up and see how much everyone loves you!?"

It used to be that companies waited for the next chip to come down the line. But now that they can use software to make any hardware dance and sing in new ways, it's software that people are waiting for and collaborating on most avidly. That is why AT&T's John Donovan said: "For us Moore's law was the good ol' days. Every twelve to twenty-four months we could plan on a new chip and we knew it was coming and we could test around it and plan around it." Today it is much more about what software is coming down the pike. "The pace of change is being driven by who can write the software," he added. "You know that something is up when the guys with all the trucks and ladders who climb up telephone poles tell you, 'Donovan, we're a software company now.' Software used to be the bottleneck and now it is overtaking everything. It has become a compound multiplier of Moore's law."

## *Networking: Bandwidth and Mobility*

While the accelerating advances in processing, sensing, storage, and software have all been vital, they would never have scaled to the degree they have without the accelerating advances in connectivity—that is, the capacity and speed of the world's network of overland and undersea fiber-optic cables and wireless systems, which are the backbone of the Internet, as well as mobile telephony. Over the last twenty years, progress in this realm also has been moving at a pace close to Moore's law.

In 2013, I visited Chattanooga, Tennessee, which had been dubbed "Gig City" after it installed what was at the time the fastest Internet service in America—an ultra-high-speed fiber-optic network that transferred data at one gigabit per second, which was roughly thirty times the average speed in a standard U.S. city. According to a February 3, 2014, report in *The New York Times*, it took a mere "33 seconds to download a two-hour, high-definition movie in Chattanooga, compared with 25 minutes for those with an average high-speed broadband connection in the rest of the country." When I visited, the city was still buzzing about an unusual duet heard on October 13, using video-conference technology with super-low latency. The lower the latency, the less noticeable are the delays when two people are talking to each other from across the country. And with Chattanooga's then-new network, the latency was so low that a human ear could not pick it up. To drive home that point, T Bone Burnett, a Grammy Award winner, performed "The Wild Side of Life" with Chuck Mead, a founder of the band BR549, for an audience of four thousand. But Burnett played his part on a screen from a Los Angeles studio, and Mead on a stage in Chattanooga. The transcontinental duet was possible, reported Chattanoogan.com, because the latency of Chattanooga's new fiber network was sixty-seven milliseconds, meaning the audio and video traveled 2,100 miles from Chattanooga to Los Angeles in one-fourth of the blink of an eye—so fast no human ear could pick up the slight delay in sound transmission.

That duet was also a by-product of accelerating breakthroughs—just in the last few years—in the science of fiber optics, explained Phil Bucksbaum, a professor of natural science in the physics department at Stanford University. Bucksbaum specializes in the laser science that is the foundation of optical communications and is the former president of the Optical Society. Early in his career, in the 1980s, he worked at Bell Labs. In those days, computer scientists would use a command called "ping" to find out if a computer they wanted to communicate with in another part of the Bell Labs building was "awake." Ping would send out an electronic message that would bounce off the other computer and indicate if it was awake and ready for a two-way conversation. Ping also had a clock that would tell you how long it took for the electric pulse to go down the wires and back.

"I hadn't used ping in more than a decade," Bucksbaum told me over breakfast in September 2015. But for the fun of it "I sat down at my computer in my house in Menlo Park and pinged a bunch of computers around the world the other day," just to see how fast the pulse could get there and back. "I started pinging computers in Ann Arbor, Michigan; Imperial College London; the Weizmann Institute in Israel; and the University of Adelaide in Australia. It was amazing—the speed was more than half as fast as the speed of light," which is three hundred million meters per second. So that pulse went from a keystroke on Bucksbaum's computer, into his local fiber-optic cable, then into the terrestrial and undersea fiber cable, and then into a computer half a world away at more than half the speed of light.

"We are already half as fast as the laws of physics will allow, and trying to go faster runs into diminishing returns," he explained. "In twenty years," he added, "we went from maybe this is a good idea to there's no turning back to hitting the physical limits . . . With ping I found out how close to the physical limits we were, and it was pretty startling. It is a way big revolution."

This revolution happened, Bucksbaum explained, thanks to a kind of Moore's law that has been steadily quickening the transmission speeds of data and voice down fiber-optic cables. "The speed at which we can transmit data over undersea cables just keeps accelerating," said Bucksbaum. The short version of the story, he explained, goes like this: We started out sending voice and data using a digital radio frequency over coaxial cable made primarily of copper wire. That is what your first cable/phone company sent into your house and into the box on your television set. They also used the same coaxial cable to carry voice and data under the ocean to the four corners of the globe.

And then scientists at places like Bell Labs and Stanford started playing around with using lasers to send voice and data as pulses of light through optical fibers—basically long, skinny, flexible glass tubes. Starting in the late 1980s and early 1990s that evolved into the new standard. The original fiber-optic cables were made of chains of cables that only went so far. After traveling a certain distance, the signal would weaken and have to stop at an amplifier box, where it would be turned from light into an electronic signal, amplified, and then converted back to light and sent on its way again. But over time the industry discovered novel ways of using chemicals and splicing the fiber cables to both increase capacity for voice and data and transmit a light signal that would never weaken.

"That was a huge breakthrough," explained Bucksbaum. "With all this internal amplification they could get rid of the electronic amplifier boxes and lay continuous end-to-end fiber-optic cables" from America to Hawaii or China to Africa or Los Angeles to Chattanooga. "That enabled even more nonlinear growth," he said—not to mention the ability to stream movies into your home. It made broadband Internet possible.

"Once you no longer had to break up the laser light signal to amplify it, the speed that you could transmit information was no longer limited by the properties and constraints of electricity but only by the properties of light," he explained. "Then us laser guys really got to do cool stuff." They found all sorts of new ways to push more information using lasers and glass. These included time division multiplexing—turning the light on and off, or pulsing the lasers to create more capacity. And it included wavelength division multiplexing, using different colors of light to carry different phone conversations at once—and then combinations of the two.

They are not done accelerating. "The history of the last twenty years is that we just keep finding faster, better ways to divide the different properties of light to pack ever more information," said Bucksbaum. "The rate of data transfer for an undersea cable today is now trillions of bits per second." At some point, you end up "bumping up against the laws of physics," he added, but we are not there yet. Companies are now experimenting not just with ways to change the pulse or the color of light to create more capacity, but also with new ways of shaping that light that can deliver more than one hundred trillion bits per second down their fiber lines.

"We are getting closer and closer to being able to transmit a nearly infinite amount of information at near zero cost—these are the kind of nonlinear accelerations you're talking about," said Bucksbaum. Most people right now are using this new power to stream movies, but it will infuse itself everywhere. "I ordered a book this morning at five a.m. and it's going to be delivered by Amazon today."

## *The AT&T Gamble*

As powerful as all those fiber-optic landlines and sea cables are, they are still only one part of the connectivity story. To unleash the power of the mobile phone revolution, it was also necessary to expand the speed and reach of wireless networks.

Many players had a hand in that, starting with AT&T and the huge bet it made that few people knew about. It happened in 2006 when the company's COO and soon-to-be CEO, Randall Stephenson, quietly struck a deal with Steve Jobs for AT&T to be the exclusive service provider in the United States for this new thing called the iPhone. Stephenson knew that this deal would stretch the capacity of AT&T's networks, but he didn't know the half of it. The iPhone came on so fast, and the need for capacity exploded so massively with the apps revolution, that AT&T found itself facing a monumental challenge. It had to enlarge its capacity, practically overnight, using the same basic line and wireless infrastructure it had in place. Otherwise, everyone who bought an iPhone was going to start experiencing dropped calls. AT&T's reputation was on the line—and Jobs would not have been a happy camper if his beautiful phone kept dropping calls. To handle the problem, Stephenson turned to his chief of strategy, John Donovan, and Donovan enlisted Krish Prabhu, now president of AT&T Labs.

Donovan picks up the story: "It's 2006, and Apple is negotiating the service contracts for the iPhone. No one had even seen one. We decided to bet on Steve Jobs. When the phone first came out [in 2007] it had only Apple apps, and it was on a 2G network. So it had a very small straw, but it worked because people only wanted to do a few apps that came with the phone." But then Jobs decided to open up the iPhone, as the venture capitalist John Doerr had suggested, to app developers everywhere.

Hello, AT&T! Can you hear me now?

"In 2008 and 2009, as the app store came on stream, the demand for data and voice just exploded—and we had the exclusive contract" to provide the bandwidth, said Donovan, "and no one anticipated the scale. Demand exploded a hundred thousand percent [over the next several years]. Imagine the Bay Bridge getting a hundred thousand percent more traffic. So we had a problem. We had a small straw that went from feeding a mouse to feeding an elephant and from a novelty device to a necessity" for everyone on the planet. Stephenson insisted AT&T offer unlimited data, text, and voice. The Europeans went the other way with more restrictive offerings. Bad move. They were left as roadkill by the stampede for unlimited data, text, and voice. Stephenson was right, but AT&T

just had one problem—how to deliver on that promise of unlimited capacity without vastly expanding its infrastructure overnight, which was physically impossible.

"Randall's view was 'never get in the way of demand,'" said Donovan. Accept it, embrace it, but figure out how to satisfy it fast before the brand gets killed by dropped calls. No one in the public knew this was going on, but it was a bet-the-business moment for AT&T, and Jobs was watching every step from Apple headquarters.

"We were expected to deal with some exponentials," said Donovan. "And I knew that I could not get there with Moore's law on the hardware alone. It would take too long to deploy at that scale. I had to get a faster solution—hence *software*. We pioneered software-enabled networking. We put everyone in the company we could muster into software development and we went to our [infrastructure] vendors and told them, 'We are moving to software.'"

I asked Prabhu to explain software-enabled networking, which he did with a simple example: "Think of the calculator on your phone," he said. "It creates the virtual effect of hardware—a desk calculator—by using software. Or think of the flashlight on your iPhone. That is software using the underlying hardware to create a virtual flashlight."

In networking, Prabhu explained, it means minting massive amounts of new capacity for transmitting data, text, and voice by taking the same networking switches, wires, chips, and cables and getting them to work better and faster by virtualizing different operations with the magic of software. The best way to understand it is to think of telephone wires as a highway, and then imagine that the only cars on this highway were self-driving vehicles controlled by computers, so they could never crash into one another. If that were the case, you could pack so many more cars on that highway, because they could drive bumper-to-bumper at one hundred miles per hour six inches apart. When you take the electric energy passing through a copper wire or fiber cable or a cellular transmitter and you apply software to that electronic signal, you can manipulate that energy in so many more ways and create so much more capacity beyond the traditional limits and safety margins built into the original hardware.

And just as you can set up a highway with automated cars driving at a hundred miles an hour six inches apart, said Donovan, you can "take the same copper wire designed to carry a two-ringy-dingy voice phone call and make it carry eight streams of video by maximizing how the bits perform. Software adapts and learns. Hardware can't. So, we blew apart the hardware components, and we forced everyone to think anew. We basically turned the hardware into a commodity and then created a baseline operating system for every router, and called it ONOS, for Open Network Operating System." Users could write programs on it to keep improving the performance.

Software, concluded Donovan, "has power and flexibility greater than anything materials can offer. Software better captures new wisdom than materials." Basically what we have done "is amplify Moore's law with software. Moore's law was viewed as the magic carpet we were riding, and then we discovered we could use software and literally accelerate Moore's law."

## *Irwin: The Cell Phone Guy*

It was wonderful for consumers for all these networking breakthroughs to occur, but someone had to pack them into a phone you could carry in your pocket to get the full frontal revolution—and no individual was more responsible for this mobile phone revolution than Irwin Jacobs. In the pantheon of the great innovators who launched the Internet age—Bill Gates, Paul Allen, Steve Jobs, Gordon Moore, Bob Noyce, Michael Dell, Jeff Bezos, Marc Andreessen, Andy Grove, Vint Cerf, Bob Kahn, Larry Page, Sergey Brin, and Mark Zuckerberg—save a few lines for Irwin Jacobs, and add Qualcomm to the list of important companies you've barely heard of.

Qualcomm is to mobile phones what Intel and Microsoft together were to desktops and laptops—the primary inventor, designer, and manufacturer of the microchips and software that run handheld smartphones and tablets. And all you have to do is walk through Qualcomm's museum at its San Diego headquarters and see its first mobile phone—basically a small suitcase with a phone on it made in 1988—to appreciate the Moore's law journey it's been on. Because Qualcomm today does not sell its products to consumers, only to phone manufacturers and service providers, most people don't know about Jacobs and the role he played in launching mobile telephony. It's worth a short reprise.

As he explained to me in an interview in the coffee shop in the lobby of Qualcomm headquarters, Jacobs had and still has one overriding goal in life: "I want everyone on the planet to have their own phone number."

Now eighty-two years old, Jacobs still has that steely stubborn streak, disguised by a grandfatherly smile and warm demeanor, which is common to great innovators whom people initially dismissed as crazy: *It is so great to meet you—now get out of the way while I disrupt your whole business. Oh, and have a nice day!*

We forget today that thinking you could get a phone in the palm of everyone's hand—with their own unique phone number—back in the 1980s was not exactly an everyday dream. And it was certainly not as inevitable as it now feels. Jacobs had been an engineering professor at MIT, where he coauthored a textbook on digital communications. In 1966, he was lured west by the great weather to take up a position at the University of California, San Diego. Soon after he got there he created a telecom consulting start-up with some colleagues, called Linkabit, which opened in 1968, and which he later sold.

In the 1980s, the mobile phone business was just emerging. The first generation, or 1G phones, were analog devices that received and transmitted over FM radio. Each country developed its own standards, and for a place like Europe—the original leader in this technology—that made it hard to roam from country to country. The next generation, 2G phones, were based on the emerging European standard for digital cellular networks, which was called GSM (Global System for Mobile) and used TDMA (Time Division Multiple Access) as its communication protocol. All European common market governments mandated the GSM standard in 1987, enabling users to roam and use their phones, and receive calls, in any western European country. The EU then tried to lobby the rest of the world to use that standard, propelled by European companies such as Ericsson and Nokia.

Around the time all that was happening, in 1985, Jacobs and his colleagues founded a new telecom start-up called Qualcomm. One of their first customers was Hughes Aircraft. "Hughes Aircraft had approached us with a project," recalled Jacobs. "They had submitted a proposal to the FCC for a mobile satellite communication system and they came to Qualcomm and asked if there were any technical improvements we could come up with for their proposal."

On the basis of his previous research, Jacobs thought that a protocol called Code Division Multiple Access, or CDMA, might be the best way to move forward, because it could vastly increase wireless capacity and therefore make mobile telephony available to many more people—and support more subscribers per satellite—than the TDMA protocol being mandated in Europe.

At the time, though, Europe's GSM and its TDMA-based U.S. equivalents were in their initial growth phase, and almost every investor asked Jacobs the same question: "Why do we need another wireless technology when GSM and TDMA seem good enough?"

Both CDMA and TDMA, Jacobs explained, worked by sending multiple conversations over a single radio wave. CDMA, however, could also take advantage of natural pauses in the way people speak to allow more conversations simultaneously. This is known as "spread spectrum," whereby each call is assigned a code that is scrambled over a wide frequency spectrum and then reconstructed at the receiving end, thus allowing multiple users to occupy the same spectrum simultaneously, using very complicated software coding and other techniques. Spread spectrum reduces the interference generated by other conversations from other cell sites. With TDMA, by contrast, each phone call took up its own slot. That limited its ability to scale, because eventually a mobile network operator would run out of slots, if too many people tried to make calls at the same time. Every network can overload, but TDMA would overload sooner with many fewer users. All in all, CDMA promised much more efficient use of the spectrum—later, it would also support the transmission of broadband data over wireless networks. In short, TDMA was the key to a finite room. CDMA was the key to an almost unlimited room. And Jacobs had an inkling that might be very important one day.

Jacobs and his colleagues, back in the Linkabit days, had worked on one of the three networks that participated in the first demonstration of the Internet in 1977. So he could already imagine that one day cellular phones might be used to connect to the Internet. When Jacobs and his colleague Klein Gilhousen floated their alternative approach, the phone industry said it was too complicated and too expensive, and might not yield the additional capacity. And furthermore, in the early 1990s, how many people thought you would use your cell phone to access the Internet? People were just happy if their calls weren't dropped. Meanwhile, Hughes scrapped their project with him and let Qualcomm, then an infant start-up, *keep the intellectual property and patents they had developed for mobile telephony.*

Bad move by Hughes—because Jacobs would not give up.

"So we issued the interim standard for CDMA in the summer of 1993, and we could not convince any other handset maker to make a CDMA phone," said Jacobs. "We made the chips, software, phones, and base station infrastructure all by ourselves—because no one else would." In September 1995, however, Jacobs persuaded the Hong Kong phone company Hutchison Telecom to adopt Qualcomm's CDMA protocol and phones, making it the world's first big commercial operator of this technology.

"Before then, everyone was very skeptical that CDMA could work in a commercial setting," he said. "That was October 1995, and in 1996 South Korea came on using our phones made in San Diego. The voice quality was better, there were fewer dropped calls, and it could carry both voice and data at a scale that TDMA could not."

And that set the stage for a decisive struggle between the CDMA and TDMA protocols. While 2G phones did voice and a little text, as the popularity of the Internet grew, operators and manufacturers recognized the need for efficient wireless access to the Internet and therefore proposed a third generation, or 3G, of cellular communications that would enable you to transmit large amounts of data and voice efficiently. There was a global phone war over whose standard would prevail.

The short story is that Jacobs won and the European GSM/TDMA-based standard lost. They lost because their technology had a finite amount of spectrum and CDMA enabled you to do so much more with the same amount of spectrum—and there was soon so much more to carry, thanks to the Internet. We don't remember these wars today, but they were bloody. The U.S.-invented standard prevailed not only because it was better but also because, unlike in Europe, where the governments mandated a standard, in the United States the government allowed the market to choose, and many chose the Jacobs CDMA pathway. Again, you probably missed most of this, but it had huge implications. The vast majority of the world's population when they access the Internet today do it through a phone and not a laptop or desktop. And the reason that happened at the speed and price that it did—making smartphones the fastest-growing technology platform in history—was Jacobs's early recognition that CDMA would efficiently support Internet access as well as voice.

Sure, you could say that in the end everything gets invented and someone would have found their way to CDMA as the foundation of mobile Internet. Perhaps. But it was due to Jacobs's titanic stubbornness in pushing the CDMA standard, when no one else thought it necessary and Europe was pushing the other way, that it happened faster and farther and cheaper. And one by-product was that American phone companies took the lead on 3G and 4G. Meanwhile, once its protocol and software were taken up for mass adoption, Qualcomm got out of the business of making phones and transmission platforms and just focused on the chips and software.

Today, said Jacobs, "people everywhere in the world have both voice and efficient access to the Internet and that supports education, economic growth, health, and good governance." One key "reason we won," he added, "was that even though CDMA was more complicated to implement, people were just thinking about the capacity of chips *at that moment in time*. They were not taking into account Moore's law that would allow the technology to improve every two years and enable the greater efficiency that could be achieved with CDMA." People say that in hockey you don't go where the puck is, you go where the puck is going, and Qualcomm went where the puck was going: to Moore's law, which was on a hockey stick–like curve upward. "Somewhere in the early 2000s when we were trying to expand to India and China," said Jacobs, "I made the outlandish prediction that one day we would see hundred-dollar phones. Now they're below thirty dollars in India."

But the Jacobs family inventions did not stop there. In late 1997, Paul Jacobs, who later succeeded his father as CEO, had a brainstorm. One day he came into a staff meeting in San Diego, took a Qualcomm cell phone and taped it together with a Palm Pilot, and told his team: "This is what we're going to do." The idea was to try to create a device that combined the Palm Pilot—at that time basically a combination calendar, Filofax, address book, and day planner, with note-taking capabilities and a wireless Web-based text browser—with a 3G cell phone. That way when you called up a phone number in the Palm Pilot address book, you could just click on it and the cell phone would dial it. And with the same device you could surf the Internet. Jacobs approached Apple to see if they were interested in partnering with Qualcomm on this, using the Apple Newton, their Palm competitor.

But Apple—this was just before Steve Jobs came back—turned them down and eventually killed the Newton. So Jacobs went to Palm and together they ended up making the first "smartphone"—the Qualcomm pdQ 1900—in 1998. It was the first phone designed not just to relay text messages, but to combine digital wireless mobile broadband connectivity to the Internet with a touchscreen and an open operating system that eventually ran downloadable apps. Qualcomm later created the first mobile telephone–based app store, called Brew, which was marketed by Verizon in 2001.

Paul Jacobs recalls the exact moment when he knew a revolution was about to happen. It was Christmas 1998 and he was sitting on the beach in Maui. "I took out a prototype of the pdQ 1900 they had sent me and I typed in 'Maui sushi' into the AltaVista search engine. I was wirelessly connected using Sprint. Up came a sushi restaurant in Maui. I don't remember the name of the restaurant, but it was good sushi! I knew viscerally right then that what I had theorized—having a phone with the connectivity of a Palm organizer connected to the Internet—would change everything. The day of the disconnected PDA was over. I searched for something I cared about that had nothing to do with technology. Today it seems obvious, but back then it was a novel experience—that you could sit on the beach in Maui and find the best sushi."

Paul Jacobs doesn't mince words: "We made the smartphone revolution." But Jacobs is quick to add that they were ahead of their time—and behind it. The early device they created was rather clunky: it had none of the easy user interfaces and beautiful design that Steve Jobs's Apple iPhone would eventually offer in 2007, and it came out before there was the Internet bandwidth to do many things.

So Qualcomm went back to concentrating on making everything inside the smartphone. Qualcomm gets its improvements by using software and hardware techniques to more densely pack and compress bits, and Jacobs believes it can improve further—maybe another thousandfold—before it reaches its limit. Most people think that they can watch *Game of Thrones* on their cell phone because Apple came out with a better phone. No, Apple gave you a larger screen and better display, but the reason it is not buffering is because Qualcomm and AT&T and others invested billions of dollars in making the wireless network and phones more efficient.

To review this acceleration: 2G was voice and data, with simple texting but not through the Internet; 3G was connecting to the Internet but at a level of speed and clumsiness that recalled the days when you needed a dial-up modem to get online; 4G wireless, the current standard, is as seamless as broadband connectivity over a landline, with particularly seamless access for data-hungry applications such as video. What will 5G be like? The Qualcomm engineers describe it as the stage when the pronouns go away—"you," "me," "I"—and the phone learns who you are and where you like to visit and who you like to connect with, and then can anticipate much of that and just do it all for you.

As Chris Anderson, the technology writer, told *Foreign Policy* magazine on April 29, 2013:

> It's hard to argue that we're not in an exponential period of technological innovation. The personal drone is basically the peace dividend of the smartphone wars, which is to say that the components in a smartphone—the sensors, the GPS, the camera, the ARM core processors, the wireless, the memory, the battery—all that stuff, which is being driven by the incredible economies of scale and innovation machines at Apple, Google, and others, is available for a few dollars. They were essentially "unobtainium" ten years ago. This is stuff that used to be military industrial technology; you can buy it at RadioShack now. I've never seen technology move faster than it's moving right now, and that's because of the supercomputer in your pocket.

And as far as Irwin Jacobs is concerned, you have not seen anything yet. Before I left he told me: "We're still in the era when cars had fins."

## *The Cloud*

The fact that all of the technologies detailed above continue to accelerate at an exponential rate owes much to the fact that they started to meld together into something that came to be called "the cloud."

The term "cloud" evokes the image of a magical energy source somewhere in the sky. The cloud, though, is not a distinct place. The term refers to a collection of computers running just about every possible software program and providing vast storage and processing capacity that any user can connect to via the Internet from their cell phone, tablet, or desktop computer.

Many innovators had a hand in shaping the cloud, but probably none were more important than the small band of technologists—Diane Greene, Mendel Rosenblum, Scott Devine, Ellen Wang, and Edouard Bugnion—who in 1998 created VMware, the unique software program referenced above which went public in 2007.

Without VMware there could be no cloud. Why? It has to do with the way computers were configured and sold, dating back to the 1980s: the hardware, operating system, and applications were packaged together and treated as a single entity. So each computer ran a single operating system, and its associated applications. This meant computers were often underutilized.

"What VMware's founders did was to break this stack of hardware, operating system, and applications apart, and that was a powerful and critical enabler," Diane Greene, VMware's cofounder and its first CEO, explained to me.

VMware did so by creating a "virtualization layer" that interfaced between any operating system—Linux or Microsoft or Apple, for instance—and any computer's hardware. As a result, on a single computer or, more importantly on a collection of computers, multiple users could run multiple different operating systems from multiple different companies and then each of those operating systems could run their own applications. As a result, many more people could share the resources and tap the power of the same physical computer or computer server farm while keeping all their operations separate. This decoupling of a computer's hardware from its software—its specific operating systems and applications—vastly expanded the amount of computing that could be done on any single computer or single collection of them, Greene explained. Users immediately leapt on it.

VMware "created a translator for any language hardware to speak to any language software, so the hardware and software did not have to be born and raised in the same village," explained John Donovan, AT&T's Chief Strategy Officer. "It created a world of any-to-any." And that created scale "because any operating system and software could work with any computer. It allowed for intertribal marriages."

This also meant that when the owner of the hardware upgraded, VMware could magically keep everyone's operating system and applications working with the new hardware.

"This is what made the cloud possible," said Greene. "We multiplexed all of these resources," and in the process "VMware revolutionized how people thought about doing their computing."

This drove the cost of computing way down and the ease of, and access to, computing way up. For the user, it created the sense that there was a giant "cloud" of computers out there (actually, a collection of server farms) that were being constantly upgraded, and you could just tap into this cloud with whatever operating system and applications you were working with and, presto, they would run alongside everyone else's and everything worked seamlessly, virtually and magically.

So let's add it all up: VMware made access to cloud services more seamless than ever from any computing device. Google and Hadoop's innovations around GFS and MapReduce ensured that once you were inside the cloud you could not only store unimaginable amounts of unstructured data, but also search that data and find whatever needle or pattern you were looking for with a level of accuracy and speed unknown in human history. Moore's law ensured that processing power and data storage kept growing exponentially, so the cloud's power and size grew exponentially. Steve Jobs's smartphone enabled more people than ever to tap into the cloud with a handheld computer, connected to the Internet, which doubled as a cell phone and a camera. And the network creators like Irwin Jacobs and the fiber-optic scientists ensured that those Internet-enabled cell phones would be connected to the cloud with digital pipes that got bigger and faster every year, so you could access the cloud, while mobile or stationary, in milliseconds.

The net result was one of the most remarkable amplifiers of human and machine power in history. With access to the cloud, potentially everyone on the planet gained access to their own virtual brain, file cabinet, and tool box, where they could find the answers to any questions, store their favorite apps, photos, health records, books, speech drafts, stock purchases, and mobile games, as well as design anything they could imagine at a cost that was unimaginably low. And APIs permitted each component to readily mesh with the others in the cloud—whether that cloud was operated by Google, Amazon, Microsoft, or Alibaba. Yes, indeed, this thing we call "the cloud" is a real force multiplier.

And you have not seen anything yet. Because the cloud is already starting to "move to the edge," as the Internet of Things and 5G telecommunications technology becomes diffused. If you are riding in a self-driving car with hundreds of sensors and cameras looking around 360 degrees and deciding every millisecond where to turn—and whether to avoid the pedestrian or hit the garbage can—that car does not have time to make all those decisions by going back and forth to the cloud. It will just reach out to the nearest 5G antenna/processor and do the computing and storage locally. We would chew up all the bandwidth in the world if we tried to send everything back and forth to the central cloud. Intel's Brian Krzanich has pointed out that each averagely driven autonomous car—with all of its sensors, radar, cameras, and computing systems—will generate as much data as about three thousand people with their individual computers, phones, and tablets. So just a million autonomous cars will generate three billion people's worth of data. It all cannot go back and forth to the big cloud. You will need mini-clouds that will work by creating "mesh networks" that share the processing and storage power on all the phones and Internet-enabled devices in their area.

That is why Microsoft today talks about building "an intelligent cloud and an intelligent edge." For instance, if you have a smart insulin pump on your body, you want it to operate on a short frequency, making decisions very fast. But you will also want some overall supervising control. The short, quick decisions will be done on the edge, depending on what's happening in your body, and the overall supervising will be done in the cloud. If the device is running okay, the cloud will leave it alone; and if it starts operating outside the model, the cloud will intervene.

That is what is meant by the cloud moving to the edge. But that is for the next book! For now, it is understandably difficult for many people to conceive how all this power can be accessed from this thing we call the cloud somewhere out there in the ether. That's why a 2012 national survey by Wakefield Research, commissioned by Citrix, found that "most respondents believe the cloud is related to weather . . . For example, 51 percent of respondents, including a majority of Millennials, believe stormy weather can interfere with cloud computing," *Business Insider* reported on August 30, 2012. Only 16 percent understood that it was a "network to store, access and share data from Internet-connected devices."

Precisely because I know exactly what the cloud is, I don't like to use the term anymore. Not because it is confusing but because it connotes something so soft, so light, so fluffy, so passive, so benign. It reminds me of a Joni Mitchell song: "I've looked at clouds from both sides now / from up and down, and still somehow / it's cloud illusions I recall / I really don't know clouds at all."

That imagery in no way captures the transformational nature of what has been created. When you combine AI, robots, big data, sensors, synthetic biology, and nanotechnology and seamlessly integrate them into and power them off the cloud, it starts to feed on itself—pushing out boundaries in multiple fields at once. And when you combine the power of the cloud with the power of wireless or fixed-line broadband connectivity, the resulting mix of mobility, connectivity, and steadily increasing computational power is without precedent. It creates a tremendous release of energy into the hands of human beings to compete, design, think, imagine, connect, and collaborate with anyone anywhere.

If you look back over human history, only a few energy sources fundamentally changed everything for most everyone—fire, electricity, and computing. And now, given where computing has arrived with the cloud, it is not an exaggeration to suggest that it's becoming more profound than fire and electricity. Fire and electricity were hugely important sources of mass energy. They could warm your home, power your tools, or transport you from place to place. But in and of themselves they couldn't help you think or think for you. They could not connect you to all the world's knowledge or all the world's people. We have simply never had a tool like this that could be accessed by people all over the world at the same time via a smartphone.

Twenty years ago, you needed to be a government to access this kind of computing power in the cloud. Then you needed to be a business. Now you need only a Visa card, and it's yours for the renting. Today, there are already more connected mobile devices on the planet than there are people, though that is partly due to the fact that many people in the developed world own two. About half the world's population still has no cell phone, smartphone, or tablet. But that number is shrinking every day. Once everyone is connected, and we will see such a day within a decade, I am sure, the collective brainpower that will be generated will be staggering.

*This ain't no cloud, folks!*

And so, instead of calling this new creative energy source "the cloud," this book will henceforth use the term that Craig Mundie, the computer designer from Microsoft, once suggested. I will call it "the supernova"—a computational supernova.

The National Aeronautics and Space Administration (NASA) defines a supernova as "the explosion of a star . . . the largest explosion that takes place in space." The only difference is that while a star's supernova is a one-time incredible release of energy, this technological supernova just keeps releasing energy at an exponentially accelerating rate—because all the critical components are being driven down in cost and up in performance at a Moore's law exponential rate. "And this release of energy is enabling the reshaping of virtually every man-made system that modern society is built on—and these capabilities are being extended to virtually every person on the planet," said Mundie. "Everything is getting changed, and everyone is being impacted by it in positive and negative ways."

No, no, no: *This ain't no soft, fluffy cloud.*

FOUR

# *The Supernova*

*I sense a disturbance in the Force.*

*—Luke Skywalker to Kyle Katarn in the video game* Star Wars: Jedi Knight

*You always sense a disturbance in the Force. But yeah—I sense it, too.*

*—Katarn to Skywalker*

Yeah—and I sense it, too.

On February 14, 2011, a turning point of sorts in the history of humanity was reached on—of all places—one of America's longest-running television game shows, *Jeopardy!* That afternoon one of the contestants, who went by just his last name, Watson, competed against two all-time great *Jeopardy!* champions, Ken Jennings and Brad Rutter. Mr. Watson did not try to respond to the first clue, but with the second clue he buzzed in first to answer.

The clue was: "Iron fitting on the hoof of a horse or a card-dealing box in a casino."

Watson, in perfect *Jeopardy!* style, responded with the question "What is 'shoe'?"

That response should go down in history with the first words ever uttered on a telephone, on March 10, 1876, when Alexander Graham Bell, the inventor, called his assistant—whose name, ironically, was Thomas

Watson—and said, "Mr. Watson—come here—I want to see you." In my mind, "What is 'shoe'?" is also up there with the first words uttered by Neil Armstrong when he set his foot down on the moon, on July 20, 1969: "That's one small step for man, one giant leap for mankind."

"What is 'shoe'?" was one small step for Watson and one giant leap for computers and mankind together. Because Watson, of course, was not a human but a computer, designed and built by IBM. By defeating the best human *Jeopardy!* champions in a three-day competition, Watson demonstrated the solution to the problem that "artificial intelligence researchers have struggled with for decades": to create "a computer akin to the one on *Star Trek* that can understand questions posed in natural language and answer them" in natural language, as my colleague John Markoff put it in his February 16, 2011, *New York Times* story summing up the competition.

Watson, by the way, won handily, showing great facility with some pretty complex clues that might easily stump a human, such as this one: "You just need a nap. You don't have this sleep disorder that can make sufferers nod off while standing up."

Watson buzzed in first—in less than 2.5 seconds—and replied, "What is 'narcolepsy'?"

Reflecting on Watson's performance and its advances since that day, John E. Kelly III, IBM's senior vice president for cognitive solutions and IBM Research, who oversaw that Watson project, put it to me this way: "For many years there were things I could imagine but I never thought were possible in my lifetime. Then I started to think, well, maybe I will see them after I retire. Now I realize I am going to see them before I retire."

Craig Mundie put it even more succinctly, in words that called to mind Astro Teller's graph: "We've jumped to a different curve."

What Kelly and Mundie are talking about is how this thing we call the cloud, and that I call the supernova, is creating a release of energy that is amplifying all different forms of power—the power of machines, of individual people, of flows of ideas, and of humanity as a whole—to unprecedented levels.

For instance, *the power of machines*—whether they are computers, robots, cars, handheld phones, tablets, or watches—has crossed a new line. Many are being endowed with all five of the senses that humans have, and a brain to process them. In many cases, machines can now think on their own. But they also have sight—they can recognize and compare images. They have hearing—they can recognize speech. They have voices—they can be tour guides and interpreters and translate from one language to another. They can move and touch things on their own and respond to that touch; they can act as your chauffeur or lift your packages or even manifest the dexterity, via a 3-D printer, to print a whole human organ. Some are even being taught to recognize smells and tastes. And we humans can now summon all of these powers with a single touch, gesture, or spoken word.

At the same time, the supernova is vastly expanding and accelerating *the power of flows.* The flows of knowledge, new ideas, medical advice, innovation, insults, rumors, collaboration, matchmaking, lending, banking, trading, friendship-forging, commerce, and learning now circulate globally at a speed and breadth we have never seen before. These digital flows carry the energy, services, and tools of the supernova all across the world, where anyone can plug into them to power a new business, participate in the global debate, acquire a new skill, or export their latest product or hobby.

All of that, in turn, is vastly amplifying *the power of one.* What one person—one single, solitary person—can now do constructively and destructively is also being multiplied to a new level. It used to take a person to kill a person; now it is possible to imagine a world where one day one person could kill everyone. We certainly learned on 9/11 how nineteen angry men, super-empowered by technology, could change the whole direction of American history, maybe world history. And that was fifteen years ago! But the flip side is also true—one person can now help so many more people—one person can educate millions with an Internet learning platform; one person can entertain or inspire millions; one person can now communicate a new idea, a new vaccine, or a new application to the whole world at once.

And, finally, this same supernova is amplifying *the power of many.* That, too, has crossed a new line. Human beings as a collective are now not just a part of nature; they have become a force of nature—a force that is disturbing and changing the climate and our planet's ecosystems at a pace and scope never seen before in human history. But again, the flip side is also true. Amplified by this supernova, the many—all of us acting together—now have the power to do good at a speed and scope we've never seen before: to reverse environmental degradation or to feed, house, and clothe every person on the planet, if we ever set our collective minds to doing so. We've never had such collective power as a species.

In sum, human beings have steadily built themselves better tools, but they have never ever built a tool like this supernova. "In the past," said Craig Mundie, "some tools had reach but were not rich in capability; others were rich in capability but had a limited number of people who could use them—i.e., they had no reach." With the emerging supernova, "we have never had this much richness with this much reach."

And people can feel it even if they can't fully understand it. It's why in researching this book, the phrase that I heard most often from engineers was "just in the last few years . . ." So many people explained to me something that they had done or that was being done to them that they never could have imagined—"just in the last few years."

This chapter will explain exactly how the supernova made that happen and, in particular, how it fueled—and is fueling—some stunning advances in what individuals and individual companies can do with technology. The next two chapters after this will discuss how this same supernova is amplifying and accelerating global flows in the Market and human impacts on Mother Nature. Together, all three chapters will show how these accelerations of technology, globalization, and the environment constitute the Machine that is reshaping everything—not just game shows.

## *Complexity Is Free*

I have found that the best way to understand how and why the supernova is amplifying the power of machines, individuals, humanity, and flows is by getting as close to the leading edge of it as possible, as if you were approaching a volcano. For me that involves getting inside big, dynamic multinationals. Unlike governments, these companies cannot go into gridlock, or just shut down out of pique like the Congress, or miss a single technology cycle. If they do, they die—and they die fast. As a result, they stay very close to the edge of the supernova. They draw energy from it and they also drive it forward. They feel its heat first and they wake up every morning and read the financial obituaries to make sure that they aren't being melted down by it. So you can learn an enormous amount about what is coming in terms of new technologies and services, and about what is already here and how it is changing things, by interviewing the engineers, researchers, and leaders of these companies.

Indeed, when I visit their labs I feel like James Bond going to visit "Q" in his British Secret Service research lab at the start of every Bond movie, where 007 gets outfitted with the latest poison pen or flying Aston Martin. You always see things you had no idea were possible.

I had that experience in 2014, when I decided to write a column about General Electric's research center in Niskayuna, New York. GE's lab is like a mini United Nations. Every engineering team looks like one of those multiethnic Benetton ads. But this was not affirmative action at work; it was a brutal meritocracy. When you are competing in the global technology Olympics every day, you have to recruit the best talent from anywhere you can find it. On that trip I was given a tour of GE's three-dimensional manufacturing unit by its then director, Luana Iorio. In the old days, explained Iorio, when GE wanted to build a jet engine part, a designer would have to design the product, then GE would have to build the machine tools to make a prototype of that part, which could take up to a year, and then it would manufacture the part and test it, with each test iteration taking a few months. The whole process, said Iorio, often took "two years from when you first had the idea for some of our complex components."

Now, Iorio told me, engineers using three-dimensional, computer-aided software could design the part on a computer screen, then transmit it to a 3-D printer filled with a fine metal powder and a laser device that literally built, or "printed," the piece out of the metal powder before your eyes, to the exact specifications. Then you immediately tested it—four, five, six times in a day, adjusting each iteration with the computer and the 3-D printer—and when it was just perfect, presto, you had your new part. To be sure, more complex parts required more time, but this was the new system, and it was a fundamental departure from the way GE had built parts since it was founded by Thomas Edison back in 1892.

"The feedback loop is so short now," explained Iorio, that "in a couple days you can have a concept, the design of the part, you get it made, you get it back and test whether it is valid" and "within a week you have it produced . . . It is getting us both better performance and speed." In the past, performance worked against speed: the more tests you did to get that optimal performance, the longer it took. What only a few years earlier had taken two years was being reduced to *a week.* That is the amplified power of machines.

Then, summing up all that was new, Iorio told me that today, "complexity is free."

I said to her: "What did you say?"

"Complexity is free," she repeated.

I thought that was a real insight. I never forgot it. But only in writing this book did I fully understand the importance of what she'd said. As we've noted, over the last fifty years microprocessors, sensors, storage, software, networking, and now mobile devices have been steadily evolving at this accelerating rate. At different stages they coalesce and create what we think of as a platform. With each new platform, the computing power, bandwidth, and software capabilities all meld together and change the method, cost, or power and speed at which we do things, or pioneer totally new things we can do that we never imagined—and sometimes all of the above. And these leaps are now coming faster and faster, at shorter and shorter intervals.

Before 2007, the previous leap forward in our technology platform happened around the year 2000. It was driven by a qualitative change in connectivity. What happened was that the dot-com boom, bubble, and then bust in that time period unleashed a massive overinvestment in fiber-optic cable to carry broadband Internet. But bubbles are not all bad. The combination of that bubble and then its bursting—with the dot-com bust in the year 2000—dramatically brought down the price of voice and data connectivity and led, quite unexpectedly, to the wiring of the world to a greater degree than ever before. The price of bandwidth connectivity declined so much that suddenly a U.S. company could treat a company in Bangalore, India, as its back office, almost as if it were located in its back office. To put it another way, all of these breakthroughs around 2000 made connectivity *fast, free, easy for you, and ubiquitous.* Suddenly we could all touch people whom we could never touch before. And suddenly we could be touched by people who could never touch us before. I described that new sensation with these words: "The world is flat." More people than ever could now compete, connect, and collaborate on more things for less money with greater ease and equality than ever before. The world as we knew it got reshaped.

I think what happened in 2007—with the emergence of the

supernova—was yet another huge leap upward onto a new platform. Only this move was biased toward easing complexity. When all the advances in hardware and software melded into the supernova, it vastly expanded the speed and scope at which data could be digitized and stored, the speed at which it could be analyzed and turned into knowledge, and how far and fast it could be distributed from the supernova to anyone, anywhere with a computer or mobile device. The result was that suddenly complexity became *fast, free, easy for you, and invisible.*

Suddenly, all the complexity that went into getting a taxi, renting someone's spare bedroom in Australia, designing an engine part, or buying lawn furniture online and having it delivered the same day was abstracted into one touch via applications such as Uber, Airbnb, and Amazon or by innovations in the labs of General Electric. No technology innovation more epitomizes this leap forward than Amazon's invention of "one-click" checkout from any e-commerce site. As Rejoiner.com, which tracks e-commerce, noted, thanks to its one-click innovation, "Amazon achieves extremely high conversion from its existing customers. Since the customer's payment and shipping information is already stored on Amazon's servers, it creates a checkout process that is virtually frictionless."

The two graphs on the following page help demonstrate how complexity became free. The first shows how the maximum speed of data transmission dramatically rose—expanding the capabilities of what you could do with a mobile device and thus attracting more users—just as the cost to users of consuming each megabyte of all that data dramatically fell, so many more people could access the power of the supernova more often. Those lines crossed around 2007–2008. The second graph shows how the supernova/cloud emerged right after . . . 2007.

If you read Apple's original announcement of the iPhone in 2007, it was all about how Apple had abstracted away the complexity of so many complex applications, interactions, and operations—from e-mailing, to map searching, to photographing, to phoning, to web surfing—and about how the company had used software to neatly condense so much into one touch on the "iPhone's remarkable and easy-to-use touch interface." Or, as Steve Jobs put it at the time: "We are all born with the ultimate pointing device—our fingers—and iPhone uses them to create the most revolutionary user interface since the mouse."

# Consumer Cost of Data per Megabyte and Data Speed

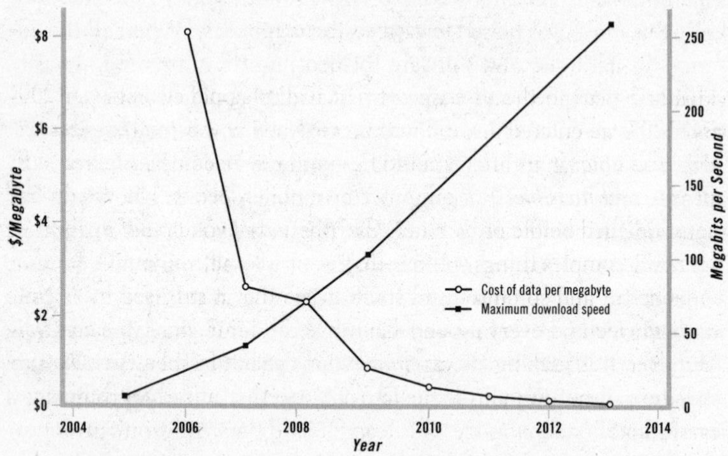


Note: Data speed indicates the maximum downlink speed, not average observed speeds. The average observed speeds depend on many factors, including infrastructure, subscriber density, and device hardware and software.

Courtesy of the Boston Consulting Group (BCG), from its report "The Mobile Revolution: How Mobile Technologies Drive a Trillion-Dollar Impact" (2015). Sources: Cisco Visual Networking Index; International Telecommunication Union; IE Market Research; Motorola; Deutsche Bank; Qualcomm

# Total Size of the Public Cloud-Computing Market, 2008–2020

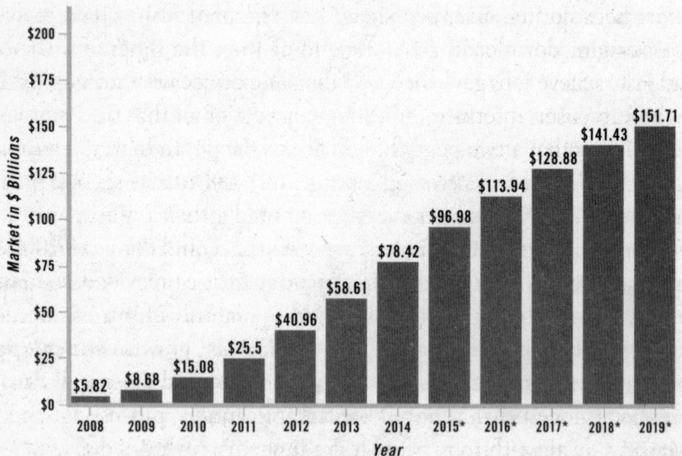


*Indicates projection

Courtesy of Statista

## *The Phase Change*

This brings us to the essence of what really happened between 2000 and 2007: we entered a world where connectivity was *fast, free, easy for you, and ubiquitous* and handling complexity became *fast, free, easy for you, and invisible.* Not only could you touch people whom you had never touched before or be touched by them, but you could do all these amazing, complex things with one touch. As a result, computing became so powerful and so cheap and so effortless that it suffused itself "into every device and every aspect of our lives and our society," said Craig Mundie. "It is making the world not just flat but fast. Fast is a natural evolution of putting all this technology together and then diffusing it everywhere."

It is taking the friction out of more and more businesses and industrial processes and human interactions. "It is like grease," added Mundie, "and is seeping into every nook and cranny and pore and everything is getting very slippery and leveraged, and so you can move it with less force"—whether it is a boulder, a country, a pile of data, a robot, the paging of a taxi, or the renting of a room in Timbuktu. It was quite a historical intersection, when you think about it. The price of sensing, generating, storing, and processing data collapsed just as the speed of uploading or downloading that data to or from the supernova soared, and just as Steve Jobs gave the world a mobile device with such an amazingly easy user interface, Internet connectivity, and rich software applications that a two-year-old could navigate it. To look at it another way, science fiction writers have long imagined machines that could think faster and store and retrieve more knowledge than any humans—not to mention robots that could talk and cars that could drive themselves. In recent years, we've come to understand the basic programming math that would make all of these possible. But it was only in this first decade of the twenty-first century when the exponential growth in microprocessing chips, storage chips, networking, software, and sensors all hit tipping points together that we got the storage capacity, processing speeds, and software algorithms to begin to make them a reality of daily life and commerce. When all those technologies merged—when connectivity became fast, free, easy for you, and ubiquitous and when handling complexity became fast, free, easy for you, and invisible—there was an energy release into the hands of humans and machines the likes of which we have never seen and are only beginning to understand. That is the inflection point that happened around 2007.

"Mobility gives you mass market, broadband gives you access to the information digitally, and the cloud stores all the software applications so you can use them anytime anywhere and the cost is zero—it changed everything," said Hans Vestberg, former CEO of the Ericsson Group.

It is the equivalent of a "phase change" in chemistry from a solid to a liquid. What is the feature of something solid? It is full of friction. What is the feature of a liquid? It feels friction-free. When you simultaneously take the friction and complexity out of more and more things and provide interactive one-touch solutions, all kinds of human-to-human and business-to-consumer and business-to-business interactions move from solids to liquids, from slow to fast, from their complexity being a burden and full of friction to their complexity becoming invisible and frictionless. And so whatever you want to move, compute, analyze, or communicate can be done with less effort.

Quite often, the reason a problem is complex and therefore expensive to solve is that the information you need is not accessible or consumable, making it difficult to gather the relevant data and turn it into applicable knowledge. But when sensing, gathering, and storing data and beaming it to the supernova and processing and analyzing it through software applications becomes virtually free, we are in a new place: as noted earlier, now any pile of data can be *analyzed* to find needles in the haystack or previously unseen patterns. Now any system can be *optimized* for peak performance—with much less effort. Now we can *prophesize.* We can understand a machine's operating life at such a granular level that we can predict when each of its parts will wear out, and replace them before they fail and cause costly delays. Now any piece of clothing, medicine, service, or computer program can be *customized* just for you. And now so many machines in our daily lives—from cars to machine tools to musical instruments—can be *automized* and *roboticized* to run themselves without human direction.

As a result, the motto in Silicon Valley today is: everything that is analog is now being digitized, everything that is being digitized is now being stored, everything that is being stored is now being analyzed by software on these more powerful computing systems, and all the learning is being immediately applied to make old things work better, to make new things possible, and to do old things in fundamentally new ways.

Think of three examples from the transportation and energy industries where all this is happening: The invention of the Uber taxi service did not just create a new competitive taxi fleet; it created a fundamentally new and better way to summon a taxi, to gather data on riders' needs and desires, to pay for a taxi, and to rate the behavior of the driver and the passenger.

In researching this book, I visited the control room at Devon Energy, the oil and gas producer in Oklahoma City that took fracking to scale. That control room is half a floor of computer screens displaying data coming out of every well Devon is drilling around the world. At the bottom of each screen are two boxes that blew my mind. One box displays how much money was budgeted to drill that particular well per foot, and the other box displays—in real time—how much the drilling of that individual well is actually costing, as it bores through different rocks. The numbers are updated every foot—depending on the nature of the rock that the sensor at the tip of the drill encounters. If the rock is softer than expected, the actual cost may come in below the predicted cost, and if it is harder rock the actual cost may be higher.

Think of the historic problem with wind-generated electricity. Because the wind blows intermittently and the electricity it generates cannot be stored at scale, and thus a utility could never be totally assured of sufficient supply, the ability of wind to replace coal-fired power has always been limited. But now, weather-prediction software using big data analytics has become so intelligent it can tell you the exact hour when the wind will blow or the rain will come or the temperature will rise. And so a utility in a city such as Houston can know twenty-four hours in advance that the next day is going to be a particularly hot day and demand for air-conditioning will spike in those exact hours, meaning that demand for wind-generated electricity could exceed supply. That utility can now notify buildings in Houston to automatically turn up their air-conditioning between 6:00 a.m. and 9:00 a.m., before employees arrive, and when the wind is generating the most electricity. Buildings are good storehouses of cooling. So that stored cooling keeps the building comfortable most of the day. As a result, the amount of wind power that utility generates, rather than being insufficient, perfectly matches the demand—without having to worry about storing it on batteries or needing to call in coal-generated power. An incredibly complex demand-response challenge was solved at a cost of . . . zero—just by bringing intelligence to all the machines and optimizing the whole system. All the complexity was abstracted away by the software, and it is starting to happen everywhere today.

## *Show Me the Money*

But if these transformations are real, why is it taking so long for them to show up in the productivity figures, as economists define them—the ratio of the output of goods and services to the labor hours devoted to the production of that output? Since productivity improvements drive growth, that is an important and now a hotly debated subject among economic writers. The economist Robert Gordon has made a compelling case in his book *The Rise and Fall of American Growth: The U.S. Standard of Living Since the Civil War* that the days of steadily rising growth are probably behind us. He believes all the big gains were made in the "special century" between 1870 and 1970—with the likes of automobiles, radio, television, indoor plumbing, electrification, vaccines, clean water, air travel, central heating, women's empowerment, and air-conditioning and antibiotics. Gordon is skeptical that today's new technologies will ever produce another leap forward in productivity comparable to that special century.

But MIT's Erik Brynjolfsson has countered Gordon's pessimism with an argument I find even more compelling. As we transition from an industrial-age economy to a computer-Internet-mobile-broadband-driven economy—that is, a supernova-driven economy—we are experiencing the growing pains of adjusting. Both managers and workers are having to absorb these new technologies—not just how they work but how factories and business processes and government regulations all need to be redesigned around them. The same thing, notes Brynjolfsson, happened 120 years ago, in the Second Industrial Revolution, when electrification—the supernova of its day—was introduced. Old factories did not just have to be electrified to achieve the productivity boosts; they had to be redesigned, along with all business processes. It took thirty years for one generation of managers and workers to retire and for a new generation to emerge to get the full productivity benefits of that new power source.

A December 2015 study by the McKinsey Global Institute on American industry found a "considerable gap between the most digitized sectors and the rest of the economy over time and [found] that despite a massive rush of adoption, most sectors have barely closed that gap over the past decade . . . Because the less digitized sectors are some of the largest in terms of GDP contribution and employment, we [found] that the US economy as a whole is only reaching 18 percent of its digital potential . . . The United States will need to adapt its institutions and training pathways to help workers acquire relevant skills and navigate this period of transition and churn."

The supernova is a new power source, and it will take some time for society to reconfigure itself to absorb its full potential. As that happens, I believe that Brynjolfsson will be proved right and we will start to see the benefits—a broad range of new discoveries around health, learning, urban planning, transportation, innovation, and commerce—that will drive growth. That debate is for economists, though, and beyond the scope of this book, but I will be eager to see how it plays out.

What is absolutely clear right now is that while the supernova may not have made our economies *measurably more productive* yet, it is clearly making all forms of technology, and therefore individuals, companies, ideas, machines, and groups, *more powerful*—more able to shape the world around them in unprecedented ways with less effort than ever before.

If you want to be a maker, a starter-upper, an inventor, or an innovator, this is your time. By leveraging the supernova you can do so much more now with so little. As Tom Goodwin, senior vice president of strategy and innovation at Havas Media, observed in a March 3, 2015, essay on TechCrunch.com: "Uber, the world's largest taxi company, owns no vehicles. Facebook, the world's most popular media owner, creates no content. Alibaba, the most valuable retailer, has no inventory. And Airbnb, the world's largest accommodation provider, owns no real estate. Something interesting is happening."

Something sure is, and the rest of this chapter is about how makers big and small are taking advantage of all the new powers coming out of the supernova to do totally new things, and to do really old things faster and smarter. And it doesn't matter if you are a cancer doctor, a traditional retailer, a cutting-edge designer, a remote innovator in the mountains of eastern Turkey, or someone who wants to turn the tree house in your backyard into a profit center and rent it online to tourists coming from as near as New York or as far as New Guinea. In the age of the supernova, there has never been a better time to be a maker—anywhere.

## *Dr. Watson Will See You Now*

I got to meet—and get my picture taken with—the original Watson on a visit to IBM's Thomas J. Watson Research Center in Yorktown Heights, New York. He didn't say much. He's retired now. He's actually unplugged—but he fills a good-sized room with his racks of servers.

I also got to meet Watson's grandson—sort of. He's the size of a big suitcase. He's actually a mock-up, though—what today's version of Watson would look like after two generations more of Moore's law. Technically speaking, though, today's version of Watson is not even that big suitcase, because Watson now resides in the supernova.

"Watson is no longer contained in a box that is unconnected to the Internet, but rather is now of the Internet," explained David Yaun, an IBM communications vice president. IBM put together the mini-Watson mockup "to illustrate that we could cram all the computing power of the *Jeopardy!* Watson into a suitcase today. But Watson itself now is literally part of your supernova—unleashed from a twentieth-century paradigm of a box or a standalone server."

And anyway—Watson's grandson would never waste his time trying to beat humans on *Jeopardy!* That is so 2011! Today's Watson is now busy ingesting all known medical research on subjects such as cancer diagnostics and treatments. Indeed, Yaun confided to me when we sat down for lunch at Watson's home base that "we're thinking of having Watson take the radiology boards"—to get certified to read and interpret X-rays. *Ho-hum, I was thinking of doing the same thing myself.* Right! Watson could practically do that in his spare time, while taking every bar exam in America, the dental boards, the pathology boards, the urology boards—and beating the pants off you on *Jeopardy!*

The supernova offers computing power for everyone everywhere. Watson offers deep knowledge everywhere for everyone. Watson is not just a big search engine or digital assistant. He does not operate looking for keywords, per se. And he is not just a big computer that is programmed by software engineers to perform certain tasks that they design. Watson is different. You have not seen the likes of him before, except on *Star Trek*. Watson represents nothing less than the dawn of "the Cognitive Era of computing," said John E. Kelly III, who divides the history of computing into three distinct eras.

The first era, he says, is the "Tabulating Era," which lasted from the early 1900s to the 1940s and was built on single-purpose, mechanical systems that counted things and used punch cards to calculate, sort, collate, and interpret data. That was followed by the "Programming Era"—the 1950s to the present. "As populations grew, and economic and societal systems got more complex, [the] manual, mechanical-based systems just couldn't keep up. We turned to software programmed by humans that applied if/then logic and iteration to calculate answers to prescribed scenarios. This technology rode the wave of Moore's law and gave us personal computers, the Internet, and smartphones. [The] problem is, as powerful and transformational as these breakthroughs have been—and for a very long time—programmable technology is inherently limited by our ability to design it."

And so, from 2007 onward, we have seen the birth of the "Cognitive Era" of computing. It could happen only after Moore's law entered the second half of the chessboard and gave us sufficient power to digitize almost everything imaginable—words, photos, data, spreadsheets, voice, video, and music—as well as the capacity to load it all into computers and the supernova, the networking ability to move it all around at high speed, and the software capacity to write multiple algorithms that could teach a computer to make sense of unstructured data, just as a human brain might, and thereby enhance every aspect of human decision making.

When IBM designed Watson to play *Jeopardy!*, Kelly explained to me, it knew from studying the show and the human contestants exactly how long it could take to digest the question and buzz in to answer it. Watson would have about a second to understand the question, half a second to decide the answer, and a second to buzz in to answer first. It meant that "every ten milliseconds was gold," said Kelly. But what made Watson so fast, and eventually so accurate, was not that it was actually "learning" per se, but its ability to self-improve by using all its big data capacities and networking to make faster and faster statistical correlations over more and more raw material.

"Watson's achievement is a sign of how much progress has been made in machine learning, the process by which computer algorithms self-improve at tasks involving analysis and prediction," noted John Lanchester in the *London Review of Books* on March 5, 2015. "The techniques involved are primarily statistical: through trial and error the machine learns which answer has the highest probability of being correct. That sounds rough and ready, but because, as per Moore's law, computers have become so astonishingly powerful, the loops of trial and error can take place at great speed, and the machine can quickly improve out of all recognition."

That is the difference between a cognitive computer and a programmable computer. Programmable computers, Kelly explained in a 2015 essay for IBM Research entitled "Computing, Cognition and the Future of Knowing," "are based on rules that shepherd data through a series of predetermined processes to arrive at outcomes. While they are powerful and complex, they are deterministic, thriving on structured data, but incapable of processing qualitative or unpredictable input. This rigidity limits their usefulness in addressing many aspects of a complex, emergent world in which ambiguity and uncertainty abound."

Cognitive systems, on the other hand, he explained, are "probabilistic, meaning they are designed to adapt and make sense of the complexity and unpredictability of unstructured information. They can 'read' text, 'see' images and 'hear' natural speech. And they interpret that information, organize it and offer explanations of what it means, along with the rationale for their conclusions. They do not offer definitive answers. In fact, they do not 'know' the answer. Rather they are designed to weigh information and ideas from multiple sources, to reason, and then offer hypotheses for consideration." These systems then assign a confidence level to each potential insight or answer. They even learn from their own mistakes.

So in building the Watson that won on *Jeopardy!*, Kelly noted, they first created a whole set of algorithms that enabled the computer to parse the question—much the way your reading teacher taught you to diagram a sentence. "The algorithm breaks down the language and tries to figure out what is being asked: Is it a name, a date, an animal—what am I looking for?" said Kelly. The second set of algorithms is designed to do a sweep of all the literature Watson had been uploaded with—everything from Wikipedia to the Bible—to try to find everything that might be relevant to a given subject area, person, or date. "The computer would look for many pieces of evidence and form a preliminary list of what might be the possible answers, and look for supporting evidence for each possible answer—[such as if] they are asking for a person who works at IBM and I know that Tom works there."

Then, with another algorithm, Watson would rank what it thought were the right answers, assigning degrees of confidence to all of them. If it had a high enough degree of confidence, it would buzz in and answer.

The best way to understand the difference between programmable and cognitive computers is with two examples offered to me by Dario Gil, IBM's vice president of science and solutions. When IBM first started to develop translation software, he explained, it created a team to develop an algorithm that could translate from English to Spanish. "We thought the best way to do that was to hire all kinds of linguists who would teach us grammar, and once we understood the nature of language we would figure how to write a translation program," said Gil. It didn't work. After going through *a lot* of linguists, IBM got rid of them all and tried a different approach.

"This time, we said, 'What if we took a statistical approach and just take two texts translated by humans and compare them and see which one is most accurate?" And since computing and storage power had exploded in 2007, the capacity to do so was suddenly there. It led IBM to a fundamental insight: "Every time we got rid of a linguist, our accuracy went up," said Gil. "So now all we use are statistical algorithms" that can compare massive amounts of texts for repeatable patterns. "We have no problem now translating Urdu into Chinese even if no one on our team knows Urdu or Chinese. Now you train through examples." If you give the computer enough examples of what is right and what is wrong—and in the age of the supernova you can do that to an almost limitless degree—the computer will figure out how to properly weight answers, and learn by doing. And it never has to really learn grammar or Urdu or Chinese—only statistics!

That is how Watson won on *Jeopardy!* "The programmable systems that had revolutionized life over the previous six decades could never have made sense of the messy, unstructured data required to play *Jeopardy!*," wrote Kelly. "Watson's ability to answer subtle, complex, pun-laden questions with precision made clear that a new era of computing was at hand."

That is best illustrated by the one question Watson answered incorrectly at the end of the first day's competition, when the contestants were all given the same clue for "Final Jeopardy!" The category was "U.S. Cities," and the clue was: "Its largest airport was named for a World War II hero; its second largest, for a World War II battle." The answer was Chicago (O'Hare and Midway). But Watson guessed, "What is Toronto?????" With all those question marks included.

"There are many reasons why Watson was confused by this question, including its grammatical structure, the presence of a city in Illinois named Toronto, and the Toronto Blue Jays playing baseball in the American League," said Kelly. "But the mistake illuminated an important truth about how Watson works. The system does not answer our questions because it 'knows.' Rather, it is designed to evaluate and weigh information from multiple sources, and then offer suggestions for consideration. And it assigns a confidence level to each response. In the case of "Final Jeopardy!," Watson's confidence level was quite low: 14 percent, Watson's way of saying: 'Don't trust this answer.' In a sense, it knew what it didn't know."

Because it is so new, a lot of scary stuff has been written about the Cognitive Era of computing—that cognitive computers are going to take over the world from humans. That is not IBM's view. "The popular perception of artificial intelligence and cognitive computing is far from reality—this whole idea of sentient computer systems that become conscious and aware and take their own direction by what they learn," said Arvind Krishna, senior vice president and director of IBM Research. What we can do is teach computers about narrow domains—such as oncology, geology, geography—by writing algorithms that enable them to "learn" about each of these disciplines through multiple and overlapping systems of pattern recognition. "But if a computer is built to understand oncology, that is the only thing it can do—and it can keep learning as new literature comes out in the narrow domain that it was designed for. But the idea that it would then suddenly start designing cars is zero."

By June 2016, Watson was already being used by fifteen of the world's leading cancer institutes, had ingested more than twelve million pages of medical articles, three hundred medical journals, two hundred textbooks, and tens of millions of patient records, and that number is increasing every day. The idea is not to prove that Watson could ever replace doctors, said Kelly, but to prove what an incredible aid it can be to doctors, who have long been challenged to keep current with medical literature and new findings. The supernova simply heightens the challenge: estimates suggest that a primary care physician would need more than 630 hours a month to keep up with the flood of new literature that is being unleashed related to his or her practice.

The bridge to the future is a Watson that can make massive amounts of diagnostic complexity free. In the past, when it was determined that you had cancer, the oncologists decided between three different forms of known treatment based on the dozen latest medical articles they might have read. Today, the IBM team notes, you can get genetic sequencing of your tumor with a lab test in an hour and the doctor, using Watson, can pinpoint those drugs to which that particular tumor is known to best respond—also in an hour. Today, IBM will feed a medical Watson 3,000 images, 200 of which are of melanomas and 2,800 are not, and Watson then uses its algorithm to start to learn that the melanomas have these colors, topographies, and edges. And after looking at tens of thousands and understanding the features they have in common, it can, much quicker than a human, identify particularly cancerous ones. That capability frees up doctors to focus where they are most needed—with the patient.

In other words, the magic of Watson happens when it is combined with the unique capabilities of a human doctor—such as intuition, empathy, and judgment. The synthesis of the two can lead to the creation and application of knowledge that is far superior to anything either could do on their own. The *Jeopardy!* game, said Kelly, pitted two human champions against a machine; the future will be all about Watson and doctors—man and machine—solving problems together. Computer science, he added, will "evolve rapidly, and medicine will evolve with it.

This is coevolution. We'll help each other. I envision situations where myself, the patient, the computer, my nurse, and my graduate fellow are all in the examination room interacting with one another."

In time, all of this will reshape medicine and change how we think about being smart, argues Kelly: "In the twenty-first century, knowing all the answers won't distinguish someone's intelligence—rather, the ability to ask all the right questions will be the mark of true genius."

Indeed, every day we read about how artificial intelligence is being inserted into more and more machines, making them more supple, intuitive, human-like, and accessible with one touch, one gesture, or one voice command. Soon everyone who wants will have a personal intelligent assistant, their own little Watson or Siri or Alexa that learns more about their preferences and interests each time they engage with it so its assistance becomes more targeted and valuable every day. This is not science fiction. This is happening today.

That is why it was no surprise to me that Kelly, at the end of our interview at Watson's home at IBM, mused: "You know how the mirror on your car says 'Objects in your rearview mirror are closer than they appear'?" Well, he said, "that now applies to what's in your front windshield, because now it's the future that is much closer than you think."

## *The Designers*

It is fun to be around really, really creative makers in the second half of the chessboard, to see what they can do, as individuals, with all of the empowering tools that have been enabled by the supernova. I met Tom Wujec in San Francisco at an event at the Exploratorium. We thought we had a lot in common and agreed to follow up on a Skype call. Wujec is a fellow at Autodesk and a global leader in 3-D design, engineering, and entertainment software. While his title sounds like a guy designing hubcaps for an auto parts company, the truth is that Autodesk is another of those really important companies few people know about—it builds the software that architects, auto and game designers, and film studios use to imagine and design buildings, cars, and movies on their computers. It is the Microsoft of design. Autodesk offers roughly 180 software tools used by some twenty million professional designers as well as more than two hundred million amateur designers, and each year those tools reduce more and more complexity to one touch. Wujec is an expert in business visualization—using design thinking to help groups solve wicked problems. When we first talked on the phone, he illustrated our conversation real-time on a shared digital whiteboard. I was awed.

During our conversation, Wujec told me his favorite story of just how much the power of technology has transformed his work as a designer-maker. Back in 1995, he recalled,

> I was a creative director of the Royal Ontario Museum, Canada's largest museum, and my last big project there before joining the private sector was to bring to life a dinosaur called a Maiasaura. The process was complicated. It began by transporting a two-ton slab of rock, double the size of a table, from the field to the museum. Over the course of many months, several paleontologists carefully chiseled out the fossils of two specimens, an adult and an infant. It was thought the dinosaurs were a parent and child: Maiasaura means "mother lizard." As the fossilized bones emerged, it was our job to scan them. We used hand-digitizing tools to precisely measure the three-dimensional coordinates of hundreds and thousands of points on the fossil surfaces. This took forever and strained our modest technology. We realized that we needed high-end tools.
>
> So, we upgraded. We got a grant for two hundred thousand dollars for software and three hundred forty thousand dollars for hardware. After the fossils were fully exposed, we hired an artist to create a three-foot-long scale physical model of the adult, first from clay, then from bronze. This sculpture became an additional reference for our digital model. But creating the digital model wasn't easy. We spent more months painstakingly measuring tiny features and hand-entering the data into our computers. The software was unstable, forcing us to do the work over and over each time the system crashed. Eventually, we ended up with decent digital models. With the help of more experts, we rigged, textured, lit, animated, and rendered [these models] into a series of high-resolution movies. The effort was worth it: museum visitors would be able to press buttons on an

exhibit panel and watch life-sized dinosaurs—the size of big SUVs—move in ways our paleontologists thought they would behave. "Here's how they would walk, here's how they would feed, here's how they might stand on their hind legs." After the exhibit opened, I thought, "Oh, my God, that was a lot of work."

From start to finish, it was a two-year project, costing more than $500,000.

Now fast-forward. In May 2015, roughly twenty years later, Wujec found himself at a cocktail party at the same museum, where he had not worked for many years, and saw that they had put out on display the original bronze cast of the scale model of the Maiasaura dinosaur that he had built. He recalled:

> I was surprised to find the sculpture there. I wondered what the digitizing process might be like using modern tools. So, on a Friday night, with a glass of wine in my hand, I took out my iPhone and walked around the model, took twenty or so photographs over maybe ninety seconds, and uploaded them to a free cloud app our company produces called 123D Catch. The app converts photos of just about anything into a 3-D digital model. Four minutes later, it returned this amazing, accurate, animatable, photorealistic digital 3-D model—better than the one we produced twenty years ago. That night, I saw how a half-million dollars of hardware and software and months and months of hard, very technical, specialized work could be largely replaced by an app at a cocktail party with a glass of wine in one hand and a smartphone in the other. In a few minutes I reproduced the digital model for free—except it was better!

And that is the point, concluded Wujec, with the advances in sensing, digitization, computation, storage, networking and software: all "industries are becoming computable. When an industry becomes computable, it goes through a series of predictable changes: It moves from being digitized to being disrupted to being democratized." With Uber, the very analog process of hailing a cab in a strange city got digitized. Then the whole industry got disrupted. And now the whole industry has been democratized—anyone can be a cab driver for anyone else anywhere, and anyone can now pretty easily start a cab company. With design, the analog process of rendering a dinosaur got digitized, then thanks to the supernova it got disrupted, and now it is being democratized, so anyone with a smartphone can do it, vastly enhancing the power of one. You can conceive an idea, get it funded, bring it to life, and scale it at an ease, speed, and cost that make the whole process accessible to so many more people.

And that is why Wujec likes to say that "the twentieth century was all about getting you to love the things we make. And the twenty-first is all about how to make the things you love."

We are entering a maker's paradise. You know what the next generation of kids' toys will be? Make your own—make the toy you love. The system will also soon enable you to make the drug you need for your particular DNA. Or as Andrew Hessel, distinguished research scientist at Autodesk, put it to me: "The gap between science fiction and science is getting really narrow now, because as soon as someone has that idea and articulates it, it can be manifested in a very short period of time."

Autodesk's business involves abstracting more and more of the complexity involved in different aspects of design into one-touches to amplify the power of one designer. Carl Bass, the CEO of Autodesk, showed me how their latest software for architects has evolved from a digital drawing tool to one in which the software works in partnership with the designer or the architect, through a concept called "building information modeling."

For starters, the design process moves from a set of drawings to an interactive database. When the designer draws on the computer screen, the system can compute the properties of the building and even suggest improvements for everything from energy efficiency to people flow while costing out every conceivable option. Every variable is built into the software, so as the designer changes the shape or floors or the building as a whole, the software immediately tells him or her how much that change will cost, how much energy it will save or add, and what the impact will be on the people using the building.

"The architect is not just working with a set of drawings but with a data model that understands the whole building as a three-dimensional living system—its windows, air-conditioning, sunlight, lighting, elevators, and how they all interact," explained Bass. The different teams working on the building can also interact and collaborate, as each change they make is dynamically integrated and optimized against the others.

When technology enables such a huge leap in the prototyping process, it empowers the designer—who can immediately see all the implications of any idea he or she attempts. At the same time, the process removes so much guessing and therefore so many mistakes, and so much lost time and money. It also invites more experimentation and creativity.

And the next stage is "jaw-dropping," explains Bass. "We call it generative design." The computer becomes a real design partner. "Say I want to design a chair, and I go to any furniture designer and say, 'Please design me a chair.' If I say that to any of us, 'Please design me a chair,' it's going to look like something we understand as a chair." But if instead you use Autodesk's Project Dreamcatcher software and just say, "I need a platform at this height that can support this much weight, as lightweight as possible and using the least amount of materials but still be able to support weight at this height and with a platform this big," the computer will come up with amazing variations on its own. Autodesk has some on display at its San Francisco offices, and they are otherworldly—but you can sit in them quite comfortably!

As with Watson, when the power of machines gets amplified, the nature of the "power of one" shifts—creativity becomes, in part, about asking the best questions. "The world of the designer changes," explains Bass, "from the form maker to the person that creates the goals and the constraints of the object to be designed—[and that person] then no longer creates the designs, but selects the design from a landscape of possibilities. We're going from what was once a point solution to more of a collaboration [between man and machine], because with the computer's help, the designer is now able to understand the whole range [of any system] beyond what any human mind can comprehend on its own."

## *The Trust Makers Who Also Rent Rooms*

As we said, the supernova is enabling radical changes in the cost, the speed, and the manner in which things are done, as well as the things that can be done—and enabling individuals or small groups to emerge from nowhere to do any of those things. Or how about all of them at once? There is no better example of super-empowered makers who have remade an entire long-standing industry in the space of a few years, with no money down, than the founders of Airbnb. It is a total child of the supernova—inconceivable without it and utterly logical and impossible to stop with it.

And it all started so analog—with air mattresses.

One of the cofounders, Brian Chesky, had parents who wanted just one thing for him when he graduated from the Rhode Island School of Design—that he get a job that came with health insurance. Chesky tried that for a while with a design firm in Los Angeles, but he got fed up and packed his stuff into his Honda Civic and drove to San Francisco to crash with his pal Joe Gebbia, who agreed to split the rental of his house with Chesky.

"Unfortunately, my share came to eleven hundred fifty dollars, and I only had a thousand dollars in the bank, so I had a math problem—and I was unemployed," Chesky told me when I first interviewed him for a column. But they did have an idea. The week Chesky got to town, in early October 2007, San Francisco was hosting the Industrial Designers Society of America, and all the hotel rooms on the conference website were sold out. So Chesky and Gebbia thought: Why not turn their house into a bed and breakfast for attendees?

The problem was, they had no beds, but Gebbia did have three air mattresses. "So we inflated them and called ourselves 'Airbed and Breakfast.' Three people stayed with us, and we charged them eighty dollars a night. We also made breakfast for them and became their local guides," Chesky, thirty-four, explained. In the process, they made enough money to cover the rent. More important, though, they discovered a bigger idea that has since blossomed into a multibillion-dollar company, a whole new way for people to make money and tour the world. The idea was to create a global network through which anyone anywhere could rent a spare room in their home to earn cash. In homage to its roots, they called the company Airbnb, which has grown so large that it is now bigger than all the major hotel chains combined—even though, unlike Hilton and Marriott, it doesn't own a single bed. And the new trend it set off is the "sharing economy."

When I first heard Chesky describe his company, I confess to being a little dubious: I mean, how many people in Paris really want to rent out their kid's bedroom down the hall to a perfect stranger—who comes to them via the Internet? And how many strangers want to be down the hall?

Answer: a lot! By 2017, there were sixty-eight thousand commercial hotel rooms in Paris and more than one hundred thousand Airbnb listings.

Today, if you go to the Airbnb website you can choose to stay in one of hundreds of castles, dozens of yurts, caves, tepees with TVs in them, water towers, motor homes, private islands, glass houses, lighthouses, igloos with Wi-Fi, and tree houses—hundreds of tree houses—which are the most profitable listings on the Airbnb site per square foot.

"The tree house in Lincoln, Vermont, is more valuable than the main house," said Chesky. "We have tree houses in Vermont that have had six-month waiting lists. People plan their vacations around tree house availability!" Indeed, the top three all-time popular Airbnb listings are tree houses—two of which made the owners enough money to pay off their actual home mortgages. Prince Hans-Adam II offered his entire principality of Liechtenstein for rent on Airbnb (seventy thousand dollars a night), "complete with customized street signs and temporary currency," *The Guardian* reported on April 15, 2011. You can sleep in the homes that Jim Morrison of the Doors once owned or take your pick of Frank Lloyd Wright houses or even squeeze into a one-square-meter house in Berlin that goes for thirteen dollars a night. The most popular Airbnb of all time is a mushroom-shaped dome cabin in Aptos, California.

In July 2014, when the World Cup soccer tournament was held in Brazil, it was only thanks to Airbnb that all the visitors had a place to stay, because Brazil had not built enough hotel rooms to house all those who wanted to come and watch the games. Said Chesky: "Roughly a hundred twenty thousand people—one in five international visitors—stayed in Brazil in Airbnb-rented rooms for the World Cup; they came from over a hundred fifty different countries. Airbnb hosts in Brazil earned roughly thirty-eight million dollars from reservations during the World Cup. The average host in Rio earned roughly four thousand dollars during the monthlong tournament—about four times the average monthly salary in Rio. And one hundred eighty-nine German guests stayed with Brazilians on the night of the Brazil/Germany World Cup semifinal match."

It turns out there is an innkeeper residing in all of us! But while the insight of Chesky and his partners was profound, their timing was even better. Why? Because it coincided with 2007. Without the technologies born that year, Chesky said, there could have been no Airbnb. For starters, connectivity had to get fast, free, easy for you, and ubiquitous—from Hawaii to Hong Kong to Havana, which happened in the early 2000s, he explained. "Then people had to get comfortable giving their credit information and paying for things online and transacting online. People forget that when eBay started, people used to mail them checks and they would end each day with these huge bags of checks." There had to be both a degree of experience for a wide swath of the global population with e-commerce and a peer-to-peer payment system, like PayPal, so people could pay on Airbnb without credit cards. The globalization of flows made that possible in the early 2000s. Then people needed to be connected online with real identity profiles, which is what Facebook helped make possible when it exploded out of high schools and colleges around 2007—so both people renting their homes and people looking to rent their homes would know who the other person was with a high degree of certainty. Because you weren't just buying a book or selling a used golf club to a stranger on eBay—or even just looking to find a single roommate on Craigslist. You were going to stay in someone's spare bedroom or rent them yours.

You also needed a rating system, said Chesky, where both sides could rate each other and build reputations that became a kind of currency, which eBay and Airbnb helped to pioneer and popularize. You needed the mass diffusion of camera-enabled smartphones, so people could easily, and basically for free, photograph the room or home they offered for rent and upload the photos onto a Web-based profile—without having to hire a photographer (although many do). Steve Jobs solved that problem in 2007. And you needed a messaging system, like WhatsApp, founded in 2009, so people offering accommodations on Airbnb and those renting rooms could communicate for free about where and when to leave the key, and so many other details, and, as Chesky put it, "so they could take 'the stranger' out of the transaction and meet virtually beforehand."

And, finally, you needed to bring them "all together into a really well-designed interface—we were all design students—where you could do all of this with one touch," said Chesky. Once those pieces were in place and scaled a few years after 2007, Airbnb just took off, not only because all that complexity—someone in Minnesota renting a yurt from someone in Mongolia—could be reduced to one touch, but also because it could be done in a way that parties totally trusted.

In fact, the most interesting thing Chesky and his fellow Airbnb makers made was one of the most complex things to make at scale: *trust*.

Airbnb's founders understood that the world was becoming interdependent—meaning the technology was there to connect any renter to any tourist or traveling businessperson anywhere on the planet. And if someone created the trust platform to bring them together, huge value could be created for all parties. That was Airbnb's real innovation—*a platform of trust*—where everyone could not only see everyone else's identity but also rate them as good, bad, or indifferent hosts or guests. This meant everyone using the system would pretty quickly develop a relevant "reputation" visible to everyone else in the system. Take trusted identities and relevant reputations and put them together with the supernova and global flows and suddenly you have more than four million homes or rooms listed on Airbnb—that's more than Hilton, Marriot, and Starwood combined. And Hilton started in 1919!

"We used to only trust institutions and companies because they had a reputation and a brand," concluded Chesky. "And we also only used to trust people in your community. You knew the people in your community and everyone else from the outside was a stranger. What we did was give those strangers identities and brands that you could trust. Do you want a stranger staying in your home? No. But would you like Michelle who went to Harvard, works in a bank, and has a five-star rating as a guest on Airbnb? Sure!"

Chesky would love to apply what Airbnb has learned about the sharing economy to other realms and experiences, or, as he once put it to me: "There are eighty million power drills in America that are used an average of thirteen minutes. Does everyone really need their own drill?"

The distance between imagining something, designing it, manufacturing it, and selling it everywhere has never been shorter, faster, cheaper, and easier—for engineers and non-engineers alike.

Frankly, if it's not happening, it's because you're not doing it.

## *The Retailers*

If the supernova empowers innovators to establish radically disruptive new business models—models that can get to global scale overnight—it also allows established companies to compete with them more effectively than ever, if they are ready to disrupt themselves. If you are interested in that competition, you can't do better than study how Walmart, the ultimate brick-and-mortar company, headquartered in a little town in Arkansas, has been trying to leverage the supernova to amplify its ability to compete with a giant retailer born entirely out of the age of accelerations—Amazon. Now, I pity any retailer that has to compete with Amazon, but Walmart is not just any retailer, so I thought it would be particularly revealing to see how it was rising to this challenge.

In April 2015, Walmart's CEO, Doug McMillon, invited me to address his company's legendary Saturday-morning meeting at their headquarters in Bentonville, Arkansas—a combination variety show, corporate revival meeting, and general good fun, with an audience of some three thousand people. It's quite a production. I told him that I would be happy to do it—I was the warm-up act for Kevin Costner—but that I wanted to be paid, and I wanted to be paid "a lot"—but *The New York Times* does not allow me to accept money from a company. He asked what I wanted. I said I wanted to be paid by having Walmart engineers show me what happens behind the scenes, in the supernova, when I try to make a purchase—we settled on a 32-inch television—on Walmart's mobile app, using my iPhone. That is how they "paid" me, and it was worth the trip.

Walmart.com launched in 2000, using off-the-shelf technology to establish an online e-commerce platform. It wasn't a worthy competitor to Amazon. In 2011, Walmart got serious, and when the world's biggest retailer gets serious it is really serious. It established a major software presence in Silicon Valley, hiring several thousand engineers. It wasn't hard to recruit them, explained Neil Ashe, who was president and CEO of global e-commerce for Wal-Mart Stores, Inc., when I visited. "We told people: if you want hard problems, we got them—and if you are interested in scale we have that, too!" As a company "we have 'conversations' with between two hundred and three hundred million people a week."

What was especially striking to me was how fast and inexpensively Walmart was able to make its mobile app—thanks in large part to what had happened in 2007. Hadoop enabled them to get scale on big data. GitHub enabled them to benefit from all the retailing software invented by others, and APIs enabled them to partner with everyone. And the Moore's law advances in storage, computing, and telecommunications deep into the second half of the chessboard enabled them to be competitive overnight.

Jeremy King, chief technology officer for Walmart eCommerce, had previously been part of the tech team that built eBay's e-commerce platform—before the supernova really existed and everything had to be built from scratch. "When I was at eBay ten years ago [in 2005], we built a very similar platform, and it took two hundred software engineers to do it—there was nothing else out there like it at the time. It took several years." Not anymore. Not after 2007. "In 2011," King said, Walmart "built a similar platform thanks to the cloud, with twelve people, in under twenty-four months." The thousands of software engineers it has hired since are to infuse IT into every aspect of its business.

In the age of GitHub, said Ashe, "when we went to build our own search engine, it just relied on the best open-source option for creating an index of searchable data—called Solr—and then we wrote our own relevance engine on top of that." In the old days, the code for that was kept internal to a company, but now they are all social on GitHub. When all of these toolboxes and components are in the cloud, available through open source, and infinitely mashable thanks to interoperable APIs, "it is all about how you put them together that creates the customer value," said Ashe.

Now, back to my search for that 32-inch television. As soon as I put the number "32" into the Walmart app on my phone, its algorithms and database knew from experience that I was probably looking for a "32-inch television," even if I misspelled both "inch" and "television." Then, within milliseconds, it displayed a variety of 32-inch televisions in stock.

"The customer is looking for a frictionless experience," explained Ashe. "People are so impatient now." He said Walmart knows that every hundred milliseconds people lose patience. "They will give up [purchasing something] over a half-second [delay] . . . It takes seven milliseconds to move data from our data center in Colorado to our center in

Bentonville—and that means a fourteen millisecond round-trip. So we can't use that database in Colorado for certain transactions. We have to rely on the data in Bentonville."

Indeed, Walmart has discovered that the consumer can actually tell a difference in milliseconds—a thousandth of a single second—and when they hit a buy or send or search button they expect a response in ten milliseconds. Walmart research has found that every half second you add to the time it takes for a customer to get a response when purchasing online results in two or more percentage points in lost transactions over the millions it does every day. That's real money.

Eventually I put a Samsung 32-inch television in my virtual shopping basket and clicked "buy." The APIs linking Walmart and Visa seamlessly handled the purchase. But then I heard one of my favorite quotes in researching this book. After I pressed "buy," the system used my zip code to determine whether there was a 32-inch television in a Walmart store near me where I could drive in and pick it up, or my TV could be delivered from a Walmart store in the region, or it had to come out of one of the new mega Walmart fulfillment centers dedicated to online orders—each of which could house two cruise ships. With some products, the Walmart system has already anticipated increased demand and prepositioned stock to serve the customer at the cheapest price everywhere: that means shovels in Michigan in winter, golf balls in Florida year-round, and big-screen televisions and Doritos in the week before Super Bowl Sunday.

"So we promised you a date of delivery, when you pressed 'buy,'" said King. "We did that on the basis of probability calculations." Now, however, the system has to go through a whole other set of optimizations to get the best delivery-pickup solution or some combination of the two. It does this based on your location, other items you may have bought besides the 32-inch television, where they might be coming from, and what size and how many boxes they may require. There are myriad combinations to sort through, given Walmart's four thousand stores and multiple fulfillment centers.

"There are about four hundred thousand variables," said King. But now that you—the customer—have already made the purchase and are not waiting online, he added, "we have time, so we do it in under a second."

I started to laugh. "What did you just say?" I asked him, incredulous. "Once I press 'buy,' you have all kinds of time. *You have under a second?*"

He laughed, too.

In the Walmart supernova today, having under a second to make complexity free is what constitutes having all kinds of time for the system to sort out four hundred thousand delivery variables. When connectivity is ubiquitous and complexity is free, the world gets really fast. But the race never ends. Just when you think you've achieved escape velocity from your competitors, someone gets faster. As I was closing this book, Walmart announced that to upgrade its ability to compete in e-commerce with Amazon—which still does eight times Walmart's sales online—it was buying Jet, a year-old Internet retail startup. *The Economist* reported on August 13, 2016, that Jet's appeal to Walmart was its "real-time pricing algorithm, which tempts customers with lower prices if they add more items to their basket. The algorithm also identifies which of Jet's vendors is closest to the consumer, helping to minimize shipping costs and allowing them to offer discounts. Walmart plans to integrate the software with its own."

It turns out that "under a second" was just too damned slow.

## *The Start-Up from Batman*

In March 2016, I was visiting Sulaimaniya, in Iraqi Kurdistan, where a mutual friend introduced me to Sadik Yildiz, whose family runs a number of information-technology companies. Among them is Yeni Medya, or New Media Inc., which exemplifies just how fast a small maker can get just how big from just how remote a place by leveraging the supernova.

New Media Inc., which was founded by Yildiz's nephew, Ekrem Teymur, does big data analytics and media monitoring for the Turkish and other governments and the private sector, among many other things. They track all the media, including social media, in real time and can report to their customers what stories appear about them in the media anywhere. They can also inform the customer in real time of the top twenty subjects people are talking about and in what proportions. It is all displayed on a dashboard in colored boxes with the headline and percentage in each box.

"The Turkish presidency is a customer, and through our system they can receive a real-time poll service—every minute you can poll the public," Yildiz explained to me. "Big data is making things easy for everyone now. The software that we have developed in-house aggregates all the news sources in Turkey and the United States every five minutes—even Google News does not track every source at this pace all the time. We track all the existing stories on Twitter, and we archive all the stories that we track—one million stories a day; no one is archiving like that even in the United States—so if a news source deletes the story they publish about you after it is circulated, you can still use our system to retrieve it and use it for judiciary purposes. And then any government or company can use that to track what is said about them."

How do you make money?

"The business makes money on a subscription basis depending on how many keywords you want tracked and how many users you will have," explained Yildiz. " 'Thomas Friedman' would only be one word." (A bargain!) "They can give you content analysis, what is being said about you, break it down by geography, where it is coming from, how many people in which city are reading it, who started the story about you or the trend first—that is, who are the influencers—and how many followers used the same wording or how the original wording evolved and changed."

I was intrigued. It turns out that whispering—like guessing—is also officially over. "All the members of the Turkish parliament are using it to track about themselves," said Yildiz. "So are some news agencies, [who can] judge their reporters by whose stories are getting picked up most."

I was pretty sure I didn't want to hear everything being said about me, but I was intrigued by the tool they'd built. How much does it cost? Packages range from one thousand to twenty thousand dollars, he said, again depending on the number of keywords you want tracked.

So with all this amazing technology and reach, I asked, where did you start this company?

"Batman," he answered.

Is that a real place? I asked. "Yes it is!" Yildiz shot back. "And actually the city's mayor has sued the *Batman* movie for using the name without permission!" Yildiz is a Turkish Kurd, and so his family's company is based in the Kurdish-speaking region of eastern Turkey, in the family's hometown, called Batman. They have other businesses—construction and water treatment. But their real success came through leveraging the supernova from Batman. How did they do that? It was a family affair that started up as soon as the global flows from the supernova hit their town.

"My nephew, Ekrem Teymur, is the founder and chief engineer behind it—he is forty-two," explained Yildiz. "He was born in Batman and is the number-one data engineer in Turkey—the company was his idea." New Media Inc. has one hundred employees, and for a long time was competing from Batman with the biggest companies in the world. Most of the key positions in the company are held by family members—Ekrem and his six sisters, all born in Batman. The sisters, most of whom had only basic education, are now working as the chief editor, sales managers, and app production managers—a remarkable thing for a city where most of the women are not even allowed by their families to work.

The main business office, though, is now in Istanbul, said Yildiz, "but we still employ a lot of people in Batman." Thanks to all the connectivity today, they "can sit in their home in front of their computers and do jobs for us—and so it creates a lot of employment opportunities." Besides Batman and Istanbul, they have offices now in Dublin, Dubai, Beirut, and Palo Alto. Why the hell not?

"There is nothing called 'underprivileged' anymore," said Yildiz. "All you need is a working brain, some short training, and then put your idea into a fantastic business from any part of the world!"

Sadik Yildiz's story—and I have met so many others like him in the past decade—is a vivid example of how education plus connectivity plus supernova means that "more and more people are being empowered at lower and lower levels of income than ever before, so they think and act as if they were in the middle class, demanding human security and dignity and citizens' rights," explained Khalid Malik, former director of the U.N.'s Human Development Report Office. "This is a tectonic shift. The Industrial Revolution was a ten-million-person story. This is a couple-of-billion-person story." And we are just at the beginning of it.

I will have more to say about this later in the book. But I did have one last question for Yildiz: When did your family start this company?

"In 2007," he said.

FIVE

# *The Market*

Kayvon Beykpour is the cofounder and CEO of Periscope—the live-streaming video app launched in March 2014 that within four months had ten million users. It was quickly bought by Twitter, which understood that it offered a kind of video version of live tweeting. Periscope became popular so fast by creating a platform on which users could employ their smartphones to share with anyone in the world the live video of whatever event they were participating in or watching, be it a hurricane, earthquake, or flood, a Donald Trump rally, a thrill ride at Disney World, a confrontation with a cop, or a sit-in by Democratic lawmakers on the floor of the U.S. House of Representatives. Beykpour describes Periscope's mission as enabling everyone "to explore the world through someone else's eyes" and in doing so build "empathy and truth"—empathy by putting people into vivid contact with one another and their circumstances, and truth because live video doesn't easily lie. You can see everything raw.

Just how raw is illustrated by this story Beykpour told me:

> Last July [2015], I was flying from San Francisco to London to go to Wimbledon. I was on the United flight and was kicking myself because I had forgotten to download movies from iTunes to watch on my iPad and I was wondering what I was going to do for nine hours. So I decided to see if the United Wi-Fi was powerful enough to get onto Periscope and watch some video,

because it takes a lot of bandwidth. So I signed on to Periscope, and it worked! The first thing I did was to watch my girlfriend walking our dog, live, at Crissy Field beach [in San Francisco] near the Golden Gate Bridge. Then I thought I would see who else was on Periscope. So, when you go on the platform there is a map feature of the world and a dot indicating where anyone is live broadcasting. You just click on that dot and you watch someone's broadcast. [You can also watch a replay of live broadcasts.] I found this dot in the Hudson River. I thought, "I wonder what that is?" and I clicked on it. And it was this woman on a ferry crossing the Hudson in a storm. And she is saying "I am in a really bad storm and I am really scared." And she is talking and it was dark and she was in the front row and you could see the silhouette of the captain steering in the background moving the wheel and all this rain is coming down, beating on the window, and you can feel the turbulence. She was terrified.

There were seven other people on the site watching this, and all of us were reassuring her that it was going to be okay. I am on this plane probably somewhere over Greenland, and we're having our own turbulence, and these other people were all over the world and we're all strangers and we're all trying to comfort her. I watched for ten or fifteen minutes. Afterwards I was thinking to myself, "How is it possible that we helped to create something that enabled me to step into someone else's shoes like that? It feels like a superpower." You can't help but empathize when you can see through other people's eyes, especially folks you wouldn't have otherwise had occasion to connect with, and talk to them in real time. Imagine you're a Syrian refugee on a boat and you are broadcasting live as you are crossing the Mediterranean or walking into Serbia . . .

Beykpour's experience is a compelling illustration of how globalization—which in this book I will refer to by the umbrella term "the Market"—is also in acceleration today. For a long time many economists insisted that globalization was simply a measure of trade in physical goods, services, and financial transactions. That definition is way too narrow. Globalization, for me, has always meant the ability of any individual or company to compete, connect, exchange, or collaborate globally. And by that definition, globalization is now exploding. We can now digitize so many things, and, thanks to mobile phones and the supernova, we can now send those digital flows everywhere and pull them in from anywhere. Those flows drive the globalization of friendships and finance, hate and exclusion, education and e-commerce, news you can use, gossip that will titillate and rumors that will unsettle. Although trade in physical goods and financial products and services—the hallmarks of the twentieth-century global economy—has actually flattened or declined in recent years, globalization as measured by flows is "soaring—transmitting information, ideas, and innovation around the world and broadening participation in the global economy" more than ever, concluded a pioneering study on this subject in March 2016 by the McKinsey Global Institute, *Digital Globalization: The New Era of Global Flows*: "The world is more interconnected than ever."

Think of the flow of friends through Facebook, the flow of renters through Airbnb, the flow of opinions through Twitter, the flow of e-commerce through Amazon, Tencent, and Alibaba, the flow of crowdfunding through Kickstarter, Indiegogo, and GoFundMe, the flow of ideas and instant messages through WhatsApp and WeChat, the flow of peer-to-peer payments and credit through PayPal and Venmo, the flow of pictures through Instagram, the flow of education through Khan Academy, the flow of college courses through MOOCs, the flow of design tools through Autodesk, the flow of music through Apple, Pandora, and Spotify, the flow of video through Netflix, the flow of news through NYTimes.com or BuzzFeed.com, the flow of cloud-based tools through Salesforce, the flow of searches for knowledge through Google, and the flow of raw video through Periscope and Facebook. All these flows substantiate McKinsey's claim that the world is, indeed, more connected than ever.

Indeed, these digital flows have become so rich and powerful they are to the twenty-first century what rivers running off mountains were to civilization and cities in days of old. Back then, you wanted to build your town or your factory along a rushing river—such as the Amazon—and let it flow through you. That river would give you power, mobility, nourishment, and access to neighbors and their ideas. So it is with these digital flows into and out of the supernova. But the rivers you want to build on now are Amazon Web Services or Microsoft's

Azure—giant connectors that enable you, your business, or your nation to get access to all the computing-power applications in the supernova, where you can tie into every flow in the world in which you want to participate.

And these rivers will only get bigger and more important as more interactions between more things and more people get digitized and the benefits to companies and customers become more obvious. And obvious it is becoming. A 2014 McKinsey study, *Accelerating the Digitization of Business Process*, neatly summarized all the ways this trend is infusing our lives—and the benefits therein: "By digitizing information-intensive processes, costs can be cut by up to 90 percent and turnaround times improved by several orders of magnitude. Examples span multiple industries: one bank digitized its mortgage-application and decision process, cutting the cost per new mortgage by 70 percent and slashing time to preliminary approval from several days to just one minute. A telecommunications company created a self-serve, prepaid service where customers could order and activate phones without back-office involvement. A shoe retailer built a system to manage its in-store inventory that enabled it to know immediately whether a shoe and size was in stock—saving time for customers and sales staff. An insurance company built a digital process to automatically adjudicate a large share of its simple claims. In addition, replacing paper and manual processes with software allows businesses to automatically collect data that can be mined to better understand process performance, cost drivers, and causes of risk. Real-time reports and dashboards on digital-process performance permit managers to address problems before they become critical. For example, supply-chain-quality issues can be identified and dealt with more rapidly by monitoring customer buying behavior and feedback in digital channels. Leading organizations have come to recognize that the traditional large-scale projects to migrate all current processes to a digital world often take an extremely long time to deliver impact, and sometimes don't work at all. Instead, successful companies are reinventing processes, challenging everything related to an existing process and rebuilding it using cutting-edge digital technology. For example, rather than creating technology tools to help back-office employees type customer complaints into their systems, leading organizations create self-serve options for customers to type in their own complaints."

Surely, the world cannot get this connected in so many new realms at such profound new depths without every aspect of life and commerce being reshaped. This chapter is about how these digital global flows are doing just that: enabling so many more people around the world to access the supernova's technology toolbox to become makers *and breakers*; making the world so much more interdependent in financial terms, so every country is now more vulnerable to every other country's economy; driving contact between strangers at a pace and scale we've never seen before, so that good and bad ideas can go viral and extinguish and manufacture prejudices much more quickly; making every leader more exposed and transparent; and ensuring that the price countries pay for adventures abroad will be much higher than they expect, making these flows a new source of geopolitical restraint.

## *Interconnections or Intercourse?*

These digital rivers now coursing around the globe, tying everyone more closely together, are only going to become richer and faster as more people connect to the supernova with mobile devices. In January 2015, the Boston Consulting Group released a study, *The Mobile Revolution: How Mobile Technologies Drive a Trillion-Dollar Impact*, funded by Qualcomm. Among the impacts it studied was just how devoted people are becoming to their mobile phones. To drill down, BCG commissioned a poll that asked people in the United States, Germany, South Korea, Brazil, China, and India this headline question: "Which of the following things would you give up for a year rather than give up personal use of your mobile phone?" Dining out? Sixty-four percent said they would give it up. Having a pet? Fifty-one percent said they would give that up. Going on vacation? Fifty percent. One day off a week? Fifty-one percent. Seeing friends in person—some forty-five percent were ready to let that go. Then they really got serious and asked: What would you give up for a year first—your mobile phone or sex?

Thirty-eight percent of respondents said *they would give up sex for a year* rather than give up their mobile phone!

Broken down by country, the South Koreans led the way in a willingness to trade human intercourse for voice and data intercourse:

sixty percent! Their reasons are not hard to grasp. The Swedish telecom giant Ericsson notes:

> Mobile technologies have transformed the way we live, work, learn, travel, shop, and stay connected. Not even the industrial revolution created such a swift and radical explosion in technological innovation and economic growth worldwide. Nearly all fundamental human pursuits have been touched, if not revolutionized, by mobile. In less than fifteen years, 3G [third generation] and 4G technologies have reached three billion subscriptions, making mobile the most rapidly adopted consumer technology in history.

If a decade ago, we would have said it feels like we're all living in a crowded village, today, argues Dov Seidman, "it feels like we're all living in a crowded theater. The world isn't just interconnected, it is now becoming *interdependent.* More than ever before, we rise and fall together.

**Mobile Cellular Subscriptions (per 100 People), 1960–2014**

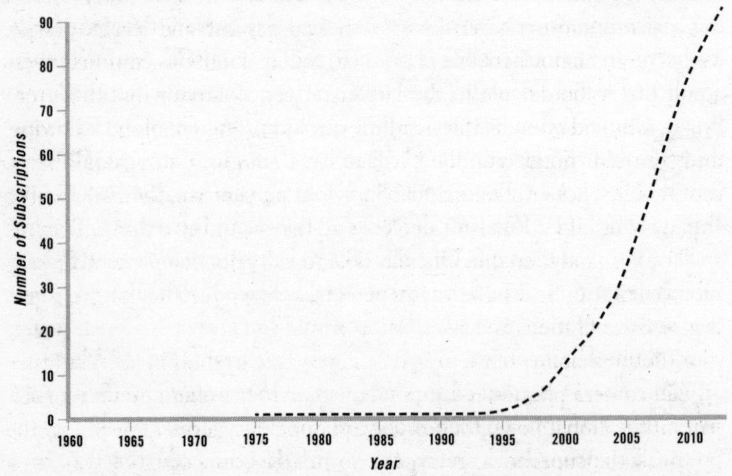


Source: International Telecommunication Union, World Telecommunication/ICT Development Report and database

So few can now so easily and so profoundly affect so many so far away . . . We are experiencing the aspirations, hopes, frustrations, plights of others in direct and visceral ways"—just the way Kayvon Beykpour did when he shared that stormy boat ride with a stranger, while flying over the ocean himself.

The French president, François Hollande, had a small breakfast for columnists during the opening session of the United Nations in September 2015, which focused heavily on the flood of refugees from the Middle East and Africa who were trying every which way to get into Europe. Afterward, one of Hollande's aides remarked to me: it is amazing how quickly information is disseminated and made operational by these refugees; they are constantly on the move, trying to cross the Mediterranean, and yet manage to stay highly informed through social networks on the things they need to know.

"One day," the French diplomat said, "we changed the regulations and said that any boat that had a handicapped person on board could not be turned back [from European shores]." Very soon, he said, boats everywhere started arriving with people in wheelchairs. "It was that fast."

In April 2016, I went to the West African state of Niger to do a documentary for the *Years of Living Dangerously* series on the National Geographic Channel. Our crew was following the route of migrants from West Africa, through Niger, across the Sahara to Libya and over to Europe. We were in the northern Niger town of Dirkou, about one hundred miles south of the border with Libya, and we were interviewing men from Niger who had gone to Libya, failed to making the crossing to Europe, and returned home penniless. They were standing alongside a large semitrailer truck overloaded with dry goods. After we finished filming them, I asked if I could take a picture of them with my iPhone. They all nodded yes. And then they all took out their cell phones and started taking pictures of me. So I have a picture of me taking pictures of them taking pictures of me.

I doubt that any of them had much money in their pocket, but they all had camera phones and they were going to use them now to participate in the global flows, even at just a rudimentary level. Drawing on the power of the supernova, everyone, no matter how poor, now can be a subject, not just an object, not just an ornament to some Westerner's trip to Africa, but the author of their own narrative to a global audience. And that is a good thing—impossible just a decade ago.

Heck, even beggars are now getting into the digital game. On a visit to Beijing in June 2017, I discovered that China has moved so fast into a cashless society, where everyone pays for everything with their mobile phones, that giving alms has been transformed. Here is how an April 24, 2017, story in Ibtimes.com described it:

"Don't have any spare change? No problem—beggars in China now accept alms transferred via mobile payments by scanning QR codes with smartphones . . . The beggars place a printout of a QR code in their begging bowls. The QR codes enable anyone with a mobile payment app like Alibaba Group's Alipay or Tencent's WeChat Wallet to scan the code and send a certain sum to the beggar's mobile payment account . . . According to Chinese state media, this is not as uncommon as you'd think . . . Chinese digital marketing firm China Channel claims that the practice of QR code begging is not merely altruistic. The firm claims that many of the beggars they encountered in Beijing are actually being paid by local businesses and startups to promote QR codes and entice passersby to scan them. The scans are used by the businesses to harvest user data on each person's WeChat IDs. When compiled, the lists of WeChat IDs can be sold for a fair amount of money to small businesses, who use them to send out unsolicited advertisements in the app—the same way in the past companies used email addresses and phone numbers."

My Chinese friends told me they don't carry purses or wallets anymore, only a mobile phone, which they use for everything—including the purchase of vegetables from street vendors. "America has been dreaming of a cash-less society," Ya-Qin Zhang, President of Baidu, China's main search engine, remarked to me, "but China is already there."

When you look at how this diffusion of these digital flows keeps accelerating, it boggles the mind to think about how interdependent the world will be in another decade. Consider just a few indices. The McKinsey *Digital Flows* study noted that back in 1990, "the total value of global flows of goods, services, and finance amounted to $5 trillion, or 24 percent of world GDP. There were some 435 million international tourist arrivals, and the public Internet was in its infancy. Fast-forward to 2014: some $30 trillion worth of goods, services, and finance, equivalent to 39 percent of GDP, was exchanged across the world's borders.

International tourist arrivals soared above 1.1 billion." But here's what's even more interesting:

> Cross-border bandwidth [terabits per second] has grown 45 times larger since 2005. It is projected to grow by another nine times in the next five years as digital flows of commerce, information, searches, video, communication, and intra-company traffic continue to surge . . .
>
> Thanks to social media and other Internet platforms, individuals are forming their own cross-border connections. We estimate that 914 million people around the world have at least one international connection on social media, and 361 million participate in cross-border e-commerce . . . On Facebook, 50 percent of users now have at least one international friend. This share is even higher—and growing faster—among users in emerging economies.

As a result, all of this connectivity is vastly expanding "instantaneous exchanges of virtual goods":

> E-books, apps, online games, MP3 music files and streaming services, software, and cloud computing services can all be transmitted to customers anywhere in the world there is an Internet connection. Many major media websites are shifting from building national audiences to global ones; a range of publications, including *The Guardian*, *Vogue*, BBC, and BuzzFeed, attract more than half of their online traffic from foreign countries. By expanding its business model from mailing DVDs to selling subscriptions for online streaming, Netflix has dramatically broadened its international reach to more than 190 countries. While media, music, books, and games represent the first wave of digital trade, 3-D printing could eventually expand digital commerce to many more product categories.

And forget the fact that so many "friends" are connecting on Facebook. How about all the "things" getting to know one another? You want to see flows—wait until the "Internet of Things" gets to scale and machines start talking to machines everywhere and always! "Only 0.6 percent of things are connected today," Plamen Nedeltchev, distinguished IT engineer at Cisco, wrote on Cisco.com in an essay entitled "It is inevitable. It is here. Are we ready?" on September 29, 2015. "There were 1,000 Internet-connected devices in 1984," said the article, a million in 1992, and ten billion in 2008. Fifty billion devices "are expected to be connected by 2020. In 2011, the number of new things connected to the Internet exceeded the number of new users connected to the Internet."

Today, data flows are "exerting a larger impact on growth than traditional goods flows," McKinsey found. "This is a remarkable development given that the world's trade networks have developed over centuries but cross-border data flows were nascent just fifteen years ago." This is sure to grow, it noted, because originally "the largest corporations built their own digital platforms to manage suppliers, connect to customers, and enable internal communication and data sharing for employees around the world," but now "a diverse set of public Internet platforms has emerged to connect anyone, anywhere," via mobile phones—including Facebook, YouTube, WhatsApp, WeChat, Alibaba, Tencent, Instagram, Twitter, Skype, eBay, Google, Apple, and Amazon.

Some messaging apps—such as Facebook Messenger and WeChat—are not only exploding in popularity but also replacing e-mail as the preferred means for communicating and becoming the preferred carriers of more and more interactive capabilities. They are becoming platforms for e-commerce, e-banking, reservations, and rapid-fire communication. The phenomenon has been dubbed "conversational commerce," and it promises to weave the world together even tighter and faster by simplifying and accelerating more and more complex interactions. With Venmo, for instance, young people today not only seamlessly split the bill for a dinner through their banks using their cell phones but also share thoughts on the food and conversation, with the same billing message.

Eleonora Sharef, a McKinsey consultant, says that in her office messaging apps such as Slack and HipChat have taken off so fast because they are like having "a live dashboard that sends you all the relevant info on your business throughout the day, while also allowing you to talk about work in a fun environment . . . All these chat tools are available on your smartphone, too, so you can always be in fast contact with your employees and see metrics anytime day or night—and be a slave to the job!"

These messaging apps are going to make conventional e-mail seem to our kids what conventional mail seemed to the first generation of e-mail users. Mobile messaging apps are "the next platform and they are going to change a lot of things," said David Marcus, who runs Facebook Messenger and used to run PayPal. "If we are successful, a lot of your life will be running on a messaging app. It is becoming the hub for everyday interactions with people and business and services. E-mail will stick around for less immediate connections." When we talked in May 2016, Facebook Messenger was about to cross one billion users a month. When one billion people are using anything, you should pay attention.

"Think about it," Marcus elaborated in a blog post about the rise of these messaging platforms:

> SMS and texting came to the fore in the time of flip phones. Now, many of us can do so much more on our phones; we went from just making phone calls and sending basic text-only messages to having computers in our pockets. And just like the flip phone is disappearing, old communication styles are disappearing too. With Messenger, we offer all the things that made texting so popular, but also so much more. Yes, you can send text messages, but you can also send stickers, photos, videos, voice clips, GIFs, your location, and money to people. You can make video and voice calls while at the same time not needing to know someone's phone number.

Messaging apps, of course, are phone number–based, but Marcus's vision for Facebook Messenger is to make phone numbers disappear. You will just click on the names of people and companies in your Facebook graph and never have to remember a phone number again. "Over time," he remarked, "it will make phone numbers obsolete." Imagine what that will do to intensify the flow of flows.

As all of these tools scale, the cost of cross-border communications and transactions keeps declining, so starting a business that is global from day one is now incredibly cheap. McKinsey noted that by 2016, there were fifty million small businesses on Facebook. "That's twice the number of two years ago . . . Alibaba in China has ten million small and medium-sized enterprises that sell products to the rest of the world through its platform. Amazon has two million small businesses . . . Some 900 million people have international connections on social media, and 360 million take part in cross-border e-commerce."

For the same reason, added McKinsey, "products can go viral on a scale that has never been seen before. In 2015, Adele's song 'Hello' racked up fifty million views on YouTube in its first forty-eight hours, and her album 25 sold a record 3.38 million copies in the United States in its first week alone, more than any other album in history. In 2012, Michelle Obama wore a dress from British online fashion retailer ASOS in a photo that was re-tweeted 816,000 times and shared more than four million times on Facebook; it instantly sold out."

Meanwhile, all of these macro and micro flows are fundamentally shifting how we think about economic power—what it consists of and who has it.

## *The Big Shift*

Which is why the management experts John Hagel III, John Seely Brown, and Lang Davison coined the term "the Big Shift." The Big Shift, they argue, is that we're moving from a long period of history in which *stocks* were the measure of wealth and the driver of growth—how much of every resource imaginable you could stock up on and then draw down and exploit—to a world in which the most relevant source of comparative advantage will be how rich and numerous are the *flows* passing through your country or community and how well trained your citizen-workers are to take advantage of them.

"Business for the past several centuries at least has been organized around stocks of knowledge as the basis for value creation," Hagel elaborated to me in a follow-up email. "The key to creating economic value has been to acquire some proprietary knowledge stocks, aggressively protect those knowledge stocks and then efficiently extract the economic value from those knowledge stocks and deliver them to the market."

Stocks were what gave us security and wealth. Hagel, Seely Brown, and Davison elaborated in a coauthored essay in the January 27, 2009, *Harvard Business Review* entitled "Abandon Stocks, Embrace Flows":

> "If you knew something valuable, something nobody else could access, you had, in effect, a license to print money. All you needed to do was to protect and defend that knowledge and then deliver products or services based on that knowledge as efficiently and as broadly as possible. Think of the proprietary formula for Coca-Cola, or the patents protecting blockbuster drugs in the pharma industry. The power, simplicity, and success of this model explain why it is so deeply engrained in the minds of executives."

But the challenge in the age of accelerations, added Hagel, "is that knowledge stocks depreciate at an accelerating rate. In this kind of world, the key source of economic value shifts from stocks to flows. The companies that will create the most economic value in the future will be the ones that find ways to participate more effectively in a broader range of more diverse knowledge flows that can refresh knowledge stocks at an accelerating rate. So, knowledge flows become the key to creating economic value. The good news is that we're in a world where knowledge flows are expanding and accelerating at an exponential pace on a global scale. This is happening for many reasons. A key factor is the deployment of rapidly improving digital technology infrastructures. Another important factor is the rapid urbanization of the world's population. As we move into more and more dense urban settlements, there's a significant increase in knowledge flows shaped by the growing interactions of people in these urban areas."

As we will discuss in more detail in chapter seven, this Big Shift from stocks to flows has enormous implications for learning and education. "As individuals," wrote Hagel, Brown, and Davison, "we expect to go through structured educational programs in the early stages of our lives. Then we enter the workforce secure in the belief that the skills and knowledge we have acquired will serve us well throughout our careers. Sure, we will acquire new knowledge as we work, but the key is to effectively leverage the knowledge stocks we acquired as we passed through the educational system."

But what if the rise of the supernova has made that whole model obsolete? What if, as the authors put it,

> a different source of value is becoming more powerful? We believe there's good reason to think that value is shifting from knowledge stocks to knowledge flows. Put more simply, we believe that *flows trump stocks* [italics added] . . .
>
> As the world speeds up, stocks of knowledge depreciate at a faster rate. As one simple example, look at the rapid compression in product life cycles across many industries on a global scale. Even the most successful products fall by the wayside more quickly as new generations come through the pipeline faster and faster. In more stable times, we could sit back and relax once we had learned something valuable, secure that we could generate value from that knowledge for an indefinite period. Not anymore.
>
> To succeed now, we have to continually refresh our stocks of knowledge by participating in relevant flows of new knowledge.

This Big Shift calls on us to keep learning for as long as we are in the workforce. And that is going to be a big challenge for individual workers, who must continuously keep up, and for government and business, which must help enable their workers to keep up. Because, the authors add, "You can't just tap flows one time. You have to contribute to them as well to really be 'in the flow.'"

"We can't participate effectively in flows of knowledge—at least not for long—without contributing knowledge of our own. This occurs because participants in these knowledge flows don't want free riding 'takers'; they want to develop relationships with people and institutions that can contribute knowledge of their own."

You can see this clearly in the open-source software communities, such as GitHub, but it is true more widely. "While there are certainly risks associated with knowledge sharing, the damage from IP theft diminishes as the rate of obsolescence increases," they argued. "At the same time, the rewards from knowledge sharing go up substantially."

Indeed, digital crowds are a new form of "flow" enabled by the age of accelerations, and more and more companies are shifting their attention to them as well. There is madness in crowds, but there is also wisdom in crowds, and in a world of increasing flows crowdfunding, crowd-designing, crowd-innovating, and crowd-correcting are not only more possible, they become the only way to keep up with the pace of change.

In the old days, argue MIT's McAfee and Brynjolfsson in their book *Machine, Platform, Crowd: Harnessing Our Digital Future*, the core of managers and engineers inside a company usually drove all the innovation and strategy. But now, with so many rich digital flows traveling in all directions, companies can and increasingly must tap the crowds to contribute ideas and inventions and to conquer new markets—and the most successful organizations are doing just that.

Case in point—General Electric. When GE is looking to invent a new product part, it no longer calls on just its own engineers in India, China, Israel, and the United States—now it is increasingly supplementing those engineers by tapping the flows, by running "contests" to stimulate the best minds anywhere to participate in GE's innovations.

Every aircraft engine has key components that hold it in place—such as hangers and brackets. Making those components both stronger and lighter is the holy grail, because the lighter they are, the less fuel the plane consumes. So in 2013 GE took one bracket, described the conditions under which it worked and the particular function it performed, and posted online the GE engine-bracket challenge. GE offered a reward to anyone in the world who could design that component with less weight, using 3-D printing. They advertised it in June 2013. As I wrote in a column, within weeks they had received 697 entries from all over the world—from companies, individuals, graduate students, and designers.

According to the GE website:

> In September [2013], the partners picked 10 finalists who received $1,000 each.
>
> Aviation 3D printed the 10 shortlisted designs at its additive manufacturing plant in Cincinnati, Ohio. GE workers made the brackets from a titanium alloy on a direct metal laser melting (DMLM) machine, which uses a laser beam to fuse layers of metal powder into the final shape.
>
> The team then sent the finished brackets to GE Global Research (GRC) in Niskayuna, New York, for destruction testing. GRC engineers strapped each bracket to an MTS servo-hydraulic

> testing machine and exposed it to axial loads ranging from 8,000 to 9,500 pounds.
>
> Only one of the brackets failed and the rest advanced to a torsional test, where they were exposed to torque of 5,000 inch-pounds.

None of the finalists were Americans, and none of them were aeronautical engineers. The best design, GE told me, actually came from Ármin Fendrik, a third-year university student from Hungary. This entry was among his first 3-D printing designs. But it turned out that he had been interning at GE's office in Budapest and therefore could not take the prize. So the first-prize money, seven thousand dollars, went to M. Arie Kurniawan, a twenty-one-year-old engineer from Salatiga in Central Java, Indonesia. Kurniawan's bracket, said GE, "had the best combination of stiffness and light weight. The original bracket weighed 2,033 grams (4.48 pounds), but Kurniawan was able to slash its weight by nearly 84 percent to just 327 grams (0.72 pounds)." GE officials noted to me that the manager who ran the challenge had worked at GE longer than that kid had been alive.

Kurniawan was quoted by GE as saying, "3D printing will be available for everyone in the very near future." Kurniawan, who, GE tells us, "runs a small engineering and design firm called DTECH-ENGINEERING with his brother," added: "That's why I want to be familiar with additive manufacturing as soon as possible."

GE ended up offering the Hungarian intern a job. Although he clearly had enormous talent, that Hungarian student had failed his engineering structural analysis class, said Bill Carter, a senior mechanical engineer in GE's Additive Manufacturing Lab: "So it shows that if you get young people excited about something, they feel and can relate to, they get excited—and instead of being in class and studying, he went out and [entered our contest]. And he went and learned from people he never would have talked to."

Discussing this whole project two years later, Prabhjot Singh, manager of the Additive Manufacturing Lab, explained to me just how much these global flows are being leveraged by a company such as GE today: "When you are looking for new ideas, you can now bring in a diversity of responses worldwide, and you engage the community to drive speed. I can rapidly scale and descale my team depending on how much I want to leverage the community. This helps us to stay up-to-date on things."

## *Globalization Redefined (Again)*

When I wrote *The World is Flat* in 2005, the argument was that globalization was once driven by countries—Spain discovering the "New World." Then it was driven by companies—think the Dutch East India Company two centuries ago or Apple today. And now, thanks to all of these digital flows, globalization is being driven by everyone and anyone—small groups, start-ups, individuals, and multinationals—and it is connecting East and West, North and South, and South and South. So many people can now take advantage of digital globalization and the flows off the supernova, to go global on their own terms. No one has observed this transition more acutely than Jeff Immelt, the longtime CEO of General Electric, who stepped down in 2017.

"I think the world's gone from macro to micro," Immelt said to me in an interview at GE's Boston headquarters in March 2017. During the last forty or fifty years, he noted, globalization was shaped by big platforms created by big governments—like the World Trade Organization or the World Bank. But as he travels the world now, Immelt said he finds that the new globalizers have never heard of the WTO and have no clue who the U.S. ambassador is in their country.

"They don't even speak those languages," Immelt said. "They are just globalizing on their own. They use platforms like Alibaba and Tencent and Amazon." These new globalizers are really "good digitally"; the ones in China are all using WeChat and on their own they figure out how to connect "Chinese funding for a power plant in Pakistan with a Canadian export credit."

In this kind of a world, said Immelt, GE now sees itself as a "multilocal" not a multinational. It is pushing down power and opportunity to its local teams all over the world and encouraging them to link up with other teams and opportunities anywhere in the world. "None of them would even know how to bring a complaint to the WTO," added Immelt.

In 2000, 70 percent of GE's revenue was in the United States. In

2017, over 60 percent came from global markets. But this was not by outsourcing. "Outsourcing is yesterday's game," Immelt said in his 2017 annual report. "During the 1980s and '90s, business looked to the emerging markets as a cheap labor source. American jobs migrated to countries that welcomed U.S. companies with open arms. American workers lost in the game of wage arbitrage. Chasing the lowest labor costs is yesterday's model."

In today's model there is no over here and over there. "We see substantial opportunity to grow around the world by investing, operating, and building relationships in the countries where we do business," Immelt said. "We partner with Chinese construction companies and leverage their funding to win contracts in Africa and Asia. Our investments have created jobs in China and the U.S., while making GE more competitive." The multinational companies that thrive in the age of accelerations will be those that digitally weave together the optimal production talents, design talents, logistics capabilities, financing, and market sales opportunities that look at every country as both a market and a source of skills. If they don't, their competitors will.

### *Google Motors, Apple Bank, Amazon Studios*

With all of these energy flows in all of these directions, competition can now come from so many more directions, individuals, and companies. Historically, noted James Manyika, one of the authors of the McKinsey report, companies kept their eyes on competitors "who looked like them, were in their sector and in their geography." Not anymore. Google started as a search engine and is now also becoming a car company and a home energy management system. Apple is a computer manufacturer that is now the biggest music seller and is also going into the car business, but in the meantime, with Apple Pay, it's also becoming a bank. Amazon, a retailer, came out of nowhere to steal a march on both IBM and HP in cloud computing. Ten years ago neither company would have listed Amazon as a competitor. But Amazon needed more cloud computing power to run its own business and then decided that cloud computing was a business! And now Amazon is also a Hollywood studio.

On January 12, 2016, CNNMoney.com ran a story about the Golden Globes award ceremony that began:

> "I want to thank Amazon, Jeff Bezos . . ."
>
> Those words were spoken at a Hollywood awards show [by the director Jill Soloway] for the first time on Sunday as Amazon's comedic television series *Transparent* picked up two Golden Globe awards, beating shows from HBO, Netflix, and the CW.
>
> The awards were an affirmation of the widening television landscape, as streaming services like Netflix and Amazon Prime Instant Video are beginning to play host to award-worthy programs just like television networks . . .
>
> A little while later, the star of *Transparent*, Jeffrey Tambor, won the award for best actor in a television comedy. He called Amazon "my new best friend."

I wonder how HBO feels about that?

For all of these reasons, McKinsey has created its own measure of globalization that basically asks a country or a company or a citizen: Are you in the flow? It's called the "MGI Connectedness Index." It ranks different countries on how much they are participating in all the different kinds of global flows, and it is a pretty good indicator of prosperity and growth. Singapore tops the list—followed by the Netherlands, the United States, and Germany.

But there is a message in the bottle: Singapore invested in both the infrastructure to make sure it was participating in every digital flow as well as the education of its workforce, to make sure they could take advantage of those flows if the government made them available. Individual cities that now do the same thing can reap the benefits. So this isn't complicated: the most educated people who plug into the most flows and enjoy the best governance and infrastructure win. They will have the most data to mine; they will see the most new ideas first; they will be challenged by them first and able to respond and take advantage of them first. Being in the flow will constitute a significant strategic and economic advantage.

A research study published in the *International Journal of Business,*

*Humanities, and Technology* in February 2013 found a correlation between countries with high GDP and "high Internet penetration," not only in the highly developed and Internet-saturated Nordic countries, but also beyond. "A pattern begins to emerge; ICT [information and communications technology] growth occurs, and as the population becomes more comfortable with the technology and more productive, the GDP level begins to increase as well."

It's a *big shift*—but that is what this era of globalization is all about.

## *The Big Shift Goes Everywhere*

What is most exciting about this acceleration in the globalization of flows is that these digital rivers now run everywhere with equal energy, and with mobile phones and tablets people anywhere can now tap into them to compete, connect, collaborate, and invent. I was lucky enough on a visit to India, in November 2011, to actually see and write a column about how the poorest people in the world can join the flow. I was on a reporting trip and was invited by Prem Kalra, then the director of the Indian Institute of Technology in Rajasthan, one of the elite MITs of India, to give a talk on his campus, meet with his students, and take a look at a project he was working on at his institute in Jodhpur that was specifically designed to connect the poorest person in India to the global flows.

Kalra explained that there is a concept in telecommunications called "the last mile"—the section of any phone system that is the most difficult to connect—the part that goes from the main lines into people's homes. He said he was dedicating his IIT to overcoming a parallel challenge: connecting to "the last person." If you want to overcome poverty, he argued, you have to answer that question today: How will we reach the last person—meaning the poorest person in India? Can "the financially worst-off person" in India "be empowered?" he asked. That is, be given the basic tools to acquire enough skills to overcome dire poverty? In a country where 75 percent of the people live on less than two dollars a day, what bigger question is there?

Specifically, India's Human Resources Development Ministry put out a challenge that Kalra and his technology institute back then decided to take up: Could anyone design and make a stripped-down, iPad-like, Internet-enabled, wirelessly connected tablet that the poorest Indian family, saving about $2.50 a month for a year, could afford if the government subsidized the rest? More specifically, could they make a simple tablet usable for distance learning, teaching English and math, or just tracking commodity prices for under fifty dollars, including the manufacturer's profit, so millions of poor Indians living on the edge could also join the global flows?

Kalra's team—led by two IIT Rajasthan electrical engineering professors, one of whom came from a village that still had no electricity—won the competition and unveiled the Aakash tablet. *Aakash* is Hindi for "sky." The original version was based on the Android 2.2 operating system, with a seven-inch touchscreen, three hours of battery life, and the ability to download YouTube videos, PDFs, and educational software. If Indians could purchase only tablets made in the West, the price points would be so high they'd never reach the last person, said Kalra, so "we had to break the price point" in a big way. They did it by taking full advantage of today's globalization of flows: pulling commodity parts mainly from China and South Korea, using open-source software and collaboration tools, and employing the design/manufacturing/assembly abilities of two companies in the West—DataWind and Conexant Systems—and Quad in India.

But what I actually remember most about my visit was a story that Kalra's wife, Urmila, told me about a chat she had had with their maid after the Aakash was unveiled in Indian newspapers on October 5, 2011. Urmila said that her maid, who had two young children, came to her one day and said that she had heard "from the night watchman that Mr. Kalra has made a computer that is very cheap, and is so cheap that even she could afford to buy it. The watchman had given her a picture of it from the paper, and she asked me if it was true."

Urmila told her maid that, yes, it was true and that the machine was meant for people who could not afford a big computer.

"She asked, 'How much will it cost?'" Urmila told me.

"I said, 'It will cost you around fifteen hundred rupees [$30].'"

The maid, dumbfounded by the low price, asked, "Fifteen thousand or fifteen hundred?"

Urmila answered, "Fifteen hundred." But, Urmila added, the maid "was sure that if the government was doing something so good for the poor, it had to have a catch. 'What can you do on it?' she asked me. I said, 'If your daughter goes to school, she can use it to download videos of class lessons,' just like she had seen my son download physics lectures every week from MIT's website."

Urmila's son was already watching lectures on MIT's OpenCourseWare platform—a predecessor to MOOCs, or massive open online courses, which MIT had put up on the Internet for free; it consisted simply of video lectures and course guides. Urmila told her maid: "You have seen our son sitting at the computer listening to a teacher who is speaking. That teacher is actually in America."

The maid's eyes "just kept getting wider and wider," Urmila recalled. "Then she asked me will her kids be able to learn English on it. I said, 'Yes, they will definitely be able to learn English,' which is the passport for upward mobility here. I said, 'It will be so cheap you will be able to buy one for your son and one for your daughter!'"

Urmila's son was already drawing from the global flows to effectively study at his home in Jodhpur on MIT's platform, and the maid's kids would not be far behind. The farther you get from the connected capitals of the developed world, the more you can see how today's globalization is rapidly distributing energy through these flows right out to "the last person."

That is not an exaggeration. And that is, for me, an enormous source of optimism.

The early stages of modern digital globalization tended to be all about "outsourcing," another name for American and European companies leveraging the fact that connectivity was becoming fast, free, easy for you, and ubiquitous, so they could hire enormous numbers of relatively cheap engineers anywhere in the world to solve *American problems.* When this first became possible at scale in the late 1990s, the big problem most people wanted solving was Y2K—the fear that many computers would stop working because of a bug that would kick in on their internal clocks on January 1, 2000. Millions of computer systems needed to be remediated, and India had hundreds of thousands of low-wage engineers to do it. Presto, problem solved.

What happened, though, with the emergence of the supernova, when complexity became fast, free, easy for you, and invisible, and globalization meant that everyone anywhere with an Internet connection could access the digital flows, was something very exciting: Indian, Mexican, Pakistani, Indonesian, and Ukrainian engineers, and many others, began tapping in to solve *their* problems. And now some of those low-cost innovations are coming back in our direction and to our benefit. India always had a strong tradition of educating people in math, science, and engineering, and America was once the big beneficiary—in the 1950s, 1960s, and 1970s, when global flows were in many countries either nonexistent or barely a trickle, those Indian graduates could not get jobs in India, so they flocked to America and helped America overcome deficiencies in its workforce. Now, thanks to the digital flows being pushed out to them from the supernova, they can stay home and act globally more than ever. As a result, many more people are now working on the world's biggest opportunities and biggest problems.

I see this everywhere I travel. Every time I visit India to write columns, I visit NASSCOM, the high-tech association, to meet with the newest crop of Indian innovators. They account for only a tiny fraction of India's 1.2 billion people, most of whom remain painfully poor, but I focus on these innovators because so many of them today are focused on making India unpoor.

In 2011, the NASSCOM team introduced me to Aloke Bajpai, who, like others on his young team, cut his teeth working for Western technology companies but returned to India on a bet that he could start something—he just didn't know what. The result was Ixigo.com, a travel search service that can run on the cheapest cell phones and helps Indians book the lowest-cost fares, whether it is a farmer who wants to go by bus or train for a few rupees from Chennai to Bangalore or a millionaire who wants to go by plane to Paris. Ixigo is today the biggest travel search platform in India, with millions of users. To build it, Bajpai leveraged the supernova, using free open-source software, Skype, and cloud-based office tools such as Google Apps and social media marketing on Facebook. They "enabled us to grow so much faster with no money," he told me.

It can be incredibly uplifting to go to a place such as Monterrey, Mexico, that country's technology hub, and meet a critical mass of young people who just didn't "get the word"—they didn't get the word that their government is a mess or that China is going to eat their lunch or that the streets are too dangerous. Instead, they take advantage of how connecting to global flows can now enable them to start stuff and collaborate on stuff really cheaply—and they just do it. Monterrey has tens of thousands of poor living in shantytowns. They've been there for decades. What is new, though, is that it now also has a critical mass of young, confident innovators trying to solve Mexico's problems, by leveraging technology and globalization.

I went to Monterrey in 2013 and wrote a column about a few of the young people I met: There was Raúl Maldonado, founder of Enova, which had created an after-school program of blended learning—teacher plus Internet—to teach math and reading to poor kids and computer literacy to adults. "We've graduated eighty thousand people in the last three years," he told me. "We plan to start seven hundred centers in the next three years and reach six million people in the next five." There was Patricio Zambrano from Alivio Capital, who had created a network of dental, optical, and hearing aid clinics to provide low-cost alternatives for all three, plus loans for hospital care for people without insurance. There was Andrés Muñoz, Jr., from Energryn, who demonstrated his solar hot-water heater that also purified water and could cook meat. There was the administrator from CEDIM, a start-up university offering a "master's in business innovation." And there was Arturo Galván, founder of Naranya, a mobile Internet company that created a range of services, including micropayments for consumers at the bottom of the pyramid. "We've all been here for many years, but I think that the confidence is starting to happen," explained Galván. "You start to see the role models who started from zero and are now going public. We are pretty creative. We had to face a lot of challenges." As a result, he added, "we are strong now, we believe, and the innovation ecosystem is happening." "Naranya" is based on the Spanish word for "orange," or *naranja*. Why that name? I asked Galván. "'Apple' was already taken," he said.

But connecting to the flows is not just a story of how easily these developing countries can now innovate new goods and services for themselves and then push them out to the world as exports. It is also a story of what the poorest of the poor can now easily *pull down* from the global flows. Consider the 3-2-1 Service in Madagascar, founded by David McAfee, CEO of Human Network International. He explained:

> At a moment of need, callers use their own simple mobile phones to proactively retrieve information across a range of topics. Callers dial a toll-free number anytime, anywhere and listen to a menu of options: "*Would you like to know about: Health? Press one. Agriculture? Press two. Environment? Press three. Water and sanitation? Press four. Land tenure? Press five. Micro finance? Press six. Family planning? Press seven.*"
>
> We use the same out-of-the-box software that every 1-800 number uses—"*Press one to continue in English. Press two to switch to Spanish.*" But we repurpose it so illiterate audiences can use their telephone keypad to select and listen to prerecorded messages free of charge and on demand. The innovation here is the "pull" aspect. Callers can "pull" the information at a moment of need . . . Up until now, development and humanitarian organizations have struggled to meet this "moment of need." Development workers managing projects with a behavior change component—like encouraging moms to have their children sleep under a mosquito net—use mass media channels—radio, television—or interpersonal communications—knocking on doors—to get their key messages out. But these "push" channels aren't adapted to meet people's individual moment of need. It sounds dumb and obvious . . . but people need access to information when they need it: at their own moment of need. And they can't "pull" the information out of their radios! . . . In the six years since the launch, more than five million callers have made sixty million information requests . . . all free to the end user.

The 3-2-1 Service is currently live in twelve countries, and the plan is to extend the service to three additional African and Asian countries by the end of 2017. Once the 3-2-1 Service is launched in these countries, 120 million subscribers will have free, on-demand access to key public service messages. In the first six months of 2017, six hundred thousand people contacted the 3-2-1 Service on average each month and listened to 1.5 million key messages. That's a lot of flows being pushed and pulled. McAfee's team then mines these flows, because they are digital, as a resource to improve the service. Unlike radio and television stations in Africa, noted

McAfee, "we know exactly how many people have listened to our key messages. We collect metadata on each call: telephone number, time/date stamp, decision at each menu, and key message selected."

What cannot be stressed enough is how early we are in this acceleration of flows. You can already see the next stage forming—the creation of clearinghouse platforms that will efficiently match the flows coming out of the developing world with those that want to come into it, weaving the world together ever more tightly. One of the most interesting start-ups I came across in this space was Globality.com, co-founded in March 2015 by Joel Hyatt and Lior Delgo with a mission to create a platform that uses artificial and human intelligence to enable small and midsize companies to become "micro-multinationals" by participating in the global economy as easily as the big guys.

Say you are a small manufacturer in America who needs a law firm and marketing firm in Lima, Peru, or you are an Indian data services company buying a three-person start-up in Houston. You would go to Globality's platform and create a project brief, using their technology dashboard. "Then we take the project brief and, using artificial intelligence and human curation, bring back to your company—for free—a small selection of the best-qualified firms to meet your needs, which we determine on the basis of our industry expertise, research, and matching algorithms," explained Hyatt.

Globality then connects you, on its site, to the firm or firms you choose, providing the video technology for the two parties to forge deal parameters and hammer out legal frameworks, check references, conclude contracts, and do all the billing—along with a star system for both sides to rate each other the way Uber, Airbnb, and eBay do. Everything a company would need to act globally "from the first moment to the last is on the platform in one easy and uniform format," said Hyatt. Globality—which makes its money by charging the service provider (that is, the seller) a commission based on the value of the transaction—aims to do for small companies wanting to work globally what Airbnb did for small homeowners wanting to rent rooms globally and for individual tourists who wanted to travel globally and have a homestay experience. It is creating a platform of trust among companies that do not know each other that will allow more global commerce to flow among much smaller players.

Some large multinational corporations are now using the Globality platform to identify small and midsize firms that do high-quality work at lower fees than the large international firms. When the big guys stop transacting only with the other big guys—and start including more and more small players in the global game—another accelerator kicks in for globalization.

## *When the Big Shift Hits Financial Flows*

Globalization was always driven by financial flows, but thanks to the supernova, these now digitized financial flows are happening at almost unfathomable rates. As a result, the interdependence of markets—particularly major markets—is becoming tighter and tighter every day. When China's government took some very questionable financial steps in the summer of 2015 and rattled its own markets, Americans immediately felt it in their retirement accounts and stock portfolios. On August 26, 2015, CNN.com reported:

> The American stock market has surrendered a stunning $2.1 trillion of value in just the last six days of market chaos.
>
> The enormous losses reflect the deep fears gripping markets about how the world economy will fare amid a deepening economic slowdown in China.
>
> The Dow, S&P 500 and Nasdaq have all tumbled into correction territory, their first such 10% decline from a recent high since 2011.
>
> The S&P 500—the best barometer for the biggest U.S. companies—has lost trillions of market value in the six-day sell-off through Tuesday, according to S&P Dow Jones Indices . . .
>
> It's like erasing almost the entire value of the British version of the S&P 500. Known as the S&P BMI U.K. . . .
>
> The dramatic retreat on Wall Street has been fueled by serious concerns about the fallout of China's economic slowdown.

With more ways being created every month to digitize money—loans, deposits, withdrawals, checking, trading, and bill-paying—this interdependence will only tighten faster. That subject is worth a book on its own, so I can only give a taste of it here, and the best time and place to take a nibble is May 6, 2010, at 9:30 a.m.

The Dow Jones Industrial Average had just opened that morning and stood at 10,862. It seemed like a nondescript day. But five hours later, it would make history. Starting at 2:32 p.m. the Dow began to tumble. By 2:47 p.m., it had fallen by 9 percent—the largest intraday point drop from the opening ever, declining 998.5 points to 9,880. One hour and thirteen minutes later, at 4:00 p.m., it closed the day at 10,517, recovering most of its losses. Depending on whether you bought or sold within those ninety minutes you could have either made or lost the gross national product of a good-sized country: the sudden drop created more than $1 trillion in losses in the space of thirty minutes.

How could market sentiment have changed so much, so fast? What were people thinking?

People weren't thinking—machines were. This was computer-driven algorithms misfiring in the age of accelerations and interdependence.

It took a while to figure out what happened, but on April 21, 2015, British authorities arrested Navinder Singh Sarao, age thirty-six, on the request of U.S. prosecutors, who alleged that he'd helped cause the crash and profited from it to the tune of $875,000. What's amazing is that Sarao operated with a computer and network connection *out of his parents' home* in Hounslow, West London. But in a hyperconnected world, he managed to use computer algorithms to manipulate the market by creating fictitious orders that "spoofed" the Chicago Mercantile Exchange and, authorities maintain, set off a chain reaction.

"Spoofing," explained Bloomberg.com on June 9, 2015, is "an illegal technique that involves flooding the market with bogus buy or sell orders to drive prices one way or another. The idea is to fool other traders, both humans and computers, to enable the perpetrator to buy low or sell high . . . Authorities say Sarao developed his computer algorithms in June 2009, to alter how his orders would be perceived by other computers . . . [He created] an algorithm that gave a misleading impression of the volume of sell orders."

His methods were different from those used by high-frequency trading firms, but the outcome he allegedly helped produce was amplified by the presence of so many such firms in the market, as well as by the advances in computer-driven, high-speed global trading. Spurred by

Moore's law, these firms have created an arms race for who can execute more trades faster. Indeed, the speeds being sought now are so fast that in researching this aspect of globalization, I found some of the most helpful background materials not in financial journals, but in a science/physics journal.

For instance, *Nature*, the international weekly journal of science, published a piece on February 11, 2015, entitled "Physics in Finance: Trading at the Speed of Light," in which it observed:

> [Financial traders] are in a race to make transactions ever faster. In today's high-tech exchanges, firms can execute more than 100,000 trades in a second for a single customer. This summer, London and New York's financial centers will become able to communicate 2.6 milliseconds (about 10 percent) faster after the opening of a transatlantic fiber-optic line dubbed the Hibernia Express, costing $300 million. As technology advances, trading speed is increasingly limited only by fundamental physics, and the ultimate barrier—the speed of light . . .
>
> High-frequency trading relies on fast computers, algorithms for deciding what and when to buy or sell, and live feeds of financial data from exchanges. Every microsecond of advantage counts. Faster data links between exchanges minimize the time it takes to make a trade; firms fight over whose computer can be placed closest; traders jockey to sit closer to the pipe. It all costs money—renting fast links costs around $10,000 per month.

The jockeying is so intense, *Nature* reported, that traders figured out that "fiber-optic cables carry the most data, but do not give the speed required. The fastest links carry information over a geodesic arc—the shortest path on Earth's surface between two points. So line-of-sight microwaves are a better option; millimeter waves and lasers are better yet, because they have higher data densities." Fast trading does keep markets liquid, *Nature* noted, which can "benefit trade in the same way that free-flowing traffic helps transport. Such markets tend to have low 'spreads'—the difference between the prices at which one can buy or sell a stock, which reflects the fee that dealers demand and thus transaction costs for investors."

But there are real downsides, it added: "The algorithms they use to trade profitably make more errors and are programmed to get out of the market altogether when markets get too volatile. The problem is exacerbated by the similarity of the algorithms used by many high-frequency trading firms—they all bail out at the same time. That is what happened in the 2010 flash crash." Humans can do the same but machines can do it bigger and faster and, arguably, can be more easily spoofed into huge losses. "In 2012, a flaw in the algorithms of one of the largest US high-frequency trading firms, Knight Capital, caused losses of $440 million in forty-five minutes as its system bought at higher prices than it sold."

But my favorite line in the *Nature* article was still to come. The story pointed out that "in the United States, some large trading firms have set up private trading spaces to eliminate the timing edge for high-frequency traders. For example, the alternative trading system IEX, launched in 2013 . . . has introduced a trading 'speed bump'—an automatic delay of 350 microseconds—which makes it impossible for traders to benefit from the faster feeds."

*Really?* So 350 microseconds in today's market constitutes a "speed bump." I was immediately reminded of the Walmart engineer telling me that once I pressed "buy," their computers had all kinds of time to figure out how to deliver my television—they had under a second.

No wonder *Nature* concluded that "finance research suggests that there may be an optimal speed for trading that today's markets have already far surpassed." Either way, there is little sign that "speed bumps" can reverse the fact that global markets have never been more interdependent. Moore's law just keeps driving innovations to weave buyers and sellers, savers and investors, into a tighter web, explained Michael L. Corbat, Citigroup's chief executive officer, who offered one of my favorite examples.

If you were a U.K. pensioner living in Australia, he told me, the Treasury used to cut you a check and put it on a mail truck to Heathrow Airport, where it got sorted, put on a flight to Sydney, dumped into a sorting bin, and distributed by the Post Office in Australia, eventually arriving in your mailbox somewhere between the seventh and tenth of the month. You would then deposit it and ask that it be converted to Aussie dollars. On the twentieth or so of the month, those Aussie dollars would show up in your account, minus a fee.

Citibank came along, though, said Corbat, and said: "We can put good money in their account the next day and do it cheaper—transferring it electronically in local currency." So the U.K. entrusted Citibank with that task, and then others in Europe and Asia did the same. But then one day, recalled Corbat, "Italy said to us: 'We have some pensioners who are over one hundred years old' "—living in some really remote places. " 'How do we wire them funds?' To serve them electronically we needed to have proof of life. This used to be done with forms and notaries. Now we are going paperless." Fortunately, there was a solution. Elderly pensioners can now confirm their identity through Web portals and claim their pension, and the money gets deposited in their accounts. How? It turns out, Corbat explained, that a person's voiceprint is actually more accurate than their fingerprint, iris scan, or any other means of identification. And as more consumers use their smartphones to pay for things, access data, and check on their accounts, passwords and PINs are less workable. So your unique voice now becomes the key that opens all doors. "Now when a credit card customer calls a service center, you have the option of no longer entering a code, a PIN, or a Social Security number," said Corbat. "You just say, 'Hi, this is Tom Friedman,' and we know from your voice that it is you. So the system says, 'Hi, Tom, would you also like to check your balance?' And it knows it is you and it starts to learn what you like to do." All that traffic got digitized and automated, and some of it has now become voice-activated, said Corbat, "and that gives us the time and resources to really deal with the stuff that creates dissatisfaction."*

One of the most important drivers of the digitization of finance today is PayPal, the digital payments platform that got its start as part of eBay and specializes in the secure, high-speed digital transmission of all

*It is not only the recognition of voices and irises that's improving rapidly. Digitized facial recognition software has gotten so good that the Chinese search company Baidu has dispensed with ID cards for workers entering their main Beijing campus. At the entry plaza, employees and guests just pause a moment before cameras, which instantaneously compare their visages with facial recognition images stored in a data bank; only those who don't match are stopped. In the not-too-distant future, you will probably never need to carry a ticket to a play, concert, or sporting event. You will just register your face online with the ticket company, it will be digitally transferred to the event venue, and cameras will then match you when you arrive—no waiting, no muss, no fuss.

kinds of financial transactions, involving everyone from the most remote buyers and sellers to the most connected.

Dan Schulman, PayPal's CEO, explained that the company's goal "is democratizing financial services and making the opportunity to move and manage money a right and possibility for every citizen—not just for the affluent." Banks, he explained, "were established in an era dominated by physical presence, not digital flows, and the physical world had expensive infrastructure. A branch needs thirty million dollars of deposits to be profitable. And so where are banks closing? In all the neighborhoods where the average income level is below the national median." They cannot attract enough deposits.

"What happened with the explosion of mobile and smartphones," said Schulman, "is that all the power of a bank branch is now in the palm of a consumer's hand. And the incremental cost to add a customer when software is at scale approaches zero. All of a sudden the transaction of cashing a check or paying a bill or getting a loan or sending money to someone you love—which were simple and easy things for us in America already—becomes simple and easy"—and nearly free—for the three billion people around the world who have been underserved. These are people who for decades have been "standing in line for three hours to get their currency changed and then go to another line to pay a bill and then get charged ten percent [in fees]. The technology is shifting the possibilities for them dramatically."

For instance, PayPal created a global loan platform called Working Capital that can underwrite loans for PayPal users in a matter of minutes, instead of the weeks that banks might take. This makes a huge difference to a small business that needs to purchase inventory or has a growth opportunity. The product has already extended $2 billion in loans in less than three years. How can PayPal do that?

Big data.

Schulman explained:

> The key is the amount of data you can now analyze. I can take all the data exhaust from our platform—six billion transactions a year and rising exponentially—and it allows us to make better decisions. You want a loan? If you are a regular PayPal customer, we know you. *And we know everyone else like you.* And we know

> that you have not changed—but [maybe] your circumstances changed because you lost your job or there was a natural disaster, and we know you will find another [job]. In a second, using our algorithms, we can compare you to anyone else like you around the world, because we have all this data and modeling capabilities and we can give you a loan based on those models.

PayPal Working Capital does not rely on FICO scores, the traditional credit scoring system used by banks and credit card companies representing a person's creditworthiness and likelihood they will pay back a loan. That's because someone once may have declared bankruptcy—and therefore has a permanent stain on their FICO record, said Schulman. PayPal has found that its own big data analytics based on your actual financial transactions on their site give them a much more reliable picture of your creditworthiness than FICO. This approach allows them to give instant loans to more people around the world with a higher rate of payback.

Using the same big data analytics, PayPal is also able to guarantee every transaction on its platform. So if a small merchant in India opens a website selling Indian saris and a customer in Europe buys two saris from that Indian merchant paying through PayPal, that customer "is either going to get the sari they ordered or we will refund the money," said Schulman. "And we can make that guarantee work because, again, *we know you*, and we have all the data . . . We have one hundred ninety million customers worldwide and are adding over twenty million new ones a year." These guarantees are also driving more globalization.

Slowly but surely people are using PayPal to do away with cash.

Like all big financial players, PayPal is experimenting with the emerging technology known as "blockchain" for validating and relaying global transactions through multiple computers. Blockchain, which is most famously used by the virtual currency Bitcoin, "is a way of enabling absolute trust between two parties making a financial transaction," explained Schulman. "It uses Internet protocols to make the transaction go around any nation-state in a way that is visible to all the participants and goes beyond all middlemen and regulatory bodies—and therefore has the promise of lower costs."

## *Reinventing Medicine and Manufacturing*

As in fintech so in medtech—digital globalization is rapidly transforming medicine, making medical delivery a hybrid system. More and more diagnoses and prescriptions will be delivered half online, especially through cell phones and half in person. The result should be new opportunities to cut costs and to extend delivery into remote locations.

I met British-born Dr. Peter Yellowlees, a professor of psychiatry at University of California-Davis, at a telemedicine conference in Orlando, where he explained his partnership with a British-based charity, the Swinfen Foundation, founded by Lord Roger and Lady Pat Swinfen. The Swinfens have gone around the world and set up 150 medical health care nodes in diverse, underserved regions. They donate laptops, cameras, generators, and satellite dishes to a local clinic that may be off the grid. The charity then trains the local health providers in the basics of how to access the Swinfen hub, so when a patient comes in they can easily get a second opinion (or a first) via satellite or the Internet. When the local staff send in their question about the patient's condition or their X-ray or EKG, Swinfen shuttles it to one of their five hundred specialists, primarily in Britain, America, and Australia, who have volunteered to offer a free opinion. Most are cardiologists, pediatricians, and orthopedists. Swinfen will put the specialist directly in contact with the local healthcare worker managing the case.

"I am one of their psychiatric specialists," explained Yellowlees. "I live in Sacramento and have been on their consultant list for about ten years. About three years ago I got an email from an English resident who was on a working vacation in Nepal, in a very isolated area. This resident was a second-year, so pretty inexperienced, and he explained that he had a patient who had become very violent, very aggressive, and was breaking up their twenty-bed hospital. He was threatening to kill himself and everyone there. The resident thought the man was psychotic—either bipolar or schizophrenic. So I asked him what drugs he had on hand, and they were fifteen to twenty years old. I told him by email what to give the patient and this calmed [the patient] down and he stopped breaking up the place. As the [local medical staffer] detailed the man's mental state on email, I thought it was more likely that he had delirium—that he was medically ill but not psychotic—and so I asked him to run some extra tests that they could do locally and it turned out that the man had cerebral malaria—an infectious disease that can cause an inflammation of the brain and make people present with confusion. Malaria requires a very different treatment so we went back and forth on email and he was treated appropriately for his medical condition and recovered. It all took me half an hour. If I can do this in Nepal, I can do this anywhere in the world, in Bangladesh, Nigeria, or anywhere necessary in America, and provide really good medicine at low cost."

Thanks to the acceleration in digital globalization, concluded Yellowlees, "I think the health-care system is going to turn into this hybrid of electronic and personal medicine."

The same shift to digital flows is happening around manufacturing. In May 2017, I visited the Oak Ridge National Laboratory in Oak Ridge, Tennessee. Oak Ridge partners with local communities in a variety of ways. It offers entrepreneurial research fellowships in advanced manufacturing to top-level local technical talent who want to start their own companies in this realm. Every summer Oak Ridge's Manufacturing Demonstration Facility (M.D.F.) hosts one hundred young interns to learn the latest in 3-D printing, and its experts coach teams from local high schools for national robotics competitions. As I first learned visiting the GE labs in upstate New York, with 3-D printers, any community can now go into the manufacturing business. Lonnie Love, a corporate fellow at Oak Ridge, reiterated the point on my visit to the M.D.F., where whole car bodies and car parts are being "printed" on giant 3-D printers.

"Traditionally to make a car part you first had to build a die, and those dies cost anywhere from $500,000 to $1 million to make," Love said. Every die consists of a female and a male die, and the way you made a car part was to stamp them together. There are hundreds of dies needed to make a car, and that was why an assembly line for a new car model in Detroit could cost upward of $200 million—and take two years to build. Sadly, that die-making industry actually moved out of America to Asia over the last fifteen years, leaving only a dozen such companies in the United States.

No more. "Large-scale 3-D printing is enabling us to re-shore that industry," explained Love, who then offered this example: the Fleet

Readiness Center located at the Marine Corps Air Station, in Cherry Point, North Carolina, performs repairs to all the Navy and Marine Corps' vertical lift aircraft from all squadrons worldwide. In early 2014, the head of science and technology at the base, Robert Kestler, contacted Love with a seemingly innocuous request. It was how Love's team turned it around that he never forgot.

Kestler "called me on a Monday," Love recalled, "and asked if we could print a die mold for them and I said, 'Sure, just send me the digital model of what you want printed.' We got it by email that afternoon, and by Friday he had the mold to make the new part. And it only weighed about forty pounds because with 3-D printing we could make it stronger but lighter weight by hollowing out the inside. The following Monday he calls and asks me how much did it cost and how long did it take me to make? I told him it took me longer—and was more expensive—to ship it to him than it was to make it."

*It took me longer—and was more expensive—to ship it to him than it was to make it.* That is the sound of complexity becoming free . . .

## *When the Big Shift Hits Strangers*

On February 24, 2016, Facebook announced that as part of its "A World of Friends" initiative, it was tracking the number of relationships forged on its site by longtime foes. Facebook said that on just that *one day* it had connected 2,031,779 people from India and Pakistan, 154,260 from Israel and Palestine, and 137,182 from Ukraine and Russia. How many strong friendships emerged from those exchanges, how long-lasting they will be, and whether they will contribute to overcoming deep historical enmities—these are different matters. But you would have to be a total curmudgeon to look at those numbers and not think that they represent some pretty impressive contact between strangers and enemies.

The acceleration of flows is clearly accelerating all forms of human contact, particularly contact between strangers. Almost no matter where you are on the planet, except the most remote locations, you are likely being exposed, directly or indirectly, to more contact with more different ideas and people than ever before in human history. It's one reason that I have found myself reading the work of the late historian William H.

McNeill, author of the classic history *The Rise of the West.* On the twenty-fifth anniversary of that book, in May 1995, McNeill wrote an essay for the journal *History and Theory,* "The Changing Shape of World History," re-asking and answering one of the most profound questions for historians that animated his original work: What is the engine of history? What drives more history forward than any other factor or factors?

Is it what he described as "the sporadic but ineluctable advance of Freedom," which "allowed nationalistic historians to erect a magnificently Eurocentric vision of the human past, since Freedom (defined largely in terms of political institutions) was uniquely at home among the states of Europe, both in ancient and in modern times"? In this view, "the rest of the world, accordingly, joined the mainstream of history when discovered, settled or conquered by Europeans."

No, McNeill decided, that was not the engine of history—World War I put paid to that thought, since "freedom to live and die in the trenches was not what nineteenth century historians expected liberal political institutions to result in."

So he offered up another popular alternative: "Spengler and Toynbee were the two most significant historians who responded to . . . the strange disembowelment that Freedom suffered in World War I," McNeill wrote. Their view was that

> human history could best be understood as a more or less foreordained rise and fall of separate civilizations, each recapitulating in essentials the career of its predecessors and contemporaries . . .
>
> To many thoughtful persons, their books gave a new and somber meaning to such unexpected and distressing events as World War I, Germany's collapse in 1918, the onset of World War II, and the breakup of the victorious Grand Alliances after both wars . . . Spengler and Toynbee put European and non-European civilizations on the same plane.

McNeill then offered a third answer—his own theory of what drives history in *The Rise of the West*, a theory that he felt even more strongly about with the passage of time: "The principal factor promoting historically significant social change is contact with strangers possessing new and unfamiliar skills." The corollary of that proposition, he argued, is that centers of high skill (i.e., civilizations) tend to upset their neighbors by exposing them to attractive novelties. Less-skilled peoples round about are then impelled to try to make those novelties their own so as to attain for themselves the wealth, power, truth, and beauty that civilized skills confer on their possessors. Yet such efforts provoke a painful ambivalence between the drive to imitate and an equally fervent desire to preserve the customs and institutions that distinguish the would-be borrowers from the corruptions and injustices that also inhere in civilized life.

McNeill explained:

Even though there is no perceptible consensus about what the term "civilization" ought to mean, and no agreed word or phrase to describe the "interactive zone" . . . I think it correct to assert that recognition of the reality and historical importance of trans-civilizational encounters is on the increase and promises to become the mainstream of future work in world history . . .

When I wrote *The Rise of the West* I set out to improve upon Toynbee by showing how the separate civilizations of Eurasia interacted from the very beginning of their history, borrowing critical skills from one another, and thus precipitating still further change as adjustment between treasured old and borrowed new knowledge and practice became necessary . . .

The ultimate spring of human variability, of course, lies in our capacity to invent new ideas, practices and institutions. But invention also flourished best when contacts with strangers compelled different ways of thinking and doing to compete for attention, so that choice became conscious, and deliberate tinkering with older practices became easy, and indeed often inevitable.

## *Contact on Steroids*

I am a deep believer in McNeill's view of history, which aligns with everything I saw as a foreign correspondent. Just as the climate is changing, and weather circulates differently, so globalization is reshaping how fast ideas circulate and change. And that is now posing some real adaptation challenges. As a result of all the accelerating flows, we are seeing *contact between strangers on steroids* today—civilizations and individuals encountering, clashing, absorbing, and rejecting one another's ideas in myriad new ways—through Facebook, video games, satellite TV, Twitter, messaging apps, and mobile phones and tablets. Some cultures, societies, and individuals are predisposed to absorbing contact with strangers, learning from them, synthesizing the best, and ignoring the rest. Others, more brittle, are threatened by such contact, or easily humiliated by the fact that what they thought was their superior culture must now adapt to and learn from others.

The difference between those cultures that can handle and take advantage of this explosion of contact between strangers and their strange ideas and those that cannot will drive a lot of history in the age of accelerations, even more than in those eras McNeill wrote about. Specifically, those societies that are most open to flows of trade, information, finance, culture, or education, and those most willing to learn from them and contribute to them, are the ones most likely to thrive in the age of accelerations. Those that can't will struggle.

The benefits of being in the flow are exemplified by the work of people like Professor Hossam Haick of the Technion, Israel's premier science and technology institute. Prof. Haick is an Israeli. He is an Israeli Arab. He is an Israeli expert in nanotechnology. And he was the first Israeli Arab professor to teach a massive open online course, or MOOC, on nanotechnology—in Arabic, based out of an Israeli university.

Needless to say, he explained to me when I visited him to write a column in February 2014 in Haifa, he got some very interesting e-mails from students registering for his MOOC from all over the Arab world. Their questions included: Are you a real person? Are you really an Arab, or are you an Israeli Jew speaking Arabic, pretending to be an Arab? Haick is a Christian Arab from Nazareth and was teaching this course from his home university, the Technion.

His course was entitled Nanotechnology and Nanosensors and was designed for anyone interested in learning about Haick's specialty—"novel sensing tools that make use of nanotechnology to screen, detect, and monitor various events in either our personal or professional life." The course included ten classes of three to four short lecture videos—in Arabic and English—and anyone with an Internet connection could tune in and participate for free in the weekly quizzes and forum activities and do a final project.

If you had any doubts about the hunger for education in the Middle East today—and how it can overcome the alienation of strangers, not to mention old enemies—Haick's MOOC would dispel them. He had nearly five thousand registrations for the Arabic version, including students from Egypt, Syria, Saudi Arabia, Jordan, Iraq, Kuwait, Algeria, Morocco, Sudan, Tunisia, Yemen, the United Arab Emirates, and the West Bank. Iranians were signing up for the English version. Because the registration was through the U.S.-based Coursera MOOC website, some registrants initially didn't realize the course was being taught by an Israeli Arab scientist at the Technion, said Haick, and when they discovered that fact some professors and students unregistered. But most didn't.

Asked why he thought the course was attracting so much interest in the neighborhood, Haick told me: "Because nanotechnology and nanosensors are perceived as futuristic, and people are curious to understand what the future looks like." Haick, forty at the time, whose PhD was from the Technion, where his father also graduated, is a science prodigy. He and the Technion already had a start-up together, developing what he calls "an electronic nose"—a sensory array that mimics the way a dog's nose works to detect what Haick and his team have proved to be unique markers in exhaled breath that reveal different cancers in the body. In between that and teaching chemical engineering, the Technion's president, Peretz Lavie, suggested that Haick lead the school into the land of flows and MOOCs.

Lavie, Haick explained, thought there was "a high need to bring science beyond the boundaries between countries. He told me there is something called a MOOC. I did not know what is a MOOC. He said it is a course that can be given to thousands of people over the Web. And he asked if I can give the first MOOC from the Technion—in Arabic." The Technion funded the project, which took nine months to prepare, and Haick donated the lectures. Haick said, without meaning to boast, "I have young people who tell me from the Arab world: 'You have become our role model. Please let us know the ingredients of how we become like you.'"

On February 23, 2016, the Associated Press carried an interview with Zyad Shehata, an Egyptian student who completed Haick's course. "Some people told me to remove this certificate from my résumé," said Shehata. "They said that I might face some problems. I have no interest in whether it is an Israeli university or not, but I'm very proud of Professor Haick and I see him as a leader."

Translation: *Never get between a hungry student and a new flow of knowledge in the age of accelerations.*

## *The Melting of Minds*

All of this contact between strangers, together with the accelerated flow of ideas on social networks, is surely contributing to rapid changes in public opinion. Points of view, traditions, and conventional wisdom that looked to be as solid as an iceberg, and just as permanent, can now suddenly melt away in a day, in ways that used to take a generation.

The Confederate flag flew over South Carolina's statehouse grounds for fifty-four years. But on July 10, 2015, it was lowered for good by a South Carolina Highway Patrol honor guard just a few weeks after nine worshippers were gunned down at a historic black church in Charleston by a self-proclaimed white supremacist—who had posed for a picture with the symbol of the Confederacy. The killings triggered a huge social network blowback, and just like that, that Confederate flag was gone from the statehouse grounds.

Running for president, on April 17, 2008, Barack Obama declared: "I believe that marriage is the union between a man and a woman. Now, for me as a Christian—for me—for me as a Christian, it is also a sacred union. God's in the mix." A mere three years later, on October 1, 2011, President Obama, talking about one of the oldest conventions in the history of relations between men and women, told a Human Rights Campaign annual dinner that he now supported gay marriage: "Every single American—gay, straight, lesbian, bisexual, transgender—every single American deserves to be treated equally in the eyes of the law and in the eyes of our society. It's a pretty simple proposition."

When you see how fast attitudes toward lesbians, gays, bisexuals, and transgender people have changed in just five years, argued Marina

Gorbis, executive director at the Institute for the Future in Palo Alto, "you have to believe it has something to do with so many young people being immersed in what is increasingly a global dialogue—often about values." This system, she added, "amplifies everything that goes through it, so it creates feedback loops used to bully people, and it creates more points of interaction and many more opportunities for people who are homophobic to meet a gay person. And now suddenly so many more people are meeting gay people. If empathy comes about through human interaction, this system creates so many more opportunities for that."

The day I interviewed Gorbis, Bettina Warburg, a researcher at the Institute for the Future, told me this story from her recent commute in the San Francisco area: "I was riding in a Lyft the other morning—where you ride-share with others headed in the same direction. My driver chatted with me and mentioned his last [passenger] was 'voted out of the car,' because he was expressing extreme homophobic rhetoric. He said, 'You won't get a ride in San Francisco with those values—you are in the wrong city.' We were a black, a Hispanic, and a woman in the car talking about how intolerance does not jibe with an economy built on platforms that value participation."

Given the myriad new technological opportunities to have contact with strangers, "the conception of community is going to evolve," said Justin Osofsky, VP of global operations and media partnerships at Facebook. In the pre-Facebook, pre–social network era, the notion of community "was constrained around you, around that time and around that place." Now, with social networks, you have "the ability to maintain relationships through every context of your life if you choose to"—and to create new contexts for relationships that were unimaginable just a decade ago. "Without this level of connectivity you used to live your life in separate chapters," he explained, "and you grew as a person in each one, but now there is a connectivity between chapters" and it is possible to open chapters far outside your geographic context that include people with shared interests. "Our mission is to connect the world. And as that happens, 'the nature of community' will evolve. In the past you basically had two life choices—stay in a community or leave it." Today, he said, "if you grew up in the world of mobile phones with Facebook, the connectivity to a community can remain strong both for those who stay and those who leave."

Moreover, "if you are the world expert on Eritrean politics you can find an audience of like-minded people at an enormous scale," said Osofsky. "You or your child could have a rare illness, and before Facebook, you were feeling lonely and lost." Now you can instantly "find support groups going through the same thing."

That's the best part of the globalization of flows today—its ability to foster contact between like-minded strangers or transform old friends who had become strangers back into friends and a community.

Unfortunately, there is also a downside to this ease of finding the like-minded. Some people crave support groups to become neo-Nazis and suicidal jihadists. Social networks have become a godsend for extremists to connect with one another and to recruit young and impressionable strangers, and the supernova just keeps increasing their firepower. It is troubling but unavoidable. (I will discuss this further in chapter nine when we deal with "the breakers.") But for now I see many more upsides than downsides.

Indeed, it is actually quite exciting how easy it can be to summon the flow to fight bad things and promote good ones. Ben Rattray founded Change.org in 2007 to create a platform where any digital David could take on any Goliath: corporate, governmental, or otherwise. *Fast Company* described Change.org on August 5, 2013, as "the destination of choice for amateur activists and squeaky wheels of all types." It now has more than 150 million global users, a number that is growing steadily—and they launch more than a thousand petitions a day. Change.org provides both advice for how to launch an online petition and a global platform on which to publish it and draw attention and supporters alike.

Striking testimony to Change.org's ability to leverage global flows to drive change faster comes from Ndumie Funda, a South African lesbian whose fiancée was gang-raped—a so-called corrective rape—by five men because of her sexuality. As a direct result of the attack, the fiancée developed cryptococcal meningitis, an infection of the brain and spinal column, and died on December 16, 2007. "Corrective rape is a relatively new term," Funda explained in a February 15, 2011, interview on WomenNewsNetwork.net. "This 'hate-filled' form of rape is found worldwide. Based on the idea that forcing a lesbian to have sex with a man will 'cure' her of a 'deviant life,' it is accompanied most often by extreme violence."

In December 2010, Funda sat down at a Cape Town Internet café and launched a petition through Change.org that demanded government action to end the "corrective rape" of lesbians in South Africa's shantytowns. It almost immediately collected 170,000 signatures worldwide. Another petition drive was started by the digital activist site Avaaz .org, reported WomenNewsNetwork.net. Together the two petitions got nearly one million signatures worldwide and embarrassed the South African Parliament into creating a national task force to delegitimize the practice. Gay marriage has been legal in South Africa since 2007, and although "corrective rape" is still a problem, the perpetrators no longer enjoy the easy public acceptance they once did.

I asked Rattray what he and his Change.org team learned from the experience. He answered: "If you ask people about a large social problem like rape, they will tell you they're against it, yet they will rarely do anything about it. But if you tell them a personal story about someone directly affected and give them an opportunity to join a movement for change, they will often immediately respond by taking action."

## *Build Floors, Not Walls*

Globalization has always been everything and its opposite—it can be incredibly democratizing and it can concentrate incredible power in giant multinationals; it can be incredibly particularizing—the smallest voices can now be heard everywhere—and incredibly homogenizing, with big brands now able to swamp everything anywhere. It can be incredibly empowering, as small companies and individuals can start global companies overnight, with global customers, suppliers, and collaborators, and it can be incredibly disempowering—big forces can come out of nowhere and crush your business when you never even thought they were in your business. Which way it tips depends on the values and tools that we all bring to these flows.

In the face of more and more uncontrolled immigration, globalization today feels under threat more than ever. We saw that in the vote by Great Britain to withdraw from the European Union and in the candidacy of Donald Trump. But disconnecting from a world that is only getting more digitally connected, from a world in which these digital flows will be a vital source of fresh and challenging ideas, innovation, and commercial energy, is not a strategy for economic growth.

That said, people have bodies and souls, and when you feed one and not the other you always get in trouble. When people feel their identities and sense of home are being threatened, they will set aside economic interests and choose walls over Webs, and closed over open, in a second—not everyone will make that choice, but many will.

The challenge is to achieve the right balance. In retrospect, in too many ways on too many days, we have failed to do that in the big Western industrial democracies over the past decade. If many Americans or Europeans are feeling overwhelmed these days by globalization, it's because we've let the rapid expansion in digital flows and the rapid expansion of trade flows and the rapid expansion of immigration flows get way too far ahead of the social technologies—our learning, adapting, and cushioning tools. As a result, all of this contact between strangers has often felt overwhelming to people—or threatening to the things that anchor them: their jobs, local culture, and sense of home, neighborhood, and nationhood.

Warning: in the age of accelerations, if a society doesn't build solid floors under people, many will reach for a wall—no matter how self-defeating that would be. Addressing that anxiety is one of today's great leadership challenges, and I will discuss it later in the book.

We can still get this balance right—and the incentives to do so, to keep trying to get the best out of digital globalization and cushion the worst, are enormous. The job of politics is to figure out this new balance, not throw out the baby with the bath water. We dare not let the downsides of globalization that need to be cushioned blind us to the many upsides that need to be scaled.

Today's mobile-broadband supernova is creating so many flows and thus enabling so many more people to lift themselves out of poverty and participate in solving the world's biggest problems. We are tapping into many more brains and bringing them into the global neural network to become "makers." This is surely the most positive—but least discussed or appreciated—trend in the world today, when "globalization" is becoming a dirty word because it is entirely associated in the West with dislocations from trade.

That's why I want to give the last word in this chapter to Dr. Eric C.

Leuthardt, a neurosurgeon and the director of the Center for Innovation in Neuroscience and Technology at the Washington University School of Medicine in St. Louis, who asked and answered the question "Why Is The World Changing So Fast?" on his "Brains & Machines" blog:

> I would posit the reason for the accelerating change is similar to why networked computers are so powerful. The more processing cores that you add the faster a given function can occur. Similarly, the more integrated that humans are able to exchange ideas the more rapidly they will be able to accomplish novel insights. Different from Moore's Law, which involves the compiling of logic units to perform more rapid analytic functions, increased communication is the compiling of creative units (i.e., humans) to perform ever more creative tasks.

SIX

# *Mother Nature*

***God always forgives. Man often forgives. Nature never forgives.***

***—Saying***

***We are wickedly bad at dealing with the implications of compound math.***

***—Jeremy Grantham, investor***

On July 31, 2015, *USA Today* reported that in Bandar Mahshahr, a city of a hundred thousand residents in southwestern Iran that is adjacent to the Persian Gulf, the heat index soared to a staggering 163 degrees Fahrenheit:

> A heat wave continued to bake the Middle East, already one of the hottest places on earth.
>
> "That was one of the most incredible temperature observations I have ever seen, and it is one of the most extreme readings ever in the world," AccuWeather meteorologist Anthony Sagliani said in a statement.
>
> While the temperature was "only" 115 degrees, the dew point was an unfathomable 90 degrees. The combination of heat and humidity, measured by the dew point, is what makes the heat index—or what the temperature actually feels like outside.

> "A strong ridge of high pressure has persisted over the Middle East through much of July, resulting in the extreme heat wave in what many would consider one of the hottest places in the world," Sagliani said.

Reading that story, I was reminded of a new phrase I had learned a year earlier while attending the World Parks Congress in Sydney, Australia. The phrase was "black elephant."

A "black elephant," it was explained to me by the London-based investor and environmentalist Adam Sweidan, is a cross between a "black swan"—a rare, low-probability, unanticipated event with enormous ramifications—and "the elephant in the room: a problem that is widely visible to everyone, yet that no one wants to address, even though we absolutely know that one day it will have vast, black-swan-like consequences."

"Currently," Sweidan told me, "there are a herd of environmental black elephants gathering out there"—global warming, deforestation, ocean acidification, and mass biodiversity extinction, just to name four. "When they hit, we'll claim they were black swans that no one could have predicted, but in fact they are black elephants, very visible right now"—we're just not dealing with them with the scale and speed that are necessary.

A 163-degree heat index in Iran is, indeed, a black elephant: you can see it sitting in the room, you can feel it, and you can read about it in the newspaper. And like any black elephant, you also know that it is so far outside the norm that it has all the characteristics of a black swan—that it is the harbinger of some very big, unpredictable changes in our climate system that we may be unable to control. Yet somehow this has not penetrated the mass consciousness of Washington, D.C., and particularly the Republican Party. "During the Cold War we wrote a blank check to deter a low-probability event—a nuclear war—with high consequence," observed Robert Litwak, a vice president of the Wilson Center and a former adviser to President Clinton on nuclear proliferation. "Now we won't even put a nickel [tax] on gasoline to deter a high-probability event—climate change—with high consequence."

It's true that no single weather event can conclusively tell you anything one way or another about climate change, but what is striking about this moment is how many totally outside-the-norm weather and climate readings have been piling up. These readings scream to us that when it comes to climate change, biodiversity loss, and population growth, particularly in the most vulnerable countries, Mother Nature has also entered the second half of the chessboard, just like Moore's law and the Market. And in many ways, she has been driven there by the multiple accelerations in technology and globalization.

When you have more and more people on the planet and then you amplify the impact that each one person can have, the "power of many" can become incredibly constructive, if harnessed to the right objectives. But left unrestrained, untempered by any kind of conservation ethic, it can be an incredibly destructive force. And that is what's been happening. While the power of men and machines and flows has been reshaping the workplace and politics and geopolitics and the economy, and even some of our ethical choices, the power of many is driving the acceleration in Mother Nature, which is reshaping the whole biosphere, the whole global ecological system. It is altering the physical and climatic contours of Planet Earth, the only home we have.

## *Learning to Speak Climate*

You can actually hear the changes before you see some of them. Just listen to how people speak these days, the expressions they now use. They know something is up. I call this language "climate-speak." It's already spoken in a lot of countries, and our kids certainly will be fluent in it. You've probably been speaking it yourself, but you just don't know it.

I first learned climate-speak while writing columns about Greenland's ice sheet, which I toured in August 2008, with Connie Hedegaard, then Denmark's minister of climate and energy. Greenland is one of the best places to observe the effects of climate change. It's the world's biggest island, but it has only fifty-five thousand inhabitants and no industry, so the condition of its huge ice sheet—as well as its temperature, precipitation, and winds—is highly influenced by the global atmospheric and ocean currents that converge there. Whatever happens in China or Brazil gets felt in Greenland. And because Greenlanders live close to nature, they are walking barometers of climate change and therefore fluent in climate-speak.

It's easy to learn. There are only four phrases you have to master.

The first is: "*Just a few years ago, . . . but then something changed . . .*"

This is the Greenland version: Just a few years ago, you could dogsled in winter from Greenland, across a forty-mile ice bank, to Disko Island. But then the rising winter temperatures in Greenland melted that link. Now Disko is cut off. And you can put that dogsled in a museum. According to a study published in the journal *Nature* in December 2015, by fifteen scientists, Greenland is losing ice at an accelerating rate. "We find that 2003–2010 mass loss not only more than doubled relative to the 1983–2003 period, but also relative to the net mass loss rate throughout the twentieth century." NASA currently states that Greenland is losing 287 billion tons of ice per year, *The Washington Post* reported on December 16, 2015. When I visited in 2008 it was "only" 200 billion a year.

**Billion-Dollar-Damage Extreme-Weather Events, 1980–2012**

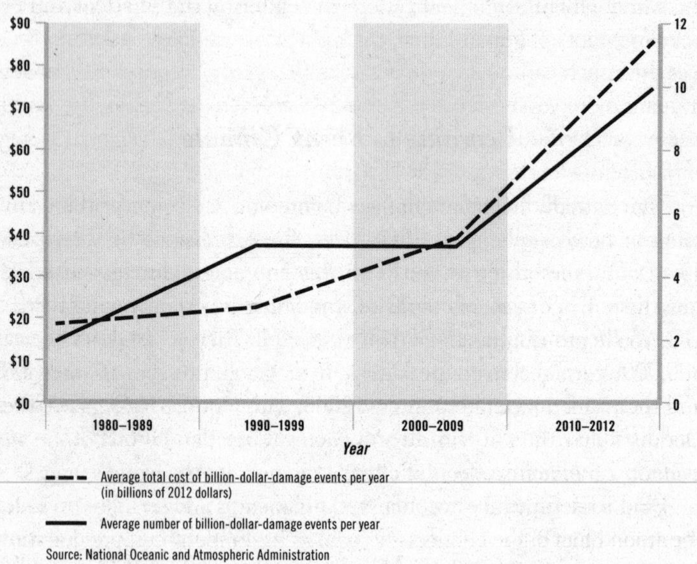

Source: National Oceanic and Atmospheric Administration

The second phrase is: "*Wow, I've never seen that before* . . ." It rained in December and January in Ilulissat, Greenland, the year I visited. This is well above the Arctic Circle! It's not supposed to rain there in winter. Konrad Steffen, then director of the Cooperative Institute for Research in Environmental Sciences at the University of Colorado, which monitors the ice, said to me on that visit: "Twenty years ago, if I had told the people of Ilulissat that it would rain at Christmas 2007, they would have just laughed at me. Today it is a reality."

The third phrase is: "*Well, usually, but now I don't know anymore . . .*" Traditional climate patterns that Greenland elders have known their whole lives have changed so quickly in some places that the accumulated wisdom and intuitions of older people are not as valuable as they once were. The river that was always there is now dry. The glacier that always covered that hill has disappeared. The reindeer that were always there when the hunting season opened on August 1 didn't show up this year . . .

And the last phrase is: "*We haven't seen something like that since . . .*" Then fill in the blank with some crazy large number of years ago. Here's Andrew Freedman writing for ClimateCentral.org on May 3, 2013, after the Mauna Loa Observatory in Hawaii reported for the first time that we had briefly hit the highest $CO_2$ concentration in the atmosphere in human history—four hundred parts per million: "The last time there was this much carbon dioxide ($CO_2$) in the Earth's atmosphere, modern humans didn't exist. Megatoothed sharks prowled the oceans, the world's seas were up to 100 feet higher than they are today, and the global average surface temperature was up to 11°F warmer than it is now."

Or consider the paragraph from a January 7, 2016, story on the environment on Bloomberg.com: "$CO_2$ is famously entering the atmosphere about *100 times faster* than it did when the planet emerged from the most recent ice age, about *12,000 years ago*. The concentration of $CO_2$ in the atmosphere is 35 percent higher than its peak for the last *800,000 years*. Sea-levels are higher than they've been *in 115,000 years*, and the rise is accelerating. A century of synthetic-fertilizer production has disrupted the earth's nitrogen cycle more dramatically than any event *in 2.5 billion years*" (italics added).

And sometimes the records being broken as Mother Nature enters the second half of the chessboard are so numerous and profound, government agencies tracking them seem to run out of even climate-speak to describe the black elephants they are seeing. Here's the National Oceanic and Atmospheric Administration report issued in April 2017: "The combined global average temperature over the land and ocean surfaces for April 2017 was 0.90°C (1.62°F) above the 20th century average of 13.7°C (56.7°F)—the second highest April temperature since global records began in 1880 . . . April 2017 also marks the 388th consecutive month that the globally-averaged temperature across the world's land and ocean surfaces was nominally above the 20th century average. Overall, April 2017 tied with March 2015, August 2016, and January 2017 as the 12th highest monthly global land and ocean temperature departure from average on record." That is the 12th highest out of 1,648 monthly records! And the extremes just keep getting more extreme. In August 2017, the National Weather Service reported that Death Valley National Park set a new high for the hottest July ever recorded, breaking the previous record that had lasted one hundred years. The average temperature was 107.4 degrees. In July 1917, it averaged 107.2 degrees.

That's all climate-speak—"surpassing," "highest," "record," "broken," "biggest," "longest." These numbers are staggering. They're telling you that something big and fundamentally different is happening, something we humans have not experienced in a long, long time. Our planet is being reshaped by the rising *power of many* as the boundaries that have defined our biosphere for millennia are being breached or nearly breached, one by one.

## *Our Garden of Eden*

To understand the importance of this moment from an environmental point of view we need to stop for a quick tutorial on geological epochs.

"The study of the Earth from the beginning of time to the present has been the task of geologists who attempt to unravel the events that have shaped our planet as it is today," explains ScienceViews.com, the history of science website. That is because "the Earth carries the history of geological events in its rock layers . . . By assembling all these layers together, scientists have worked out what is known as the stratigraphic column or record of the various ages of rock. This record spans the 4.6 billion year record of Earth's history. In order to simplify the huge amount of geological information, geologists have broken Earth's history down into sections, which are called geological eras, periods, and epochs."

The Earth was formed about 4.6 billion years ago, but the fossil record shows signs of simple life starting only about 3.8 billion years ago, and complex life forms only about 600 million years ago. Over the millennia, life forms have changed and evolved, depending on the epoch. For the last 11,500 years or so, geologists tell us, we have been in the *Holocene* epoch, which followed the *Pleistocene* epoch, also known as "the Great Ice Age."

Why should we care? Because we will miss the Holocene if it goes, and it appears to be going.

For most of the Earth's 4.6-billion-year history, its climate was not very hospitable to human beings, as it oscillated between "punishing ice ages and lush warm periods" that "locked humanity into seminomadic lifestyles," explained Johan Rockström, director of the Stockholm Resilience Centre, one of the world's leading Earth scientists, and one of my teachers on all things climate. It's only been in the last eleven thousand years that we have enjoyed the calm, stable climate conditions that allowed our ancestors to emerge from their Paleolithic caves and create seasonal agriculture, domesticate animals, erect cities and towns, and eventually launch the Renaissance, the Industrial Revolution, and the information technology revolution.

This period, which geologists named the Holocene, was an "almost miraculously stable and warm interglacial equilibrium, which is the only state of the planet we know for sure can support the modern world as we know it," said Rockström, the author of *Big World, Small Planet*. It finally gave us the ideal balance of "forests, savannahs, coral reefs, grasslands, fish, mammals, bacteria, air quality, ice cover, temperature, fresh water availability and productive soils" on which our civilization was built.

As geological epochs go, the Holocene has been our "Garden of Eden era," added Rockström. In this Holocene we've maintained just the right amount of carbon dioxide in the atmosphere, acidity in the oceans, coral in the sea, tropical forest cover along the equator, and ice at the two poles to store water and reflect the sun's rays, to support human life and a steadily growing world population. The balance between all of them determined our climate and ultimately our weather. And when any of these systems got somewhat out of balance, Mother Nature had an amazing capacity to absorb, buffer, and mute the worst impacts on the planet as a whole.

But that cannot go on indefinitely without limits. Mother Nature's bumpers, buffers, and spare tires are not inexhaustible. And right now all of our climate-speak and all of those black elephants are telling us that we are stressing to the limits and beyond many of the individual systems in our system of systems that has provided human beings with the most stable and benign geological epoch we have ever known—the Holocene.

Talk about *reshaping* the world . . .

"We are threatening to push Earth out of this sweet spot," said Rockström, and into a geological epoch that is not likely to be anywhere near as inviting and conducive for human life and civilization as the Holocene. That is what the current debate is all about.

The essential argument is that since the Industrial Revolution—and particularly since 1950—there has been a vast acceleration of human impacts on all the Earth's key ecosystems and stabilizers that have kept us balanced in the Holocene. These impacts have become so great in recent decades, and have started to transform the operations of so many individual systems, that many scientists believe they are driving us out of the relatively benign Holocene into a new, uncharted geological epoch.

This is what I mean by "the power of many." We as a species are now *a force of, in, and on nature.* That has never been said of humans before the twentieth century, but starting in the 1960s and 1970s, when the Industrial Revolution reached many new parts of the globe with full force, particularly places such as China, India, and Brazil, populations and middle classes started to expand together. In effect, many more people around the world started living an American middle-class lifestyle—cars, single-family homes, highways, air travel, high-protein diets.

Then, starting in the 2000s, the supernova created another surge of global industrial manufacturing, urbanization, telecommunications, tourism, and trade. The combination of all of these trends has begun to put pressure on each one of the Earth's major ecosystems and its plumbing to a degree never witnessed before in the history of our planet. The result: our Garden of Eden way of life is now in danger.

---

## *The Great Acceleration*

In order to establish just how profoundly that is the case, it was important for Earth scientists to try to quantify the accelerating stresses on Mother Nature, which were almost certainly pushing her out of her comfort zones and past certain normal operating boundaries. They gave these stresses a name: "the Great Acceleration." As I noted in chapter one, the Great Acceleration graphs were first assembled in a book published in

2004, *Global Change and the Earth System*, by a team of scientists led by Will Steffen—an American chemist and formerly the executive director of the Australian National University Climate Change Institute.

The graphs vividly illustrate the "power of many": just how many technological, social, and environmental forces, in the hands of more and more people, were having an accelerating impact on Mother Nature's body—the human and biophysical landscapes of the planet—from 1750 to 2000, and particularly since 1950. When Steffen and his colleagues Wendy Broadgate, Lisa Deutsch, Owen Gaffney, and Cornelia Ludwig published an updated version of the Great Acceleration graphs—bringing them from 1750 to 2010—in the *Anthropocene Review*, March 2, 2015, they were even more convinced that these accelerations were pushing us past the Holocene's planetary boundaries into an unknowable unknown.

This was how they put it:

> The Great Acceleration marks the phenomenal growth of the global socio-economic system, the human part of the Earth System. It is difficult to overestimate the scale and speed of change. In little over two generations—or a single lifetime—humanity (or until very recently a small fraction of it) has become a planetary-scale geological force. Hitherto human activities were insignificant compared with the biophysical Earth System, and the two could operate independently. However, it is now impossible to view one as separate from the other. The Great Acceleration trends provide a dynamic view of the emergent, planetary-scale coupling, via globalization, between the socio-economic system and the biophysical Earth System. We have reached a point where many biophysical indicators have clearly moved beyond the bounds of Holocene variability. We are now living in a no-analogue world.

Let's repeat that: *We are now in a no-analogue world.* That means we're somewhere that we've never been before as a human species. We have pushed all of Earth's key systems up to and maybe beyond the safe operating boundaries that defined the Holocene. "A no-analogue world" . . . I am definitely going to add that to my climate-speak dictionary.

This is what the graphs look like:

# Earth system trends

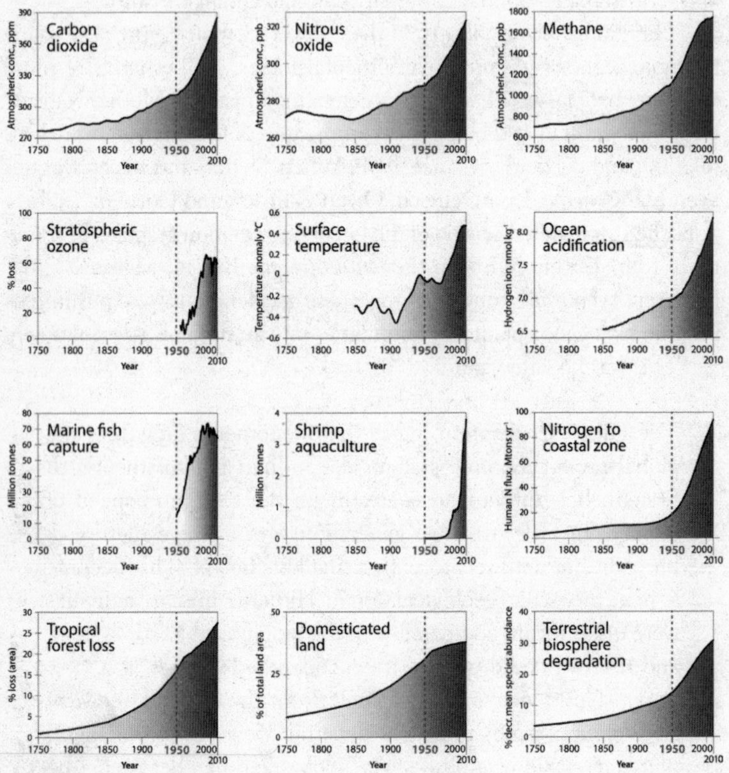


Source: Steffen, W., Broadgate, W., Deutsch, L., Gaffney, O., and Ludwig, C., "The Trajectory of the Anthropocene: The Great Acceleration." *Anthropocene Review* (vol. 2, no. 1), pp. 81–98. Copyright © 2015 by the authors. Reprinted by permission of SAGE Publications, Ltd.

# Socio-economic trends

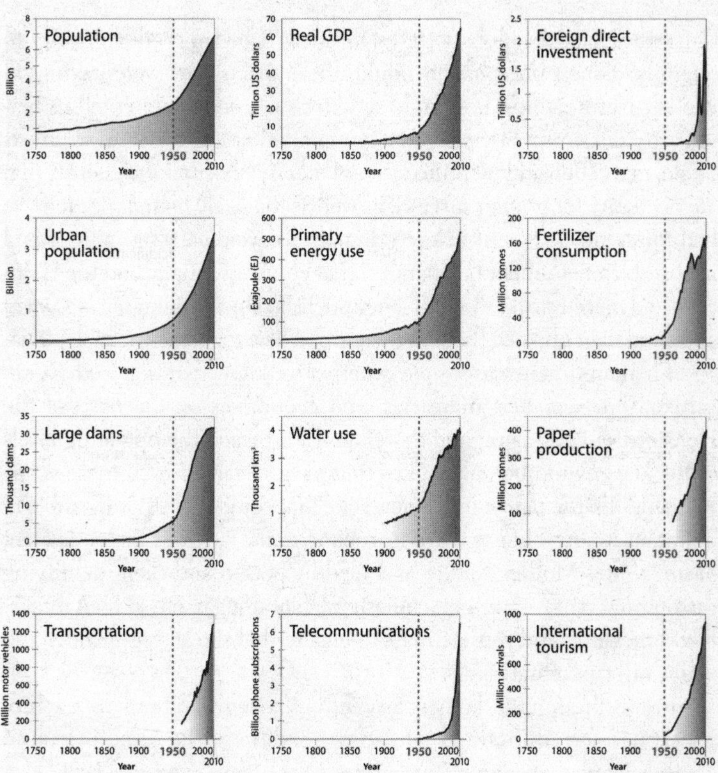

## *The Planetary Boundaries*

Once these accelerations were established, it became critical to try to quantify, as best one possibly could, the impacts they were having on Mother Nature's most important systems, since she couldn't tell us herself. So Rockström, Steffen, and a group of other Earth system scientists sat down in 2008 and identified the "planetary life-support systems" that are necessary for human survival, as well as the likely boundaries that we had to remain inside of in each domain to avoid causing "abrupt and irreversible environmental changes" that could essentially end the Holocene and make Earth unlivable. They published their findings in *Nature* in 2009 and then updated them in the journal *Science* on February 13, 2015.

Their argument was simple: whether we know it or not, we have organized our societies, industries, and economies on the basis of the Holocene environment, and therefore if we breach the operating levels of the key environmental systems that have sustained it all these years, we could flip the planet into a new state that could make it impossible to maintain modern life as we've learned to enjoy it. It was the equivalent of imagining Mother Nature as a healthy person and then identifying the optimal range of weight, cholesterol, blood sugar, fat, oxygen intake, blood pressure, and muscle mass to ensure that she stayed healthy and could still run marathons.

Just as the human body is a system of systems and organs, each of which has certain optimal operating conditions, the same is true of Mother Nature, explains Rockström. Our organs, and our body as a whole, can and do operate beyond these optimal conditions—up to a point. We don't know in every case how far beyond the optimal we can go before the body breaks down, but in some cases we do. We know that our optimal core body temperature is 98.6 degrees Fahrenheit. We've learned that average humans will die—their internal systems will fail—if their body temperature gets as hot as about 108 degrees or as cold as about 70 degrees. Those are our human health tolerance boundaries, and the closer you get to either extreme, the worse your organs and internal fluids will function.

Well, Mother Nature is a system of systems and organs—oceans, forests, atmosphere, ice caps—and Earth scientists have learned over the years what are the most stable operating levels for each of these systems and organs. True, Mother Nature is not a living being—she cannot tell us how she feels—"but she is a biogeophysical rationally functioning complex unit"—like a human body, Rockström told me. "We don't know exactly where her operating boundaries are, because we don't understand Mother Earth as accurately as a human body, but *she knows* exactly where they are. And there is no give. The Greenland ice sheet melts at a hardwired tipping point. The Amazon rain forest tips at a hardwired tipping point. And just as we would never manage our bodies at the edge of the tipping point, we should not be doing that with the planet."

So in the absence of Mother Nature being able to tell us how her most important systems are feeling, Rockström and Steffen and their planetary boundary team of scientists have attempted to make some educated estimates of where those tipping points reside, beyond which systems tip into a different state. They identified nine key planetary boundaries that we humans must make sure we do not breach (or continue to breach further, since we have already breached several). Breaching those boundaries could set in motion chain reactions that might flip the planet into a new state that could make it impossible to sustain modern civilization.

Here's their 2015 planetary boundaries health report. Warning: It doesn't look good.

The first boundary is *climate change*—and we've already breached it. The planetary boundary team, in line with the prevailing consensus among climate scientists, believes that we needed to stay below 350 parts per million of carbon dioxide in the Earth's atmosphere if we wanted to stay comfortably below the 2 degrees Celsius rise in global average temperature since the Industrial Revolution—the redline beyond which most climatologists believe we will be risking unmanageable ice melt, sea level rise, extreme temperature variations, and much more severe storms and droughts. We are now at more than 400 parts per million of $CO_2$ in the atmosphere—that blanket is now getting really thick and thickening at an accelerating rate—pushing, as noted earlier, the combined average temperature over global land and ocean surfaces to the highest levels we've seen since before the Industrial Revolution.

Mother Nature knows she's getting a fever. NASA's Vital Signs of the Planet report on global surface temperatures noted at the end of

2015: "The 10 warmest years in the 134-year record all have occurred since 2000, with the exception of 1998. The year 2015 ranks as the warmest on record." The climate system determines the growth environment for all living species, and that environment is heading into a zone well beyond the planetary boundary—threatening to make Earth into a hothouse the likes of which humans have never lived in before.

The second boundary, they argued, is *biodiversity*—which includes all the living species in the biosphere and all the nature covering the planet—that is, forests, grasslands, wetlands, coral reefs, and all the plants and animals residing within them. The planetary boundaries team determined we should maintain 90 percent of biodiversity cover from preindustrial levels. We are already down to 84 percent in parts of Africa, and going down further.

People forget, noted Rockström, that it is impossible to regulate the climate without biodiversity. If you don't have pollinators in the air and microrganisms in the soil and birds and other animals depositing seeds for new trees through their waste, you don't have a forest. If you don't have a forest, you don't have trees to soak up the carbon. If you don't have trees to soak up the carbon, it goes into the atmosphere and intensifies global warming or into the oceans and changes their composition. The natural species loss rate is one species or less per year out of every one million species. "We set the boundary at ten," Rockström explained, but with globalization that level is being regularly breached—we are now losing somewhere between ten and one hundred species per million species per year. That is as close a proxy as you can get for how much we are losing biodiversity.

The third planetary boundary we have breached, said Rockström, is *deforestation*. This concerns the minimum level of key biomes—mainly rain forests, boreal forests, and temperate forests—that we need to maintain on land to have a balanced, regulated Holocene. The scientists estimate that we must maintain around 75 percent of the Earth's original forests. We are now down to 62 percent, and some forests are showing signs of absorbing less carbon.

The fourth boundary that has already been breached is called *biogeochemical flows*. "We're now adding way too much phosphorus, nitrogen, and other elements to the world's crop systems, poisoning the Earth" with fertilizers and pesticides, said Rockström, and then those chemicals run off into the oceans and harm plant and fish life there as well. "To develop plants and animals that eat and create protein, you need a balance of nitrogen and phosphorus," he explained. "They determine the state of oceans and the landscape—too much nitrogen and phosphorus and you choke them—too little and they don't grow. It is all about how much fertilizer and pesticide we can allow ourselves to use, without choking other plants in the biosphere." Climate change can cause top-down tipping, and overuse of fertilizers and pesticides can create bottom-up tipping. Right now, said Rockström, "we have to go down to twenty-five percent of current usage."

In four other realms, we have managed to stay just inside the levels set by the planetary boundaries team, but not with much room to spare. One is rising *ocean acidification*. Some of the $CO_2$ we emit goes into the atmosphere, but a lot is actually absorbed by the oceans. This, however, is increasingly harming fish and coral reefs, which are like the tropical rain forests of the ocean. When you mix $CO_2$ with water you get carbonic acid, which dissolves the calcium carbonate that is the essential building block for all marine organisms, particularly those with shells, and for coral reefs. When that happens, "oceans, instead of playing host to marine organisms, break them down," said Rockström. "We can only ruin so much calcium carbonate before the marine system turns over and cannot host fish and coral reef as it did throughout the entire Holocene epoch before now."

Another area the planetary boundaries team says we are still managing to stay barely just inside is *freshwater use*—the maximum amount of water we can remove from the world's rivers and underground reservoirs, so our wetlands and rain forests can remain in their Holocene state and we can continue to engage in agriculture at scale.

A third boundary that we haven't quite crossed is *atmospheric aerosol loading*. These are the microscopic particles we put into the atmosphere with conventional pollution from factories, power plants, and vehicles. The inefficient burning of biomass (mostly by cooking stoves) and fossil fuels creates layers of smog that damage plant life by blocking sunlight; it also contributes to asthma and other lung diseases in humans.

And the fourth area where we are still just inside the boundary is known as the *introduction of novel entities*, namely, our invention of chemicals, compounds, plastics, nuclear wastes, and the like that are alien to nature and seep into soils and water. They do weird things we don't fully understand, and the fear is that they could one day even change the genetic code of different species, including humans.

There is one boundary we have safely retreated from after breaching it in the past. This is the appropriate thickness of the *stratospheric ozone layer* that protects us against dangerous UV radiation that causes skin cancer. Without that ozone layer, large parts of the planet would be uninhabitable. After scientists discovered an ever-widening ozone hole caused by man-made chemicals—chlorofluorocarbons—the world got together and implemented the Montreal Protocol in 1989, banning CFCs, and, as a result, the ozone layer remains safely inside its planetary boundary of losses not greater than 5 percent from preindustrial levels.

The planetary boundaries team does not claim that any of the boundaries they set are hard and fast or that if they are breached we will go straight over a cliff. Their redlines are educated estimates, beyond which we enter a "zone of uncertainty"—where no one can predict what might happen, because we have just not been there before as humans.

The one thing we've had going for us is that up to now Mother Nature has been very good at finding ways to adapt to stress, Rockström notes. Oceans and forests absorb the extra $CO_2$; ecosystems such as the Amazon adapt to deforestation and still provide rain and freshwater; the Arctic ice shrinks but does not disappear. Again, the Earth has a lot of buffers and adaptive capacities. But eventually, we can exhaust them. And that is exactly what we've been doing, particularly over the last half century.

"The planet has demonstrated an impressive capacity to maintain its balance, using every trick in its bag to stay in the current state by buffering our actions," added Rockström. But if we keep breaching these planetary boundaries, "we might shift the planet from friend to foe." That is a world where the Amazon becomes a savannah and the Arctic Circle a year-round ocean that absorbs the sun's heat rather than reflecting it away from Earth. That would almost certainly create a world for humans "that would be nowhere near as benign and friendly as the Holocene—the one steady state we know has sustained the only civilization we've ever known."

Already, many Earth scientists argue that it is no longer appropriate to describe our current geological epoch as the Holocene. They believe we've *already* left it behind and entered a new era that is being driven by . . . us. The name being given to this era is the "Anthropocene," as in *anthropo*, for "man," and *cene*, for "new." It is a fancy scientific name for the power of many.

"Human activity is leaving a pervasive and persistent signature on Earth," said Colin Waters of the British Geological Survey, coauthor of an essay in the January 8, 2016, issue of *Science* making the case that the Anthropocene deserves to be defined as a distinct new epoch from the Holocene.

The authors acknowledge that "any formal recognition of an Anthropocene epoch in the geological time scale hinges on whether humans have changed the Earth system sufficiently to produce a stratigraphic signature in sediments and ice that is distinct from that of the Holocene epoch," and they go on to make the case that we have. Everything from the hundreds of millions of tons of cement we've poured across the Earth's surface to radionuclides from atomic testing will be shaping the planet for years and years to come.

As the singer Joni Mitchell once put it in her song "Big Yellow Taxi," "They paved paradise / And put up a parking lot."

Waters and his colleagues simply put that lyric into scientific language:

> Recent anthropogenic deposits contain new minerals and rock types, reflecting rapid global dissemination of novel materials including elemental aluminum, concrete, and plastics that form abundant, rapidly evolving "technofossils." Fossil fuel combustion has disseminated black carbon, inorganic ash spheres, and spherical carbonaceous particles worldwide, with a near-synchronous global increase around 1950. Anthropogenic sedimentary fluxes have intensified, including enhanced erosion caused by deforestation and road construction. Widespread sediment retention behind dams has amplified delta subsidence.

It's weird to think of future geologists hitting our layer of sediment and trying to figure out iPods, Cadillacs with fins, and selfie sticks. Even if geologists one day can agree on this new era, there remains a dispute as to when it started. Some say it should be the dawn of agriculture—thousands of years ago; others argue it started with the onset of transoceanic Western colonialism in the early seventeenth century. "Of all the candidates for a start date for the Anthropocene," wrote Steffen and the Great Acceleration team, "the beginning of the Great Acceleration is by far the most convincing from an Earth System science perspective. It is only beyond the mid-twentieth century that there is clear evidence for fundamental shifts in the state and functioning of the Earth System that are (1) beyond the range of variability of the Holocene, and (2) driven by human activities and not by natural variability."

Because of this dispute, the International Commission on Stratigraphy, which is in charge of naming geological epochs, still has us in the Holocene. But for the purpose of this book we're in the Anthropocene, an epoch where the power of many—*that be us*—is now the dominant factor shaping and reshaping Earth systems and pushing out planetary boundaries.

But whatever era we are in, insists Rockström, "we have a responsibility to leave the planet in a state as close to the Holocene as possible." That is not going to be easy, though, because the "many" in the power of many is also in acceleration more than people realize in many places.

## *The Power of Many, Many, Many*

In April 2016, when I visited Niger to do a documentary on the impact of climate change on African migration patterns, our first stop was the northern town of Dirkou, in the middle of the Sahara. It was 107 degrees Fahrenheit in April; I was interviewing African migrants, many from Niger, who had traveled to Libya in search of work and, for a lucky few, a leaky boat ride to Europe. As I mentioned earlier, however, most found no work and no boat, only abuse from Libyans who did not want them in their country, which was experiencing its own economic and political meltdown.

So what we found in Dirkou were hundreds of men from Niger and other countries of West Africa who were marooned in a twilight zone of no jobs and no more money and unable to get north for work or south to go back home. They were being cared for by the International Organization for Migration. I interviewed several of these men under a burning hot sun next to a semitrailer overflowing with goods heading south. Most had been gone from their home villages for more than a year, so I asked one of them, Mati Almaniq, from Niger, how his family was faring.

He told me had left his three wives and seventeen children back in his village to search for work in Libya or Europe and returned deeply disillusioned. Almaniq said he left them with a store of food but knew that by now they must have eaten through it all. "They are in the hands of God now," he said. Such is life on the edge. One of his traveling companions standing next to him told me he left twelve kids back home. This was not unusual—mothers in Niger have an average of seven kids each.

I wrote all this into my *New York Times* column, and the next day I got an e-mail from my friend Robert Walker, president of the Population Institute, pointing out that "Niger's population in 1950 was just 2.5 million. Today its population is 19 million, and the U.N.'s latest population projection indicates that its population, even with declining fertility rates, will reach 72 million by 2050. Factor in climate change, along with regional conflict and instability, and you have a demonstrably unsustainable country. Making matters even more untenable is the child marriage prevalence rate in Niger: the highest in the world."

Niger is one of many countries, not all in Africa, where the other acceleration that I would include under the banner of "Mother Nature"—population growth—is still occurring. This growth will lead to more and more consumption of "natural capital," harming rivers, lakes, soils, and forests in their countries and beyond. Even though in many other parts of the world, population growth has flattened out or even reversed, the planet's total population will rise from about 7.2 billion today to about 9.7 billion by 2050, according to the latest United Nations report. That means in just over thirty years there will be another two billion people on the planet.

*Pause for a moment and think about that number: two billion more people.*

And still even more important is the fact that the impact on the planet's natural systems and climate will become exponentially more devastating, because more and more of those 9.7 billion are moving to large urban areas and up the socioeconomic ladder into their respective middle classes—where they will drive more cars, live in more and bigger homes, consume more water and electricity, and eat more protein. Their per capita impact on the planet will be much greater. Today, roughly

86 percent of Americans have air-conditioning in their homes and apartments. Only 7 percent do in Brazil and less than that in India. But once their basic needs are covered, they will want air-conditioning, too, and they are entitled to demand it as much as anyone living in Japan, Europe, or America.

I am a baby boomer, born in 1953, and that makes me part of a very unusual cohort. Not since Adam met Eve and gave birth to Cain and Abel has any generation been able to say what I and my fellow baby boomers can say: the population of the world doubled in our lifetimes. Indeed, if we eat enough yogurt, exercise well, and practice yoga, we could live long enough to see it *triple*. It was three billion in 1959 and six billion in 1999 and, as I said, is now expected to hit 9.7 billion in 2050.

I use the phrase "now expected to hit" to underscore the point emphasized by the Population Institute in its 2015 report: it's true that the world generally is undergoing a demographic transition from high mortality and high fertility to low mortality and low fertility; in many parts of the world that transition is well under way. In Europe, North America, and much of Latin America and East Asia, mortality and fertility rates have fallen so far so fast that they are now at, or below, the replacement rate, and population is actually declining in countries such as Taiwan, Germany, and Japan. But that is not the whole story.

"On the other side of the global 'demographic divide,'" the Population Institute notes, "mortality and fertility rates remain relatively high, but mortality rates have fallen faster. As a consequence, population is rising, and in some cases, rapidly. At current rates of growth, *nearly forty countries could double their population in the next thirty-five years*" (italics added).

It has not gotten much attention, but the U.N.'s population agency—the Department of Economic and Social Affairs' Population Division—keeps quietly increasing its global population projections. On July 29, 2015, it issued its "World Population Prospects: The 2015 Revision"—revising upward its projections of just two years earlier. It stated that the current world population of 7.3 billion is expected to reach 8.5 billion by 2030 (the previous projection was 8.4 billion), 9.7 billion in 2050 (up from 9.55 billion), and 11.2 billion in 2100—up from the previously estimated 10.8 billion.

Said the U.N.:

> Most of the projected increase in the world's population can be attributed to a short list of high-fertility countries, mainly in Africa, or countries with already large populations. During 2015–2050, half of the world's population growth is expected to be concentrated in nine countries: India, Nigeria, Pakistan, Democratic Republic of the Congo, Ethiopia, United Republic of Tanzania, United States of America (USA), Indonesia and Uganda . . .
>
> China and India remain the two largest countries in the world, each with more than 1 billion people, representing 19 percent and 18 percent of the world's population, respectively. But by 2022, the population of India is expected to surpass that of China.
>
> Currently, among the ten largest countries in the world, one is in Africa (Nigeria), five are in Asia (Bangladesh, China, India, Indonesia, and Pakistan), two are in Latin America (Brazil and Mexico), one is in Northern America (USA), and one is in Europe (Russian Federation). Of these, Nigeria's population, currently the seventh largest in the world, is growing the most rapidly. Consequently, the population of Nigeria is projected to surpass that of the United States by about 2050, at which point it would become the third-largest country in the world. By 2050, six countries are expected to exceed 300 million: China, India, Indonesia, Nigeria, Pakistan, and the USA . . .
>
> With the highest rate of population growth, Africa is expected to account for more than half of the world's population growth between 2015 and 2050.
>
> During this period, the populations of 28 African countries are projected to more than double.

The Population Institute notes that much of the projected population increase

> will occur in countries already struggling to alleviate hunger and severe poverty. Many countries with rapidly growing populations are threatened by water scarcity or deforestation; others

> are struggling with conflict or political instability. While progress is not precluded, rapid population growth for these countries is a challenge multiplier. Their populations are demographically vulnerable and more likely to suffer from hunger, poverty, water scarcity, environmental degradation and political turmoil.

In other words, if you go from high mortality to low mortality and don't also go from high fertility to low fertility, you create enormous strains. If a woman has twenty kids and the twenty kids all live and have twenty kids, you have four hundred grandchildren—in one family. And that is actually happening in places like Niger. The countries whose populations continue to balloon because of continued high fertility but lower mortality "are also those with the highest levels of gender inequality and child marriage," explained Walker. "Niger is number one in total fertility." Saudi Arabia, Egypt, and Pakistan are right up there, too. It is not about a lack of contraceptives. It is about a lack of modern gender norms and persistent male religious opposition to birth control. The blessing "May you have seven sons and seven daughters" is alive and well in these countries. And so is poverty and the lack of sufficient schooling and infrastructure.

That combination was never good. But when Moore's law and globalization accelerate at their current rates and your country falls behind on education and infrastructure, it falls behind at an accelerating rate as well. So you have more people who are less able to participate in the global flows. And then they have more kids as social security. And then climate change kicks in and undermines agriculture. And that can foster so much more disorder (as we will explore shortly)—when you have so many more people and governments less equipped to dig out of the hole. It is a frightening vicious cycle that is already under way in Afghanistan, the Middle East, and West Africa.

Adair Turner, a former chairman of the United Kingdom's Financial Services Authority and currently chairman of the Institute for New Economic Thinking and author of the book *Between Debt and the Devil: Money, Credit, and Fixing Global Finance*, put this problem succinctly in an August 21, 2015, essay published on *Project Syndicate*. He noted that while it is true that the U.N.'s latest population projections indicate that Europe, Russia, and Japan face considerable aging problems owing to low fertility rates, that is a manageable problem.

What is not a manageable problem is this, he wrote: "From 1950 to 2050, Uganda's population will have increased 20-fold, and Niger's 30-fold. Neither the industrializing countries of the nineteenth century nor the successful Asian catch-up economies of the late twentieth century ever experienced anything close to such rates of population growth. Such rates make it impossible to increase per capita capital stock or workforce skills fast enough to achieve economic catch-up, or to create jobs fast enough to prevent chronic underemployment." And all of this is happening before you even factor in the rising power of machines and robots to supplant bottom-rung blue-collar and white-collar jobs in these developing countries, not to mention the developed ones.

Turner also observed:

> By making it possible to manufacture in almost workerless factories in advanced economies, automation could cut off the path of export-led growth that all of the successful East Asian economies pursued. The resulting high unemployment, particularly of young men, could foster political instability. The radical violence of ISIS has many roots, but the tripling of the population of North Africa and the Middle East over the last 50 years certainly is one of them . . .
>
> With Africa's population likely to increase by more than three billion over the next 85 years, the European Union could be facing a wave of migration that makes current debates about accepting hundreds of thousands of asylum seekers seem irrelevant . . .
>
> Both increased longevity and falling fertility rates are hugely positive developments for human welfare . . .
>
> Achieving this goal does not require the unacceptable coerciveness of China's one-child policy. It merely requires high levels of female education, the uninhibited supply of contraceptives, and freedom for women to make their own reproductive choices, unconstrained by the moral pressure of conservative religious authorities or of politicians operating under the delusion that rapid population growth will drive national economic success.

Tom Burke, chairman of E3G, Third Generation Environmentalism, a green group in Britain, likes to reduce the problem to four numbers: 1, 1.5, 2.0, and 2.5. Says Burke:

> Today there are 1 billion who have arrived at middle class or above on the planet, with secure assets and high and secure incomes. There are 1.5 billion people who are in transition. They moved to the cities fifteen years ago in the emerging economies. By now they have some assets and secure incomes, but they're beginning to feel nervous because a lot of them work in the public sector and are being squeezed by globalization and technology. There are another 2.0 billion who have just recently moved to cities and they have virtually no assets and pretty insecure incomes, and you see them sitting by the roads selling stuff. And there are 2.5 billion who are the rural poor, living as subsistence farmers and on the edges of forests, who have not joined the global economy at all. If the climate changes, some migrate and the rest die.

If we cannot meet the expectations of the 1.5 and the 2.0—who are primarily in the cities in a hyperconnected world, where they can see everything they are missing, added Burke—they will destabilize the middle classes in all these countries. They will become the substrate for ISIS and other movements of the disaffected. Future growth and stability depend largely on creating rising real incomes for the bottom two quartiles of the urban population. They are the people who buy things when they get money, and they are the people who get hammered most by rising food prices and water prices and severe weather events. A significant number of those participating in the Arab Awakening starting in late 2010 emerged from the newly urbanized 1.5 and 2.0.

"Just as there are climate deniers, there will always be population deniers who refuse to acknowledge the impact that population growth is having on the planet," Robert Walker observed in a *Huffington Post* article first posted on January 30, 2015. "Population, in one form or another, touches upon a whole host of scientific concerns, including climate change . . . If world population grows as currently projected, it's hard to imagine that we will succeed in meeting the ambitious targets that must be met to avoid the worst effects of climate change."

This is truly not meant to blame the developing world, although some of these countries do have certain cultural practices related to the treatment of women in particular that they should overcome for their own benefit. When it comes to climate impacts, we in the West have been much worse for much longer. We have a much greater responsibility to invent the clean energy, efficiency, and conservation models that will allow an ever more middle-class planet to stay on the right side of every planetary boundary.

## *The Rain Room*

On November 1, 2015, NPR's *Weekend Edition* carried a story that illuminated as well as anything could the challenge posed by the Great Acceleration in Mother Nature. It was about an unusual exhibit at the Los Angeles County Museum of Art called the "Rain Room." In an interview, Hannes Koch, one of the artists who made the exhibit, said he and his fellow artists wanted to explore the relationships between art, nature, and technology.

Hence they created the Rain Room, which was described by Artnet .com on October 30, 2015, as a single, large, blackened room with artificial rain and a "bright spotlight shining in one corner." Visitors are invited to walk in and dare to believe that everywhere they stand sensors will ensure that the rain stops. Or as the article explained, they are asked "to enter a torrential downpour, trusting to science and art that they won't get wet, even as the storm continues unabated . . . Only seven people can be in the room at any given time, and visits can be no longer than fifteen minutes. As frustrating as this might be to potential visitors, this is all to their benefit: Sensors that detect the presence of visitors stop the rain above them, creating a six-foot-wide dry spot. Too many people, and there would be no rain to speak of."

I loved that line: *Too many people, and there would be no rain to speak of.*

Such is the impact of *the power of many.* While Moore's law and globalization have vastly expanded the power of machines and the power of one and the power of flows, the fact that they have also vastly expanded the power of many means that for the first time in the history of both mankind and Planet Earth, mankind has become large enough in numbers and empowered enough by the supernova to be both a force *of* nature and a forcing function *on* nature.

Our actions today, more than ever before, can turn the rain on and turn the rain off—literally. Climate change means more extremes—more torrential storms in some regions, more extended droughts in others. This power is so new that it is hard for people to get their minds around it. "Okay," some skeptics will say, "I will grant you that the climate is changing, but I don't believe that humans have anything to do with it." We are hardwired to consider nature limitless because for so many years nature seemed so limitless—and we were so relatively few and so relatively light a force upon it; how could it be that we cannot devour as much as we want? But, alas, we are now many, and the many are becoming many more and each of the many more is much more impactful and consuming much more than ever before.

As Jeremy Grantham, the well-known global investor, once observed: we humans "are wickedly bad at dealing with the implications of compound math"—another way of saying it's hard to recognize what a powerful impact we can have on the environment when the Market, Mother Nature, and Moore's law together all continue accelerating at once in the second half of the chessboard.

To be blunt, adds Adam Sweidan, "we have reaped the rewards of technological progress without due concern for its unintended consequences." All living things, he explained on his blog, "exist in and as ecosystems," which are the foundation of all life and commerce. "The degradation of that foundation will eventually cause the pyramid to crumble." And that is where the Machine is taking us, if we don't take heed of the planetary boundaries. "The system today appears to be in runaway mode," Sweidan added. "Increased demand for goods has led to the use of ever-advanced and more invasive technology to extract natural resources that keep the economy growing. These insult the land and degrade natural ecosystems while increasing inequalities, human population displacement, and social unrest."

And "it's happened so fast," writes Rockström in his book *Big World, Small Planet*. "In just two generations, humanity has overwhelmed Earth's capacity to continue supporting our world in a stable way. We've gone from being a small world on a big planet to a big world on a small planet. Now Earth is responding with environmental shocks to the global economy. This is a great turning point."

We spent the twentieth century protecting nature "from people," adds Glenn Prickett, a senior leader of The Nature Conservancy. "And we need to spend the twenty-first century protecting nature for people." Because without nature, we are not going to have the forests people need to absorb carbon and maintain watersheds, we are not going to have the mangroves that protect us from storm surges, we are not going to have the healthy oceans and coral reefs to feed all the people who will be on this planet, we are not going to have the ice caps and glaciers that regulate temperatures by reflecting the sun's rays. Nature doesn't need us, but we sure as hell need nature—especially when we become a big world on a small planet. Losing sight of that simple fact could be disastrous for the human species.

It doesn't have to end this way. The door to the Holocene does not have to be totally shut behind us. Or if it does, maybe it is still possible to have, as Rockström once put it to me, an "Anthropocene planetary equilibrium for us—for the world—without irreversibly pushing us to a hot-disaster state" of permanent disequilibrium.

But what we know for certain is that this is the moment, the turning point, when our options will be decisively shaped and determined. Much is now riding on whether we make the age of accelerations our friend or deadly foe. The supernova can amplify our powers to destroy or amplify our powers to protect and preserve.

We have to make our newfound power of one, the power of machines, the power of many, and the power of flows our friends—and our tools to create abundance within the planetary boundaries—not just our enemies. But organizing ourselves to use them that way will require a level of will, of stewardship, and of collective action the likes of which we have never seen humanity display as a whole. Every day there are new breakthroughs in solar energy, wind power, batteries, and energy efficiency that hold out the hope that we can have clean energy at a scale and price that billions can afford—provided we have the will to put a price on carbon so these technologies can rapidly scale and move down the cost-volume curve.

As environmentalists have often noted, we have been great at rising to the occasion after big geopolitical upheavals—after Hitler invaded his neighbors, after Pearl Harbor, after 9/11. But this is the first time in human history that we have to act on a threat we have collectively made to ourselves, to act on it at scale, to act before the full consequences are felt, and to act on behalf of a generation that has not yet been born—and to do it before all the planetary boundaries have been breached.

This is the challenge before humanity, now, right now, and it is for this generation. We could rebuild Europe after World War II, rebuild on the site of the World Trade Center, and even rebuild the economy after the 1929 and 2008 crashes, but if we cross Mother Nature's planetary boundaries, there are things that can never be rebuilt. We cannot rebuild the Greenland ice sheet, the Amazon rain forest, or the Great Barrier Reef. The same is true of the rhinos, macaws, and orangutans. No 3-D printer will bring them back to life.

That's why the only way to confront these compounding threats before they tip the wrong way is with a compounding commitment to stewardship, a compounding willingness to act collectively to do compounding research and make compounding investments in clean energy production and more efficient consumption, along with a willingness, at least in America, to impose a carbon tax to get compounding investments in clean power and efficiency, plus a compounding commitment to women's education and an ethic of empowerment everywhere. Without compounding, multiplicative commitments along all these fronts that are commensurate with the magnitude of the challenge we face, we stand no chance—zero—of preserving a stable planet when there will be so many more people, armed with so many more powerful tools, propelled by a supernova.

I have said it before and I will keep saying it as long as I have the breath: we are the first generation for whom "later" will be the time when all of Mother Nature's buffers, spare tires, tricks of the trade, and tools for adapting and bouncing back will be exhausted or breached. If we don't act quickly together to mitigate these trends, we will be the first generation of humans for whom later will be *too late*.

Sylvia Earle, the renowned oceanographer, puts it succinctly: "What we do right now, or fail to do, will determine the future—not just for us, but for all life on Earth."

# PART III

# INNOVATING

SEVEN

# *Just Too Damned Fast*

*We're entering an age of acceleration. The models underlying society at every level, which are largely based on a linear model of change, are going to have to be redefined. Because of the explosive power of exponential growth, the twenty-first century will be equivalent to 20,000 years of progress at today's rate of progress; organizations have to be able to redefine themselves at a faster and faster pace.*

*—Ray Kurzweil, director of engineering at Google*

*My other vehicle is unmanned.*

*—Bumper sticker on a car in Silicon Valley*

Now that we have defined this age of accelerations, two questions come to mind—one primal, one intellectual. The primal one is this: *Are things just getting too damned fast?* The intellectual one is: Since the technological forces driving this change in the pace of change are not likely to slow down, *how do we adapt?*

If your answer to the first question is "yes," then let me assure you that you are not alone. Here is my favorite story in Erik Brynjolfsson and Andrew McAfee's book *The Second Machine Age*: The Dutch chess grandmaster Jan Hein Donner was asked how he'd prepare for a chess match against a computer, like IBM's Deep Blue.

Donner replied: “I would bring a hammer.”

Donner isn’t alone in fantasizing that he’d like to smash some recent advances in software and artificial intelligence (AI). These advances are not only replacing blue-collar jobs but also supplanting white-collar skills—even those of chess grandmasters. Jobs have always come and gone, thanks to creative destruction. If horses could have voted there never would have been cars. But the disruptions do seem to be coming faster these days, as technological advances keep building on themselves, taking us from one platform to another and touching a wider and wider swath of the labor market.

I know, because as a sixty-three-year-old journalist I’ve lived through a bunch of these platform changes, and I’ve seen them coming faster and faster. I am already bracing myself for the day when I have grandchildren and one of them asks me: “Grandpa, what is a typewriter?”

Here is a short history of how I personally have felt the impact of the change in the pace of change of technology. I am sure many readers will recognize themselves.

Immediately after finishing a master’s degree in Arabic and modern Middle East studies from Oxford University, I was hired for my first job by United Press International (UPI), the wire service, in their London bureau on Fleet Street in the spring of 1978. To write my stories in that UPI London bureau, we used both desktop manual typewriters and early word processors. For those of you too young to remember, About .com explains that a “typewriter” was “a small machine, either electric or manual, with type keys that produced characters one at a time on a piece of paper inserted around a roller.” Wikipedia states that typewriters were invented “in the 1860s” and “quickly became indispensable tools for practically all writing other than personal correspondence. They were widely used by professional writers, in offices, and for business correspondence in private homes” up until the end of the 1980s, when “word processors and personal computers . . . largely displaced typewriters in . . . the Western world.”

Think about that for a moment: authors, businesses, and governments basically used the same writing machine—the typewriter—for more than a century. That’s three generations. That is how slow the pace of technological change was—although it was a whole lot faster than before the Industrial Revolution. Of course I didn’t know it at the time, but I was starting my journalism career at the very tail end of the Industrial

Revolution—the very close of the typewriter era—and on the very eve of the information technology revolution.

And once the late twentieth century arrived, that progress started cascading considerably faster still. But having started in the Industrial Revolution, I first needed to learn how to type fast on a typewriter! So, after I was hired by UPI in 1978, the first thing I did was go to a night secretarial school in London to learn how to write shorthand and how to type fast with both hands. Most of my classmates were young women looking for starting secretarial jobs.

There were no cell phones then, either. Because of that I got my first big lesson in journalism. It came on the very first real news story UPI sent me out to cover, after I joined its London bureau. And that lesson was: *Never ask your competition to hold the phone for you.*

The Islamic Revolution in Iran was just unfolding. A group of pro–Ayatollah Khomeini Iranian students in London took over the Iranian Embassy there, ousted the shah's diplomats, and then locked themselves inside the main embassy building. I managed to talk my way into the building to interview some of the student revolutionaries. I don't remember what they said, but I was so excited by whatever it was that after filling my notebook I ran directly to the phone booth next to the embassy to call my story in to the bureau. It was one of those classic red English phone booths. There was a line of six or seven reporters—all grizzled Fleet Street veterans—waiting to use the phone to call in their stories. I patiently waited my turn. When, after about twenty minutes in line, I got inside the booth, I excitedly told my editors all that I had seen and heard from the Iranian students inside, flipping through pages of my notebook so as not to miss any details. At one point, the editor who was taking my dictation asked me a detail about the embassy building that I didn't have. So, I said, "Wait a minute, I'll check."

I then opened the door of that red phone booth and said to the Fleet Street reporter who had been waiting in line behind me, "Do me a favor, hold the phone for me." Then I dashed out of the phone booth to get that minor detail for my editor.

Before I had taken two steps, the guy in line behind me slipped into the phone booth, slammed down the receiver, disconnected my call, started dialing his own newspaper, and turned to me to say two words I will never forget: "Sorry, mate."

I've never asked my competition to hold the phone for me since.

Of course, in this age of ubiquitous mobile phones, no reporter in the world will ever have to learn—or teach—that lesson.

A year later, in 1979, UPI sent me off to Beirut as their number-two correspondent, in the middle of the civil war there. Here was my technology platform: I wrote my stories on a big desktop manual typewriter. I filed those stories to our London headquarters via telex, which, again for those of you too young to recall, is defined by *Merriam-Webster's Collegiate Dictionary* as "a communication service involving teletypewriters connected by wire through automatic exchanges." The way we filed our stories was that we first typed them out on plain white typing paper, double-spaced, and only three paragraphs at a time. Then we handed those three paragraphs to a telex operator, who punched them into telex paper tape and then fed that coded tape into a big clanking telex machine in our office. It then went out over global telephone cables from that end of the world, across the ocean, and was spewed out on a telex printer machine on the other end of the world—in my case, first at UPI headquarters in London, and later at the *New York Times* headquarters in Manhattan.

Writing a story three paragraphs at a time, without the ability to move paragraphs, delete, or spell check, can be a challenge. Try it sometime! The way I did it was to type out my whole story or news analysis from start to finish, then do the whole thing again, and then, once I was satisfied that the paragraphs were right and in proper order, and I knew where the story was going, I would type it through a third time, in three-paragraph chunks, and hand it to the telex punch operator. The telex system in Beirut ran through the Lebanese PTT—Post, Telephone, and Telegraph—located in central downtown Beirut, right along the civil war dividing line.

In 1981, I went to work for *The New York Times*. I served as a business reporter in New York for a year, and in 1982 the *Times* sent me back to Beirut as its bureau chief. I went back with a portable typewriter. I remember it well. It was a German-made Adler with a white case. That Adler portable manual typewriter was the best you could buy back then, probably cost me three hundred dollars, and I remember thinking when I got it: "Now I am a real foreign correspondent!" I was so proud of that typewriter. There was a real firmness to the keys when you pressed them to create a letter.

So in writing this book, I Googled "Adler portable typewriter" to refresh my memory as to what it looked like, and the third item to come up caught my eye. It said: RARE VINTAGE ANTIQUE KLEIN ADLER PORTABLE TYPEWRITER GERMANY, for sale on eBay.

Ouch! Hard to believe that the writing device that I began my reporting career on almost four decades ago is now a "rare vintage antique." That sounds like something from 1878. I wish I could show you a picture of mine, but, alas, I don't have that typewriter any longer. It was blown up, along with the rest of my apartment in Beirut, in the first few days of the Israeli-Palestinian war in June 1982, when two groups of refugees from south Lebanon got in a fight over who would get the empty apartments in my building off Bliss Street. The group that lost destroyed the whole building, tragically killing my driver's wife and two daughters, who were sitting in my home office.

I was in south Lebanon when Israel invaded in early June 1982 and stayed in Beirut the whole summer. My deal with *The New York Times* was that I would remain there until Yasser Arafat's Palestine Liberation Organization fighters began departing by ships, which was eventually negotiated for August 21, 1982, from the Beirut port. I wanted those two six-column headlines as bookends for my scrapbook—"Israel Invades" and "Arafat Leaves." Well, the day came. It was a beautiful morning. I stood in the port with Peter Jennings of ABC News and together we watched it all—truckloads of Palestinian guerrillas, firing their Kalashnikovs in the air, showering us in shell casings, leaving Beirut for Algeria and Tunisia and a future unknown. It was a dramatic, poignant, and incredibly colorful scene, and when it was over I went to the Reuters bureau in Beirut, where I had desk space, took out my portable typewriter, and started writing it all up, three paragraphs at a time—putting a summer's worth of passion and energy into this closing chapter.

When the story was ready I handed it to the telex operator. He punched it into tape, but before he could feed it to the *New York Times* offices in New York City, all the communications between Beirut and the rest of the world were severed. Everything in those days went out through one cable switching box at the PTT, and for whatever reason it went down. I stayed up all night by the telex machine, waiting for it to come alive so that I could feed my story to New York. It never did. Yes, kids, there was a time and place when such things happened. No phone, no telex, no cell phone, no Internet, no nothing. I still have that punched telex tape in a shoe box in my basement. The next morning, August 22, 1982, *The New York Times* ran a banner headline about Arafat leaving Beirut, and the byline read: "By the Associated Press," which had filed its story several hours before mine and before the PTT had melted down.

By the time I ended my tour in Beirut in 1984, the digital IT revolution was just starting to emerge and *The New York Times* sent me something called a TeleRam Portabubble, which was a suitcase-sized box word processor, with a tiny screen and cups on the top to insert a phone that transmitted your story by sound waves back to the *Times'* first-generation computers in Times Square. From Beirut, I went to Jerusalem, from 1984 to 1988. At first I worked on TeleRam there, too, and eventually, in my last year or so, we got the first IBM desktops, with big floppy disks. The pace of change was starting to quicken a bit. My technology platforms were improving faster.

After my tour in Jerusalem, I moved to the Washington bureau, where I served as the *New York Times* diplomatic correspondent, starting in 1989. I had a front-row seat traveling with Secretary of State James A. Baker III for the fall of the Berlin Wall and the end of the Cold War. On those trips we used Tandy laptops to write and file over long-distance phone lines. We reporters became experts at taking apart telephones in hotel rooms around the world to directly fix the wires to our Tandys. You always had to travel with a small screwdriver, along with your reporter's notepad.

When I shifted to covering the White House under a new president, Bill Clinton, in 1992, no one I knew had e-mail. By the end of his second term just about everyone I knew had e-mail. My last reporting job was as the *New York Times* international economics correspondent, from 1993 through 1994. I began as a columnist in January 1995. That very same year, on August 9, 1995, a start-up company called Netscape went public, selling something called an Internet "browser," which would bring the Internet, e-mail, and eventually the World Wide Web alive on a computer screen, like nothing ever before. Netscape's public offering—its shares were priced at $28, soared to $74.75 by midday, and closed at $58.25—would launch the Internet boom and bubble.

Since then, I have gone through an increasingly rapid succession of Dell, IBM, and Apple laptops and desktops, with faster and faster connectivity to the Web. Starting a decade ago, it became obvious that careers in the newspaper business were rapidly shrinking as more and more newspapers shut down and more and more advertising went to the Web and more and more people were reading the paper on mobile devices. I watched as reporters went from writing one story a day for the print edition of *The New York Times* to having to write multiple stories a day to keep the Web edition updated, as well as file tweets and Facebook posts, and narrate videos. It reminded me exactly of my days as a wire service reporter in Beirut—filing a breaking news story, filing a picture, doing a radio spot—all the hectic things you had to do at once that made me want to be a newspaper writer with only one deadline. Now newspaper reporters, just like wire service reporters, have a deadline every second.

With every passing year, I see my own industry, my own tools, and the tools of other white-collar workers changing faster than ever, thanks to the supernova. In May 2013, I found myself standing in the passport line at Heathrow Airport in London, waiting to get stamped into the country at immigration control. At one point the man in front of me turned around, told me that he was a reader, and engaged me in a friendly conversation. I asked him what he did. He said his name was John Lord and that he was in the software business.

"What kind of software?" I asked. He said that his company's goal was to make "lawyers obsolete" wherever possible by creating software applications that enable individuals to do more and more legal work without the aid of an attorney. Indeed, Neota Logic, his company, says that its goal is to massively improve access to advice and justice for "the 40+% of Americans who can't afford an attorney when they need one"—in order to produce wills and basic legal documents and even to handle crucial life events such as home foreclosure, domestic abuse, or child protection.

Neota Logic is part of a new strain of software called "expert systems" that aims to identify a large chunk of business that clients need, and that lawyers charge for, but that actually can be done by software: think TurboTax for the legal profession. The company's website quoted one commentator complaining that Neota Logic's technology cannot "read between the lines . . . [or] hold hands and wipe away tears." To which Neota Logic responded: "You will surely see a press release when we can."

Lord later explained to me that "I have always had a special respect for trial lawyers and hope it will be a long time before algorithms replace them and juries." Alas, he added, that is "not beyond the realm of possibility of course, but not yet Neota's mission."

Suddenly I was glad my daughters were not planning to be lawyers.

But the hits just kept on coming—again and again, I found myself witnessing something I never dreamt I'd see, and it reminded me that the supernova was upending our world for good. In early 2015, I found myself reporting with my cell phone camera from the backseat of a car that had no driver! I was visiting Google's X research and innovation lab and was given a ride in a driverless Lexus RX 450h SUV. In the front seat were two X staffers. The one in the passenger seat was a Google engineer with an open laptop in her lap. The other was a staffer sitting in the driver's seat, but with no hands on the wheel. He was basically there as a prop to reassure other drivers who might pull up at a stoplight alongside us that someone was driving this car—even if he wasn't! I was seated in the back.

Off we went, driving through the neighborhoods and commercial districts of Mountain View, California. The route was preprogrammed and the car drove itself—or, rather, its software drove. We were in "autonomous mode." After five minutes watching the car calmly navigate every intersection, make perfect left turns, wait on pedestrians, and carefully pass bikers, I realized that I had crossed a line myself—something I never expected: I felt safer with the software driving than myself or any chauffeur.

And with good reason: the X website reports that thousands of minor accidents happen every day on typical American streets, 94 percent of them involving human error, and as many as 55 percent of them go unreported. Until 2016, though, Google's fifty-three vehicles had autonomously driven more than 1.4 million miles and been involved in only seventeen crashes, none that were the fault of its vehicles and none that involved fatalities. Google has acknowledged, though, that more than a dozen times its human drivers had to intervene to head off an impending crash. (Alas, on February 14, 2016, a Google self-driving car, trying to avoid a sandbag in the road, sideswiped a bus while going less than two miles an hour. That's a pretty good driving record for six years.)

So when I confessed to the Google engineer in the front seat of the autonomous vehicle driving me around how relaxed I felt, she calmly turned away from her laptop—which was tracking every move the car made—and gave me a quote I had never heard as a reporter.

"Mr. Friedman," she said, "the car has no blind spots. Almost all the accidents are drivers rear-ending us because they were not paying attention."

*This car has no blind spots!* I wrote that down in my reporter's notebook.

Google's cofounder Sergey Brin picked up the tour when we returned back to X's headquarters. There, he showed me Google's prototype of a two-person autonomous vehicle. It does not yet have a name, but it looks like a big egg on wheels or something you would ride up a mountain in on a ski lift. There were just two seats, no dashboard, and no steering wheel—nothing. But it is a self-driving vehicle.

"How do you tell it where to go?" I asked Brin.

"You will just program it on your cell phone," he answered, as if that were the most obvious thing in the world.

Of course, why didn't I think of that! My cell phone that I was using to take pictures, as a good reporter, would double as the key to my next car. Why not? Suddenly, I understood what the organizational consultant Warren Bennis meant when he once famously observed that the "factory of the future will have only two employees, a man and a dog. The man will be there to feed the dog. The dog will be there to keep the man from touching the equipment."

And then I stopped laughing at even that joke. This was getting serious and starting to get way too close to home.

On March 7, 2015, *The New York Times* ran this news story/quiz: "Did a Human or a Computer Write This? A shocking amount of what we're reading is created not by humans, but by computer algorithms. Can you tell the difference? Take the quiz:"

> 1. "A shallow magnitude 4.7 earthquake was reported Monday morning five miles from Westwood, California, according to the U.S. Geological Survey. The temblor occurred at 6:25 a.m. Pacific time at a depth of 5.0 miles."
>
> □ Human
> □ Computer

2. "Apple's holiday earnings for 2014 were record shattering. The company earned an $18 billion profit on $74.6 billion in revenue. That profit was more than any company had ever earned in history."

□ Human
□ Computer

3. "When I in dreams behold thy fairest shade
Whose shade in dreams doth wake the sleeping morn
The daytime shadow of my love betray'd
Lends hideous night to dreaming's faded form."

□ Human
□ Computer

4. "Benner had a good game at the plate for Hamilton A's-Forcini. Benner went 2–3, drove in one and scored one run. Benner singled in the third inning and doubled in the fifth inning."

□ Human
□ Computer

5. "Kitty couldn't fall asleep for a long time. Her nerves were strained as two tight strings, and even a glass of hot wine, that Vronsky made her drink, did not help her. Lying in bed she kept going over and over that monstrous scene at the meadow."

□ Human
□ Computer

6. "Tuesday was a great day for W. Roberts, as the junior pitcher threw a perfect game to carry Virginia to a 2–0 victory over George Washington at Davenport Field."

□ Human
□ Computer

7. "I was laid out sideways on a soft American van seat, several young men still plying me with vodkas that I dutifully drank, because for a Russian it is impolite to refuse."

□ Human
□ Computer

8. "In truth, I'd love to build some verse for you
To churn such verse a billion times a day
So type a new concept for me to chew
I keep all waiting long, I hope you stay."

□ Human
□ Computer

Answers: 1. Computer algorithm. 2. Human. 3. Computer poetry app. 4. Computer algorithm. 5. Computer algorithm. 6. Computer algorithm. 7. Human. 8. Computer poetry app.

Today it's poets. Tomorrow it's columnists . . .

In April 2016, as noted earlier, I had traveled to Agadez, in northern Niger, in the middle of the Sahara, with the country's minister of the environment, Adamou Chaifou, to watch the caravans of economic migrants from across the region transiting Niger for Libya and, many of them hoped, on to Europe. On April 13, 2016, I wrote a column from Niger that quoted Chaifou. It moved on NYTimes.com at 3:20 a.m. Eastern Standard Time, or 8:20 a.m. Niger time. I was departing the country that afternoon and went to the airport around 1:00 p.m. Chaifou came out to say goodbye to me and I took the opportunity to be the first to tell him: "I quoted you in my column in *The New York Times* today. It's up on the Web on NYTimes.com."

"I know," he responded. "My kids are studying in China and they already sent it to me!" So today, a minister in Niger is telling me his kids studying in a remote university in China have e-mailed him my column from Niger before my wife had woken up and read it in Bethesda.

And, finally, there was writing this book. In the two and a half years I have taken to research it, I have had to interview almost all the principal technologists at least twice and sometimes three times to make sure what I was writing remained up-to-date. I have never had that experience as an author before—it was like chasing a butterfly with a net and every time I moved to catch it, it fluttered away just outside my reach.

So there you have it: just inside of four decades, I went from writing my own stories three paragraphs at a time on a manual typewriter to riding in a self-driving car and recording it on my phone to reading poetry crafted by an algorithm to filing my story wirelessly on the Internet from Niger and having it read the next morning in China and e-mailed from there back to my host in Niger—before I could even let him know that I had quoted him—to writing a book about technological change that kept getting overtaken by . . . technological change.

Am I the one who needs a hammer now?

## *Mind the Gap*

Much as I fantasize about it some days, the answer is no. We have no choice but to learn to adapt to this new pace of change. It will be harder and require more self-motivation—and that reality is surely one of the things roiling politics all over the world today, particularly in America and Europe. The accelerations we've charted have indeed opened a wide gap between the pace of technological change, globalization, and environmental stresses and the ability of people and governing systems to adapt to and manage them. Many people seem to be feeling a loss of control and are desperate for navigational help and sense-making.

Who can blame them? When so many things are accelerating at once, it's easy to feel like you're in a kayak in rushing white water, being carried along by the current at a faster and faster clip. In such conditions, there is an almost irresistible temptation to do the instinctive thing—but the wrong thing: stick your paddle in the water to try to slow down.

*That will not work,* explained Anna Levesque, a former member of the Canadian Freestyle Whitewater Kayak Team and an Olympic bronze medalist, who has more than fifteen years of experience as an international competitor, instructor, and guide. She posted some simple strategies on her blog for how to control a kayak on a fast-rushing river that are worth keeping in mind for managing our own age of accelerations.

Her post was entitled "Why 'Keep Your Paddle in the Water' Is Bad Advice for Beginners."

> Have you ever stopped to consider what the phrase "keep your paddle in the water" actually means? If you did you wouldn't ever recommend it to a beginner whitewater paddler. The paddlers and instructors who give this advice are well intended and what they are really expressing is: "Keep paddling to maintain

> your stability through rapids." When beginners hear "keep your paddle in the water," they end up doing a bad version of a rudder dragging their paddle in the water back by their stern while using their blade to steer. This is a really bad position to be in . . .
>
> To enhance stability in rapids it's important to move as fast or faster than the current. Every time you rudder or drag your paddle in the water to steer you lose momentum and that makes you more vulnerable to flipping over.

And so it is with governing today. The only way to steer is to paddle as fast as or faster than the rate of change in technology, globalization, and the environment. The only way to thrive is by maintaining *dynamic stability*—that bike-riding trick that Astro Teller talked about. But what is the political and social equivalent of paddling as fast as the water or maintaining dynamic stability?

It's innovation *in everything other than technology.* It is reimagining and redesigning your society's workplace, politics, geopolitics, ethics, and communities—in ways that will enable more citizens on more days in more ways to keep pace with how these accelerations are reshaping their lives and generate more stability as we shoot through these rapids.

It will take workplace innovation to identify exactly what humans can do better than machines and better *with* machines and increasingly train people for those roles. It will take geopolitical innovation to figure out how we collectively manage a world where the power of one, the power of machines, the power of flows, and the power of many are collapsing weak states, super-empowering breakers, and stressing strong states. It will take political innovation to adjust our traditional left-right party platforms, born to respond to the Industrial Revolution, the New Deal, and the Cold War, to meet the new demands for societal resilience in the age of the three great accelerations. It will take moral innovation—to reimagine how we scale sustainable values to everyone we possibly can when the power of one and the power of machines become so amplified that human beings become almost godlike. And, finally, it will take societal innovation, learning to build new social contracts, lifelong learning opportunities, and expanded public-private partnerships, to anchor and propel more diverse populations and build more healthy communities.

One of my favorite thinkers on this challenge is Eric Beinhocker, the executive director of the Institute for New Economic Thinking at Oxford University and author of *The Origin of Wealth: The Radical Remaking of Economics and What It Means for Business and Society.* In an interview, Beinhocker succinctly summarized the challenge before us. He began by distinguishing between the evolution of "physical technologies"—stone tools, horse-drawn plows, microchips—and the evolution of "social technologies"—money, the rule of law, regulations, Henry Ford's factory, the U.N.:

> Social technologies are how we organize to capture the benefits of cooperation—non-zero-sum games. Physical technologies and social technologies coevolve. Physical technology innovations make new social technologies possible, like fossil fuel technologies made mass production possible, smartphones make the sharing economy possible. And vice versa, social technologies make new physical technologies possible—Steve Jobs couldn't have made the smartphone without a global supply chain.

But there is one big difference between these two forms of technology, he added:

> Physical technologies evolve at the pace of science—fast and getting exponentially faster, while social technologies evolve at the pace at which humans can change—much slower. While physical technology change creates new marvels, new gadgets, better medicine, social technology change often creates huge social stresses and turmoil, like the Arab Spring countries trying to go from tribal autocracies to rule of law democracies. Also, our physical technologies can get way ahead of the ability of our social technologies to manage them—nuclear proliferation, bioterrorism, cyber crime—some of which is happening around us right now.
>
> Our physical technologies won't slow down—Moore's law will win—so we're in a race for our social technologies to keep up. We need to more deeply understand how individual psychology, organizations, institutions, and societies work and find ways to accelerate their adaptability and evolution.

*This will be a huge ongoing challenge.*

Every society and every community must compound the rate at which it reimagines and reinvents its social technologies, because our physical technologies will not likely be slowing anytime soon. As the systems thinker Lin Wells put it in his November 1, 2014, essay, "Better Outcomes Through Radical Inclusion," if, roughly speaking,

> computing power per unit cost is doubling about every 18 months, in a year and a half we'll have 100% more power, in five years more than 900% and in ten years over 10,000% . . . Moreover, the change is not just happening in the information domain. Biotechnology is changing even faster than information, robotics and autonomous systems are becoming ubiquitous, nanotechnology is poised to affect a range of commercially useful areas, from new materials to energy storage, and energy itself is undergoing profound changes affecting all of society. Collectively, the rates of technological change in just these five areas—bio, robo, info, nano, and energy (BRINE, for short)—pose legal, ethical, policy, operational, and strategic opportunities, and risks, that no company or individual can address alone.

This is a full-on societal reinvention challenge.

America, precisely because it is decentralized into fifty states and thousands of localities that enable multiple different experiments in governing, is ideally suited for such a broad project of societal reinvention. But in 2008—right after 2007 birthed a whole new set of accelerating technologies—we entered a severe economic recession that also triggered severe political gridlock in Washington. As a result, we've seen a lot of our physical technologies hurtling along, while our social technologies—the learning, governing, and regulating systems we need to go along with these accelerations to get the best out of them while cushioning the worst—stall. As I suggested earlier, it is as if the ground under everyone's feet started shifting faster and faster just as the governing systems meant to help people adjust and adapt largely froze—and few political leaders could explain to people what was happening.

This policy gap has left way too many citizens in America and around the world feeling unmoored and at sea, prompting an increasing number to seek out candidates from the far left or far right. So many people today seem to be looking for someone to put on the brakes, or take a hammer to the forces of change—or just give them a simple answer to make their anxiety go away.

It is time to redouble our efforts to close that anxiety gap with imagination and innovation and not scare tactics and simplistic solutions that will not work. I would not even pretend to know all that is sufficient in this regard. But in this next section of the book I will offer some of the best adaptation ideas I have gleaned that are surely necessary in five key areas—the workplace, geopolitics, politics, ethics, and community building—to help people feel more anchored, resilient, and propelled in this age of accelerations. The last thing we want is for everyone to stick their paddles in the white water to slow down. That is exactly how you destabilize a kayak and a country.

EIGHT

# *Turning AI into IA*

Let's get one thing straight: *The robots are not destined to take all the jobs.* That happens only if we let them—if we don't accelerate innovation in the labor/education/start-up realms, if we don't reimagine the whole conveyor belt from primary education to work to lifelong learning.

But that has to start with an honest conversation about work—and, in America, we have not had an honest conversation about this subject for a long, long time. Ever since the early 1990s, President Bill Clinton and his successors have been telling Americans the same old, same old: if you "work hard and play by the rules" you should expect that the American system will deliver you a decent middle-class life and a chance for your children to have a better one. That was true at one time: just show up, be average, do your job, play by the rules, everything will be fine . . .

Well, say goodbye to all that.

Just as we seem to be leaving the Holocene climate epoch—that perfect Garden of Eden period when everything in nature was nicely in balance—we are also leaving the Holocene epoch for work. In those "glorious" decades after World War II, before the Market, Mother Nature, and Moore's law all entered the second half of their chessboards, you could lead a decent lifestyle as an average worker with an average high school or four-year college education, belonging to an average union or none at all. And by just working an average of five days a week at an average of eight hours a day, you could buy a house, have an average of 2.0 kids, visit Disney World occasionally, save for an average retirement and sunset to life.

So many things then were working in favor of the average worker. America dominated a world economy in which the industrial foundations of many European and Asian countries had been destroyed by World War II, so for years thereafter there were vast numbers of manufacturing jobs to be filled. Outsourcing was limited, and China had yet to join the World Trade Organization (that happened in December 2001) and its workforce was not yet much of a threat to most good blue-collar jobs. With the push and pull of globalization relatively mild, innovation slower, and the barriers to entry into different industries higher, unions were relatively stronger and could negotiate healthy and steady wage and benefits packages from employers. Companies could also afford more in-house training of their workers, who were less mobile and therefore less likely to learn and quit. Because the pace of change was slower, whatever you had learned in high school or college stayed relevant and useful much longer; skills gaps were less prevalent. Machines, robots, and, most important, software had not advanced to the point where they could abstract so much complexity so easily and cheaply—and in the process undermine the bargaining power of both industrial and service unions.

As a result of all these factors, many middle-skilled workers were able to prosper between 1945 and the 1980s. I never forget that my dad's only brother, Jerome Friedman, worked as a loan officer at the Farmers & Mechanics Bank in Minneapolis—and he only had a high school degree! There surely is not a bank loan officer in America today who only has a high school degree. But back then it was possible, which is why I call it a labor Holocene—so many workers were able to enjoy what was known as "a high-wage middle-skilled job," in the words of Stefanie Sanford, chief of global policy and advocacy for the College Board.

Well, say goodbye to all that, too.

The high-wage, middle-skilled job has gone the way of Kodak film. In the age of accelerations, there is increasingly no such animal in the zoo anymore. There are still high-wage, high-skilled jobs. And there are still middle-wage, middle-skilled jobs. But there is no longer a high-wage, middle-skilled job.

Average is officially over. When I graduated from college I got to *find* a job; my girls have to *invent* theirs. I attended college to learn skills for life, and lifelong learning for me afterward was a hobby. My girls went to college to learn the skills that could garner them their first job, and lifelong learning for them is a necessity for every job thereafter. Today's American dream is now more of a journey than a fixed destination—and one that increasingly feels like walking up a down escalator. You can do it. We all did it as kids—but you do have to walk faster than the escalator, meaning that you need to work harder, regularly reinvent yourself, obtain at least some form of postsecondary education, make sure that you're engaged in lifelong learning, and play by the new rules while also reinventing some of them. *Then you can be in the middle class.*

That's not a great bumper-sticker slogan, I know. And I say that with no delight—I liked the old world. But we terribly mislead people by saying otherwise. Thriving in today's workplace is all about what LinkedIn's cofounder Reid Hoffman calls investing in "the start-up of you." No politician in America will tell you this, but every boss will: You can't just show up. You need a plan to succeed.

Like everything else in the age of accelerations, securing and holding a job requires *dynamic stability*—you need to keep pedaling (or paddling) all the time. Today, argues Zach Sims, the founder of Codecademy, "you have to know more, you have to update what you know more often, and you have to do more creative things with it" than just routine tasks. "That recursive loop really defines work and learning today. And that is why self-motivation is now so much more important"—because so much of the learning will now have to happen long after you have left high school, college, or your parents' home—not in the discipline of a classroom. "An on-demand world requires on-demand learning for everyone, accessible to anyone around the world, anywhere on your phone or tablet, and this really changes the definition of learning," added Sims, whose platform provides an easy method to learn how to write computer code. "When I walk into a subway and see someone playing Candy Crush [Saga] on their phone, [I think] there's a wasted five minutes when they could be bettering themselves."

For more than a decade after the Internet emerged in the mid-1990s, there was much lamenting about the "digital divide"—New York City had Internet and upstate New York didn't. America had it and Mexico didn't. South Africa had it and Niger didn't. That really mattered because it limited what you could learn, how and where you could do business, and with whom you could collaborate. Within the next decade that digital divide will largely disappear. And when that happens only one divide will matter, says Marina Gorbis, executive director of the Institute for the Future, and that is "the motivational divide." The future will belong to those who have the self-motivation to take advantage of all the free and cheap tools and flows coming out of the supernova.

During the fifty years after World War II, if the world had a dial on it, that dial was set to the left, and the closer you were to the Soviet Union the more leftward the dial pointed. And what it pointed to was a sign that said "You live in a world of defined benefits: just do your job every day, show up, be average, and here are the benefits you will get." Since the emergence of the supernova, that dial has whipped sharply to the right, and the sign it points to today says "You live in a world of defined contributions—your wages and benefits will now be more and more directly correlated to your exact contribution, and with big data we will get better and better at measuring just exactly what your contribution is." It's a 401(k) world now. To paraphrase an old World War II poster, Uncle Sam wants—*to put more on*—you.

General Electric's CEO, Jeff Immelt, put it bluntly in a May 20, 2016, commencement address to graduates at New York University's Stern School of Business: "Technology has raised the competitive requirements for companies and people." John Hagel, the management expert, puts it even more bluntly: "There is mounting performance pressure on all of us—as individuals and institutions. All of this connectivity means significantly lower barriers to entry and movement, accelerating change and the increasing occurrence of extreme, disruptive events, all of which put significant pressure on our institutions . . . On a personal level, the example I use is a billboard that used to be up on a highway here in Silicon Valley which asked a simple question: 'How does it feel to know that there are at least one million people around the world who can do your job?' While we might argue whether it is one thousand or one million, it would have been an absurd question to ask twenty or thirty years ago because it didn't really matter—I'm here and they are somewhere else. Now it is increasingly a central question, and one might add, 'How does it feel to know there are at least one million robots who can do your job?' We are all feeling mounting performance pressure at a very personal level."

## *The New Social Contract*

But can everyone keep up?

This is one of the most important socioeconomic questions of our day—probably *the* most important. Here is one way to think about it: in every major economic shift, "a new asset class becomes the main basis for productivity growth, wealth creation, and opportunity," argued Byron Auguste, a former economic adviser to President Obama who cofounded Opportunity@Work, a social venture that aims to enable at least one million more Americans to "work, learn, and earn to their full potential" in the next decade. "In the agrarian economy, that asset was land," Auguste said. "In the industrial economy it was physical capital. In the services economy it was intangible assets, such as methods, designs, software, and patents."

"In today's knowledge-human economy it will be human capital—talent, skills, tacit know-how, empathy, and creativity," he added. "These are massive, undervalued human assets to unlock"—and our educational institutions and labor markets need to adapt to that." We need—at all costs—to avoid a growth model based on assets or opportunities that are accessible only to a fortunate few. The massive redistribution of wealth that would be required to support such a society is not politically sustainable.

"We need to focus on a growth model based on investment in human capital," argued Auguste. "That can produce a more dynamic economy and inclusive society, since talent and human capital are far more equally distributed than opportunity or financial capital."

So where do we begin? The short answer, says Auguste, is that in the age of accelerations we need to rethink three key social contracts—those between workers and employers, students and educational institutions, and citizens and governments. That is the only way to create an environment in which every person is able to realize their full talent potential and human capital becomes a universal, inalienable asset.

## *Let's Hire More Bank Tellers*

To understand the necessary components of such a new set of social contracts, we have to start with a clear picture of what is actually happening in the labor market, so we know what exactly we are trying to fix.

Here, I rely on the excellent work of the economist James Bessen, a researcher and lecturer at Boston University School of Law and author of *Learning by Doing: The Real Connection Between Innovation, Wages, and Wealth.* There are a lot of myths and misunderstandings surrounding this issue.

The central challenge we need to be focusing on, Bessen argues, is the issue of skills—not the issue of jobs per se. There is a huge difference, he insists, between automating tasks and completely automating a job—and dispensing with the human beings entirely. To be sure, we have jobs that have completely disappeared because the industry completely disappeared. There is probably no one in America, or anywhere for that matter, who makes their living today producing buggy whips—not since the horse and buggy gave way to the automobile. But it is critical to remember that even 98 percent automation of a job is not the same as 100 percent automation. Why? In the nineteenth century, 98 percent of the labor involved in weaving a yard of cloth got automated. The task went from 100 percent manual labor to 2 percent.

"And what happened?" asked Bessen. "The number of weaver jobs increased."

Why? "Because when you automate a job that has largely been done manually, you make it hugely more productive." And when that happens, he explained, "prices go down and demand goes up" for the product. At the beginning of the nineteenth century, many people had one set of clothes—and they were all man-made. And by the end of that century, most people had multiple sets of clothing, drapes on their windows, rugs on their floors, and upholstery on their furniture. That is, as the automation in weaving went up and the price went down, "people found so many more uses for cloth, and so demand exploded enough to actually offset the substitution of more machines for labor," explained Bessen.

Using government data, Bessen studied the impact of computers, software, and automation on 317 occupations from 1980 through 2013. In a research paper he published on November 13, 2015, he concluded: "Employment grows significantly faster in occupations that use computers more." He cited the example of cash machines, which began to be deployed in the 1990s in large numbers and now are everywhere. Everyone assumed that they would replace bank tellers. It didn't happen.

> The ATM is sometimes taken as a paradigmatic case of technology substituting for workers; the ATM took over cash handling tasks. Yet the number of fulltime equivalent bank tellers has grown since ATMs were widely deployed during the late 1990s and early 2000s. Indeed, since 2000, the number of fulltime equivalent bank tellers has increased 2.0 percent per annum, substantially faster than the entire labor force. Why didn't employment fall? Because the ATM allowed banks to operate branch offices at lower cost. This prompted them to open many more branches, offsetting the erstwhile loss in teller jobs. At the same time, teller skills changed. Non-routine marketing and interpersonal skills became more valuable, while routine cash handling became less important. That is, although bank tellers performed relatively fewer routine tasks, their employment increased.
>
> Even though the ATM automated routine cash handling tasks, the technology alone did not determine whether employment of tellers grew or fell; economics mattered. New technology can increase demand for an occupation, offsetting putative job losses. Nor is this example exceptional:
>
> - Barcode scanners reduced cashiers' checkout times by 18–19 percent, but the number of cashiers has grown since scanners were widely deployed during the 1980s.
> - Since the late 1990s, electronic document discovery software for legal proceedings has grown into a billion dollar business doing work done by paralegals, but the number of paralegals has grown robustly.
> - E-commerce has also grown rapidly since the late 1990s, now accounting for over 7 percent of retail sales, but the total number of people working in sales occupations has grown since 2000.

Bessen's point is that technology's impacts are not uniform: It can reduce demand for certain activities—routine tasks such as answering the phone and taking a message, for instance, have been largely wiped out by voice mail. But technology can also transfer tasks from one occupation to another. "There are still receptionists who answer phones and take messages," noted Bessen, "but do other things as well. So the number of telephone operators has declined dramatically (317,000 full-time equivalent in 1980 to 57,000 today) while the number of receptionists has grown by more (438,000 to 896,000); receptionists require new and different skills, of course, than switchboard operators."

At the same time, he pointed out, technology can create demand for totally new jobs—think data science engineers—even as it transforms the skills needed for some very old routine jobs, such as bank tellers and paralegals and store clerks, that would seem to be made obsolete by computers and robots but actually aren't. And it can vastly increase the skills needed to practice old jobs that have been transformed by technology—graphic designers, for instance. Which is why graphic designers who can use computer-aided design software make a lot more money than a typesetter of old.

Some economists keep telling us that there is no skills gap—because if there were, median wages would go up in those professions when the supply of skilled labor does not meet demand. They need to look under the hood more, argues Bessen.

"The wages of the median worker tell us only that the skills of the median worker aren't in short supply," said Bessen. At the same time, some workers in a given field could still have skills that are in much higher demand and the workforce could have a gap in those able to meet it. "Technology doesn't make all workers' skills more valuable; some skills become valuable, but others go obsolete," explains Bessen. When you look inside many professions what you discover is soaring demand and very high pay for those best able to leverage technology—and the opposite for those least able. That is where the real "skills gaps" show up in many occupations. Try hiring a quality data scientist in Silicon Valley today who can leverage the supernova to find needles in haystacks. Get in line!

For all of these reasons, Bessen concludes: "Jobs are not going away, but the needed skills for good jobs are going up." And with this new technology platform we're now on, it's all happening faster. For instance, new software—such as AngularJS and Node.js, both Java-based programming languages to build Web-based mobile apps—can come out of nowhere and become the industry standard overnight, far faster than any university can adjust its curriculum. When that happens, demand and pay for people with those skills soar.

So now we're defining the problem a little more accurately—what's over is not jobs. What's over is *the Holocene era for jobs*. Every middle-class job is now being pulled in four directions at once—and if we are going to train our citizens to thrive in such a world, we have to think afresh about each direction and what new skills or attitudes go with it in order to find a job, hold a job, and advance in a job.

For starters, middle-class jobs are being *pulled up* faster—they require more knowledge and education to perform successfully. To compete for such jobs you need more of the three *R*s—reading, writing, and arithmetic—and more of the four *C*s—creativity, collaboration, communication, and coding.

Consider a *New York Times* story from April 22, 2014, that reported:

> Something strange is happening at farms in upstate New York. The cows are milking themselves.
>
> Desperate for reliable labor and buoyed by soaring prices, dairy operations across the state are charging into a brave new world of udder care: robotic milkers . . .
>
> Robots allow the cows to set their own hours, lining up for automated milking five or six times a day—turning the predawn and late-afternoon sessions around which dairy farmers long built their lives into a thing of the past.
>
> With transponders around their necks, the cows get individualized service. Lasers scan and map their underbellies, and a computer charts each animal's "milking speed," a critical factor in a 24-hour-a-day operation.
>
> The robots also monitor the amount and quality of milk produced, the frequency of visits to the machine, how much each cow has eaten, and even the number of steps each cow has taken per day, which can indicate when she is in heat.

In the future, a successful cow milker may need to be an astute data reader and analyst.

Every job is also being *pulled apart* faster. For instance, being a cow milker may become disaggregated. The high-skilled part of that job may move up—now you either have to learn computing or become a veterinarian who understands the anatomy of cows or be a big data scientist who can analyze a cow's behavior. At the same time the less skilled part of that job—herding cows into and out of the milking barn and cleaning up their manure—may get pulled down so that it can be done by anyone for a minimum wage (and probably soon by a robot). This is a broad trend in the workplace, as Bessen noted: the skilled part of each job requires more skill and rewards more skill, and the routine, repetitive part, which can much more easily be automated, will pay minimum wages or just be given over to a bot.

At the same time, every job is also being *pulled out* faster—more machines, robots, and workers in India and China can now compete for all of it or a bigger part of it. That demands more self-motivation, persistence, and grit to learn new technical or social-emotional skills to keep one step ahead of the robots, Indians, Chinese, and other skilled foreigners through lifelong learning.

And, finally, every job is being *pulled down* faster—it's being outsourced to history in its present form and made obsolete faster than ever. And that demands more entrepreneurial thinking at every level: a constant searching for new niches, new opportunities to start something from which to profit and create employment.

So, at a minimum, our educational systems must be retooled to maximize these needed skills and attributes: strong fundamentals in writing, reading, coding, and math; creativity, critical thinking, communication, and collaboration; grit, self-motivation, and lifelong learning habits; and entrepreneurship and improvisation—at every level.

## *The Compounding Solution*

Fortunately, new technology tools will aid this endeavor. The new social contracts we need between government, business, the social sector, and workers will be far more feasible if we find creative ways—to borrow a phrase from Nest Labs' founder, Tony Fadell—to turn "AI into IA." In my rendering, that would be to turn artificial intelligence into intelligent assistance, intelligent assistants, and intelligent algorithms.

*Intelligent assistance* involves leveraging artificial intelligence to enable the government, individual companies, and the nonprofit social sector to develop more sophisticated online and mobile platforms that can empower every worker to engage in lifelong learning on their own time, and to have their learning recognized and rewarded with advancement. *Intelligent assistants* arise when we use artificial intelligence to improve the interfaces between humans and their tools with software, so humans can not only learn faster but also act faster and act smarter. Lastly, we need to deploy AI to create more *intelligent algorithms*, or what Reid Hoffman calls "human networks"—so that we can much more efficiently connect people to all the job opportunities that exist, all the skills needed for each job, and all the educational opportunities to acquire those skills cheaply and easily.

"When you have a compounding problem, you need a compounding solution," added Hoffman. The jobs issue "is a power law problem, and the only way to solve a power law problem is with a power law solution" for improving humanity's ability to adapt. Turning more forms of AI into more forms of IA is that solution.

## *Ma Bell's Intelligent Assistance*

I visited a lot of companies in researching this book, and none was more innovative in creating intelligent assistance to help its employees become lifelong learners than old, reliable AT&T. Don't let the moniker "Ma Bell" fool you. Don't let AT&T's CEO Randall Stephenson's good ol' boy Oklahoma accent take you in. Don't let the mild Midwest demeanor of Bill Blase, their head of human resources, get you to drop your guard. And whatever you do, don't move your eyes off their chief strategy officer, John Donovan, and Krish Prabhu, the head of AT&T Labs, because they will disrupt your business on behalf of one of your competitors before breakfast. They might even do it just for fun.

*Attention, Kmart shoppers: this is not your grandma's Ma Bell!*

Back in 2007, when AT&T found itself pioneering software-enabled networks to handle the explosion of data created by the iPhone, for which it was the original exclusive network provider, it realized that it had to up its own innovation metabolism—broadly and quickly. If you're running with Apple, you need to be able to run as fast as Apple. In 2016, AT&T was still at it. That year it opened up an "Internet of Things Foundry" in Dallas, an innovation shop full of network engineers. They invited customers in with this proposition, explained Vice Chairman Ralph de la Vega: "Tell us what problem you want us to solve, and we commit that within two weeks we'll give you a prototype solution for you that works on a real live network . . . Every time we do this, it results in a contract."

So, for instance, the global shipping giant Maersk needed a sensor that it could affix to every shipping container it owns, enabling the company to track its containers anywhere in the world. The sensor had to affix to two hundred thousand cargo refrigerator containers, it had to be able to measure their humidity, temperature, and whether they had suffered any damage, and it had to broadcast that data to their headquarters, and—this was the real catch—the sensor had to operate without batteries and be able to last ten years, because they couldn't be changing them all the time. In two weeks, AT&T engineers built a prototype of a sensor, half the size of a shoe box, that could affix to every Maersk container and was powered by a combination of solar and kinetic energy.

What happened to AT&T was that the supernova transformed its business overnight. Its decision to virtualize networks to expand its capacity made it more of a software/networking company, and then it really struck gold with the rise of big data, which meant that the data and voice traffic AT&T was carrying over its wires could be aggregated and anonymized and then mined for trends. So suddenly, as noted earlier, AT&T, using wireless cell phone data, could tell a signage company how many people who drove by their sign on the freeway ended up shopping in the store advertised on that sign—and if the sign became digital, and changed every hour, they could tell them which message was most effective. AT&T started telling some customers, heck, we'll cut you a deal on the transmission costs if we can mine the data and use it to solve customer problems or puzzles. In the blink of an eye your friendly phone company became an all-around business solutions company, also competing with IBM or Accenture.

Precisely because Stephenson understood that for his company to thrive it had to be the networking enabler and solutions provider to the most disruptive companies in the world, he knew he had to disrupt his own workforce in the process.

"We felt a fundamental obligation to reskill our workforce," said Donovan. "We needed a smaller and smarter workforce. STEM skills are now table stakes." But they also knew that when you're dealing with three hundred thousand employees, if you want to challenge them to upgrade their skills, you need a strategy of what I call "intelligent assistance" to provide the scaffolding and incentives that make a new learning journey for so many people sustainable.

AT&T's version of intelligent assistance begins, explained Blase, with the management teams becoming increasingly transparent about the company's direction and the skills it will need. Each year starts with Stephenson giving a town hall speech for all the top managers of AT&T. "The idea is to be totally transparent with our employees about where the business is heading and what the challenges will be," explained Stephenson.

That message gets filtered down through managers, so that eventually every employee has a broad understanding of the company's objectives for the next twelve to fourteen months, "and where this company is going for the next five to ten years," added Blase. "We start in January, and by July everyone has the message." Ideally, he said, most employees will say, "I get it, I want to be part of this. I am one of the three hundred thousand. How do I get to be part of this?" Others, though, will say, "You know what? I've had my thirty-five years; it's time to go. I am not ready to learn anything new." So about 10 percent of AT&T's workforce drops out each year.

Blase added:

> We do not have enough internal people with the skills to be able to lead effectively through this change or the technical knowledge of what we are selling or the technical knowledge of what is behind what we are selling. So we hire thirty thousand [employees] off the street each year. We fill another thirty thousand jobs through rotations and promotions. It costs about two thousand dollars just to hire someone, so our preference always is to use our internal employees. It is more cost-effective and will generate more employee engagement and productivity, which means employees will go the extra mile so customers will be served better and shareholder value will increase. The companies with the most highly engaged workforces earn three times those with less.

But this has meant placing a lot more lifelong learning demands on a lot more employees. Most employees "embrace what we're trying to do," said Blase. "They say, 'Just give me the tools, point me in the right direction, help me make [the transition] seamless, and make it cost-effective and make it mobile and make it Web based, so I can do it on my own time, and make it flexible and make the training in a format that I can learn quickly and effectively.'"

Added Donovan: "We have employees here who want to make the pivot—employees who built this company that they would die for—and we need to give them the opportunity to pivot. Many of these were traditionally blue-collar folks, who just finished high school, and we need to constantly retrain them to work in a networked house."

To act on that bargain, AT&T six years ago asked all of its managers to set up internal profiles similar to LinkedIn accounts, detailing their job experiences, skills, education, certifications, and specialties. Today, 90 percent of the 110,000 managers have done so. (The company categorizes as managers all its professional employees not covered by union contracts, regardless of whether they are a supervisor or an individual contributor.) Now when a new job opens up, the first thing Blase's team does is check those profiles for internal candidates who have the necessary skill sets.

At the same time, the company posts the hot new jobs and identifies where they are located, the exact skills needed to get those jobs, and how to get the training for those skills on its own in-house "Personal Learning Experience"—a one-stop training platform for employees.

The top trending tech jobs are software engineer and data analyst. On the business side, they are technical consultants who work to support sales and the customer implementation of AT&T security, network, and voice products and services. Also in demand are intelligence specialists who work with government customers on cloud computing, encryption technologies, and risk management. In the entertainment group, which manages DIRECTV and other services, the rising jobs are "content creators" and data analytics specialists, who can use their skills to improve the marketing of emerging new products.

To supplement AT&T's own learning content, the company also partnered with many universities—from Georgia Tech to Notre Dame to the University of Oklahoma system to online universities such as Udacity and Coursera—to provide affordable graduate and undergraduate degrees or just specialized training for each of the new skills it is asking its workers to acquire. AT&T's only requirement is that you take the courses *on your own time*, but the company will reimburse your tuition up to eight thousand dollars a year (or more, for certain courses) and thirty thousand dollars over your lifetime at the company.

Then, to make sure that money went as far as possible, AT&T pressed universities to create a menu of online learning opportunities that would fit its budget. This approach has driven a lot of innovation in education, most notably the partnership between Udacity, AT&T, and Georgia Tech to create an online master's degree in computer science for $6,600 for the entire course—as compared with the $45,000 it would cost for two years on campus at Georgia Tech. Coursera has partnered with Johns Hopkins and Rice to create a similar certificate in data science.

Among the Udacity and Coursera online offerings are courses on artificial intelligence providing "the foundational basics of AI," including "search and optimization, logic, planning and reasoning, probability and building models, natural language processing and computer vision," as well as a course on how to become a "Virtual Reality Developer" and a "Predictive Analytics" specialist.

AT&T's push into online learning is driving down the cost of education for everyone. The education "pie just got bigger," said Blase. "We can now assist you to get the job of your dreams."

That's intelligent assistance. "We do $250 million of training a year," said Blase.

> A lot of it is teaching people to climb poles, install services, and run retail stores, but now a lot more is in these new areas of data science, software-defined networks, Web development, introduction to programming, machine learning, and the Internet of Things. And if you want to take a general STEM [science, technology, engineering, math] course that is not part of our program, we will pay for it, too. If you want to learn, we're all in, because, again, it leads to more engaged employees; that equals better customer service, more loyal customers, and more shareholder value. We did not have anything like this when I was growing up in the company.

These supports are for jobs paying sixty thousand to ninety thousand dollars a year.

The company registers all certificates and degrees earned by employees in their company profiles and can easily search them through big data tools. And if you show that you are motivated to get these advanced degrees and certificates, said Blase, "we will give you first crack when a job opens. People need to know that if I am clearly motivated to learn, I am going to get rewarded."

The way the system works, said Blase, is:

> Say I am a manager and I have ten openings for technical jobs. I come to HR and they say you have to look [internally] first. Then you look through the online profiles and you find people who have shown or demonstrated a desire to get into one of these capacities. So HR will pull a bunch of names capable for those ten openings, [including] someone who has a majority of the skills. You pull the aspiring people and the perfect fits, and we ask the person who is doing the hiring to give these people who are trying a shot.

Because, said Blase, those employees will then tell their story to other employees: I played by the new rules and got rewarded. As he put it:

> It is a contract between the company and employees. It's a new bargain. If you want to get an A in your performance review, now you have to do the "What" and the "How." The "How" is that you get along with people, you achieve results by effectively partnering and teaming and leading change through [and with] others and don't just sit in your cubicle. The "What" is that you are not only proficient in your job but that you are reskilling to improve your capacity, continuing to learn, and that you are aspiring to go beyond where you are. Maybe you're a salesperson and you're making yourself more valuable to the company by getting [to know] the technical side as well. You're not just selling products but understanding how our network works. Our best employees have it down and they know it is the What and the How.

The new social contract, Donovan added, is that

> *you can be a lifelong employee if you are ready to be a lifelong learner.* We will give you the platform but you have to opt in . . . Everyone has a personal learning portal and they can see where the endpoint is [for whatever skill set they are aiming to acquire] and the courses that will get them there. You can pick a different future and how to get there. You can be anything you want to be in this system. But again, *you have to opt in.* The executive's role here is to define the vision for the future. The company's responsibility is to provide the tools and platforms for employees to get there, and the individual's role is to provide the selection and motivation. We need to make sure that anyone who leaves here [does not do so] because we did not provide them the platform—that it was their lack of motivation that did not make it happen.

AT&T is a big whale. When it moves down this pathway of education to employment, it creates a huge wake. As Blase put it: "Now, universities are modifying their behaviors to meet us in the marketplace. We are writing a new blueprint." If universities are paying attention, they will start creating more degrees and certificates "that are profitable for them but are cost-effective for this model."

Donovan is certain that this new social contract is raising both the company's average skill level and its morale. "What we have done is to take our best and make it our average," he said, "and our average is now right up there. Our cycle times for [new ideas] are much faster now. Anyone who finds a solution, we can scale it through the company. Our employee engagement surveys showed a 30 percent improvement in lost sick days in one year. People are calling in sick less because they are feeling more empowered, more of a sense of ownership, and more connected."

The best example of someone who understands that to be a lifelong employee you have to be a lifelong learner that I found in my research is Martin Reeves, who leads the Boston Consulting Group's in-house think tank—the Henderson Institute. Martin handed me his business card one day and instead of a title under his name it had all the ideas he was thinking about and researching at that moment.

"I change my business card every week now," he explained. Indeed, when I met him for lunch on March 13, 2017, he began with an apology: "I'm sorry I don't have this week's card." So he handed me his "old one" for the week of March 1, 2017. It read:

Martin Reeves, Director, BCG Henderson Institute, Senior Partner and Managing Director.

5 Things I am Thinking About (March 1, 2017)

1. Link between diversity and performance
2. Theories of intervention into complex adaptive systems
3. Designing supply chains for resilience
4. How to exploit math of serendipity
5. Scenarios for future of work

## *Jump-Starting the Curriculum*

The AT&T model for lifelong learning—and its supercharged version expressed in the likes of Reeves's business card—has wide-ranging implications for the entire universe of education. In short: if the nature of work is being transformed, education better not be far behind. Consider Udacity, which built the online low-cost master's degree in computer science with Georgia Tech. Today, the business it created with AT&T enables it to offer the same intelligent assistance to the world at large and plant the seeds of the real revolution in education.

Udacity was founded by German-born Sebastian Thrun, formerly an artificial intelligence professor from Stanford and an expert on robotics. Thrun likes to recall his first meeting with Randall Stephenson at AT&T headquarters in Dallas: the two of them sat on the floor together in Stephenson's executive suite so Thrun could use his laptop to sell the AT&T exec on how mini online courses, or nanodegrees, that could teach the latest technology skills could elevate the AT&T workforce. Stephenson got up off the floor after the demo and signed him up on the spot. One of the things that Thrun learned from teaming up with Georgia Tech to create the $6,600 online master's in computer science was that the course did not cannibalize Georgia Tech's far more expensive campus-based master's. It turns out there were two different markets—

one for people who want a campus experience and the other for people starved for lifelong learning that they could do in their spare time at a price they could afford. "The average age of the students for our online course is thirty-four and for the campus course is twenty-three," he explained. The demand for more lifelong learning platforms was clearly out there. People get it. So today Udacity offers nanodegree programs on building websites, introduction to programming, machine learning, developing apps for Android mobile devices, and developing apps for Apple mobile devices, among others.

But here is where things get really interesting: Udacity now develops some of its courses with the help of working Google engineers. So, for instance, in October 2015, Google released the basic algorithms for a program called TensorFlow for public consumption by the open-source community. TensorFlow is a set of algorithms that enable fast computers to do "deep learning" with big data sets to perform tasks better than a human brain.

"By January 2016 we had a course online on how to use the TensorFlow open-source platform to write deep learning algorithms to teach a machine to do anything—copyediting, flying a plane, or legal discovery from documents," explained Thrun. This is a huge new field of computer science. TensorFlow was released into the wild in October, and by January, Udacity, working directly with Google engineers, was teaching the skill on its platform. "We can now update your skills at the pace of Moore's law, at the pace of the industry," explained Thrun. The traditional "academic world cannot do that." It would probably take a year for a university to have such a course up on TensorFlow, and for many it would take a lot longer.

Udacity has developed a stable of on-demand freelancers around the world whom it employs to grade the work of its online students—and the students also grade their graders. "I can hire a thousand graders in a week from around the world," said Thrun, "try them out, find the two hundred best, and let the other eight hundred go." It is a fast way to get high quality. There are Udacity freelance graders who make several thousand dollars a month grading computing projects—like how to build a map from Google's GPS—sent in from students around the world. "We had one project grader who made twenty-eight thousand dollars one month," said Thrun. "The gig economy is moving up. It's not just about TaskRabbit errands anymore."

And Udacity is not just providing intelligent assistance for companies such as AT&T. Its platform is creating intelligent assistance for "the start-up of you"—whoever and wherever you are. In the fall of 2015, I found myself in a small conference room at Udacity's Palo Alto headquarters interviewing—via Skype—Ghada Sleiman, a thirty-year-old Lebanese woman, who was taking Udacity's online course to advance her skills in Web-page design. She explained that she was sitting at her home in Beirut taking a course from a company in Palo Alto in order to better serve her clients, most of whom she has never met in person, in Australia and the United Kingdom.

"I studied graphic design at the American University of Science and Technology in Ashrafiyya," a neighborhood of East Beirut, she explained to me. "After college I was looking for a course on Web design, and I found Udacity and decided to try it out. I started last year [in 2014]. I used to learn [on the Web] only through tutorials." But she found that with Udacity's platform "there was a sense of community, and I could communicate with other people, so it was more interesting and more interactive."

Why did you have to go on the Web for this course? I asked.

"Universities here offer graphic design and computer science, but not Web design," she explained. "It's a whole new field, and universities have not caught up . . . The course I am taking [on Udacity] is Web design and programming. I am good at the design, but I needed to get more into the programming-development part. It complements my work."

What kind of clients do you have in Australia? I asked.

"One is a publication about start-ups, one is a business-related blog, one is a new-moms-themed blog, and one is an Australian social media company," said Sleiman, who marketed herself as Astraestic.com, a combination of "artistic" and her nickname, "Astra." "At first my parents were surprised and asked, 'How did you get to know them?'—but now they like the idea and believe in a great future for me because of this, because I can reach other people in other countries. There are not so many clients here as you can find globally."

What advice would she give to other young people her age? I asked. "I would tell them that they should first of all develop their technical skills, but that is not enough. They also need to know how to market themselves. Marketing is not just for marketers—[it's] a huge part of getting work. I would say work *on yourself.*"

Sleiman's story underscores the new contract you need with yourself—more self-motivation to tap into the new global flows for work and learning—and the new contract that schools need with students. People thought that the advent of MOOCs heralded the revolution in education. It was a revolution, but it was just the tip of the iceberg, because it was still based on the old model: MOOCs essentially just used the Internet and video as a new delivery system for old-fashioned lectures. The supernova is enabling a deeper revolution that is just beginning, spurred by learning platforms such as Udacity, edX, and Coursera, that will change the very metabolism and shape of higher education and, one hopes, lift the adaptability line in the way that Astro Teller urged. When a company like Udacity can respond to a major technological leap forward, such as TensorFlow from Google, and offer a course online to teach it to anyone in the world within three months, the word is going to get out and the market will change. Who is going to wait until next year to take that course on the campus of a university—assuming that school can even change its curriculum that quickly?

But they are going to have to learn to do just that to survive. Jerry Steward, the president of Oklahoma City Community College, remarked to me during a visit there that the Oklahoma legislature wants more for its money and the customers—the students—want more for their money, "and with the advent of online education students have choices. They may live next door to our campus but if they don't like what they see they can enroll in an online college in Singapore if they want to now."

Indeed, popular learning programs can now be found everywhere.

There are even game platforms such as Foldit, the crowdsourcing computer game, that enable anyone to contribute to important scientific research. These are becoming popular learning platforms. Foldit set up an online "game" where anyone could play and win a substantial cash prize by designing proteins. "Since proteins are part of so many diseases, they can also be part of the cure. Players can design brand new proteins that could help prevent or treat important diseases," Foldit explains on its site. The game attracted thousands of contestants from all over the world, some with no formal education in biology at all, to compete for the prizes and, in so doing, win not a bachelor of science degree, but a reputational badge that may soon be more meaningful to the marketplace.

These new approaches to rapid learning are already filtering into the traditional brick-and-mortar institutions, with some radically new models popping up. Consider just one example: Olin College. In a speech, the school's president, Richard K. Miller, explained that "in 1997, the F. W. Olin Foundation established Olin College [in Needham, Massachusetts] for the specific purpose of inventing a new paradigm for engineering education that prepares students to become exemplary engineering innovators" ready to take on the biggest problems. "The role of the engineer we envision is that of 'systems architect' of complex technical, social, economic, and political systems capable of addressing the global challenges we now face," said Miller.

To produce such engineers, he said, Olin maintains a highly flexible structure that can move at Internet speed. Olin "is not internally organized into academic departments, and faculty members do not have tenure," Miller explained. "Instead, faculty members are employed with renewable-term contracts with a range of term lengths." I was the commencement speaker there in 2016, and could not help but notice that half the graduates were women—unprecedented in an engineering school.

"A particularly important aspect of Olin College," added Miller, "is the precept requiring the college to devote itself to continuous improvement and innovation." As a result, at Olin nearly everything has an "expiration date." This includes the bylaws and the curriculum. "The Olin College curriculum is continually evolving—by design," said Miller. "The current incarnation provides a snapshot of the best efforts of the Olin community to provide a new paradigm for engineering education. The curriculum currently expires every seven years and must be actively reviewed and either revised or reinstated." To graduate, Miller added, all Olin students "must complete a yearlong engineering design project in small teams with a corporate sponsor that provides financial support for each project. The projects require a corporate liaison engineer and often involve nondisclosure agreements and new product development."

Olin is small and young, but this engineering lab school demonstrates a lot of the revolutionary features that eventually will be incorporated at most schools—the end of tenure, close partnership with change agents in the working world, a constantly adapting curriculum and no departments, and a synthetic teaching approach that blends engineering and humanities—such as a course that combines biology and the history of pandemics. That's intelligent assistance at its best. That is the real revolution in education, and it will be coming to a community near you, as more and more workers need and demand intelligent assistance. Miller calls it "expeditionary learning"—creating your own knowledge and inventing your own career.

"Improvisation is what you must do constantly," he said. "It is well beyond problem-based learning or even project-based learning. You are literally marching into a forest that no one has explored in search of things you have never seen." All that he can promise, said Miller, is that you'll find jobs there that you can't imagine today that will require rapid and continuous learning.

## *Intelligent Assistants*

One of the most intriguing online intelligent assistants for the workplace that I came across in my research was LearnUp.com, cofounded by Alexis Ringwald, an adventurous young entrepreneur whom I first met in India, where she and a partner were highlighting that country's grassroots renewable energy initiatives by touring around in a solar-powered car—with a solar-powered rock band!

After doing a start-up in the States in solar energy, Ringwald got interested in the employment sector and spent six months interviewing workers looking for jobs. She discovered something she did not expect—that a majority of jobs on offer today don't require a four-year college degree, and nine out of the top ten jobs in America by volume don't require more than a high school degree. But she also discovered that the widespread assumption in America that anyone can get these entry-level customer service jobs by just showing up and demonstrating a pulse was even more wrong: these jobs do require basic skills that way too many applicants actually lacked.

As she put it: "Even a clerk at the Gap, a hamburger flipper at McDonald's, or a receptionist requires certain basic workplace skills," but "most people who apply for them don't have them—they just think,

'Hey, I like clothes. I can work here'—and their high schools or community colleges don't teach them."

"My first epiphany was to realize that the whole system is designed to weed people out, not to get people in," Ringwald explained. "The whole system is built for employers to fend off all the people flooding their career systems. And so people just throw themselves out there, applying to a hundred places at once; then they get rejected and don't know why . . . I saw employers inundated with candidates who were not qualified for basic jobs, and employees who didn't really know what they were applying for."

She also learned that once people got the job, they often faced huge challenges holding on to it—people would think that if they could not make it to work one day, because they were sick, or their car broke down, or they had to stay home with the kids, that meant they had to quit, not just try to explain it to their manager.

Ringwald thought all of these problems were fixable and cofounded LearnUp in 2012, to do just that: job seekers go to its website and find an online platform with a mini course they can use to learn about the actual requirements and skills needed for a job before they apply. It offers modules on how to prepare for an interview, as well as the specific skills needed for different open positions, including how to build a customer relationship at AT&T, how to sell clothing at Old Navy, and how to solve a customer problem at the Fresh Market—as well as how to help a customer find the right fit, how to make the store look good, and how to operate basic office equipment—copying machines, for instance, if you're applying at a retailer such as Staples. The trainings are set up to take just one to two hours, but that is enough for job candidates to learn about the company, gain skills needed for the role, and become qualified to apply. For the companies it also identifies those who have the persistence to learn the basics and those who don't. Once you complete the course, LearnUp actually sets up a job interview for you with the company of your choice.

"LearnUp is linked to a specific job opening with a real open interview slot," explained Ringwald. "Job seekers who are about to apply online to a job at one of our partners—like Old Navy, the Fresh Market, AT&T—can access LearnUp by clicking a button on an employer's career site that says 'prepare before you apply.'" LearnUp does not screen candidates out—rather, it tries to train them and coach them in for a particular job opening. As they learn more about the job through LearnUp, candidates choose to go forward with the application or opt out by clicking a button that says: "I don't want this job."

Most important, in my view, LearnUp also provides an online "coach" who proactively gives you encouragement, interview reminders, and advice, and who will answer your questions. It is so easy to forget that many, many people in America don't have a professional network, an alumni network, two parents, or in some cases anyone around them with a job, to consult about how to get one. Ringwald was surprised to learn how many people would ask their coach something as simple as: "What should I wear to my job interview? What do I do if I am going to be late?" Some candidates will text the coach with a photo of what they are wearing to a job interview and ask: "Is this OK?"

These questions may sound elementary to you, but, said Ringwald, you'd be amazed how many people need this kind of advice: "All the people we talk to are grateful for it." She explained that the coach button was

> inspired by the power of real human coaches we encountered at workforce development offices. Their enthusiasm and support make a huge difference in a job seeker's success. That's why we built the coach into the platform. There's so much friction to getting a job—while trying to manage your life and your family's lives in the meantime—from deciding where to apply, filling out an application, figuring out if you live close enough, making sure you're qualified, preparing, getting yourself to the interview with transportation, wearing the right thing, saying the right thing, following up. And then imagine doing it a thousand times over in your search for a job. People suffer not just from decision fatigue, but also lack of hope and confusion. In a world with so many choices, it's hard to know what to do. And it's a thousand times harder when you don't have anyone around you who has done it. For seventy percent of the workforce, in jobs without degrees, that is their world. There is *no* support. If you don't have it in your family, in your community, then it's tough . . . What's powerful about the LearnUp coach is its accessibility and ease of use. Most of the nondegree demographic in America may not even think to reach out to an adviser or mentor.

In fact, I would say there is even a stigma for people to go to the unemployment offices and ask for help. It's really hard to do.

I asked Ringwald for examples of what their coaches offer in the hiring process. She sent back the following list:

- Tell you what to wear & provide the weather forecast for interview day
- Where to go with Google street map view of job location & public transit route to job location
- Send interview reminders about the time and how long you should prepare to get there
- Have you dial-in to a practice interview line, record your answers, then hear "best practices" answers
- Provide tips from previously hired job seekers or managers at each step
- Provide more transparency of what and why at each step of a job search so that the benefits are clear
- Show other previously hired job seekers at the job location
- Share interesting facts about the location and the manager with job seekers
- Provide more info about the hiring manager whom they will meet
- Ask job seekers to share interesting facts about themselves with the hiring managers
- Auto schedule a Lyft or Uber to take them to their interview
- Remind you to send a thank-you note to the interviewer

Concluded Ringwald: "Everyone needs someone who says, 'I believe in you' . . . There is not just a skills gap—there's a confidence gap."

And you can't sustainably fill one without the other.

## *You Need Work on Fractions*

Maybe the most popular intelligent assistant in the world today is Khan Academy, which was started in 2006 by the educator Salman "Sal" Khan and offers free, short YouTube video lessons in English on subjects ranging from math, art, computer programming, economics, physics, chemistry, biology, and medicine to finance, history, and more. Anyone anywhere can go there to learn or brush up on any subject. Not only has it become the most important intelligent assistant for generalized learning in the world, but in 2014 it formed a partnership with the College Board, which administers the SAT college entry exams and the PSAT practice exams. Together, they created an intelligent assistant for anyone who wants to improve their SAT scores to get into college. They not only offer free SAT prep—so you don't have to pay a small fortune to some private test prep service to get your kid into college—but they also created an amazing practice platform to help students fill their knowledge gaps.

The system works like this, explained Stefanie Sanford of the College Board: In tenth or eleventh grade you take the practice SAT, known as the PSAT. And let's say, for instance, you scored 1060 out of 1600 on English and math. Your results are fed into a computer, which, using AI and big data, then spits out a message: "Tom, you did really well, but you need some work on fractions. You have a real opportunity to grow here. Click here for customized lessons just for you on fractions."

Suddenly, I not only know exactly what I need to work on but am also intelligently assisted into a practice program that addresses my exact weaknesses. I don't need to practice everything and drown in review problems. I can focus precisely where the artificial intelligence of the College Board platform points out that I need help. So far more than 3 million kids have signed up for free SAT prep from Khan Academy online. This represents four times the total population of students who use commercial test prep classes in a year. In fact, more kids now are using Khan Academy than paying for test preparation at every level of income. That tells you what a valuable intelligent assistant it has become. Meanwhile, more than 500,000 have linked their results on the College Board's PSAT exams with Khan Academy to get tailored tutoring on the questions they missed, which they can then practice on their own time wherever they are—including through their cell phones.

This is one of the quietest but most important intelligent-assistant education tools being made available for free in America today. Practice for the SAT—and advice for getting into college—have long, rightly, been thought of as areas where privilege rather than merit matters, where the wealthy have special access.

"We are trying to change that so that many more students have the tools of ownership," explained David Coleman, president of the College Board. "We are providing personalized learning at a time when students need to take far greater command of the cultivation of their talents and their career trajectory. The College Board used to just give tests to measure and mark progress, now we are actually trying to provide the tools of practice and coaching to change trajectories."

Doing so, though, requires some important attitude changes in line with where the world of work is going. "You have to own your own performance," said Coleman, "and realize that it is not something that is given, but achieved through practice." Coleman has worked to change all aspects of the SATs, to make clear that the test does not measure IQ or general aptitude, but a focused set of learning skills you use over and over again in high school and college. The test is steadily and subtly being reshaped to encourage practice and lifelong learning. "That's why we partnered with Khan Academy—to provide the best in test prep," Coleman added. "Now all students can own their performance, because they have access to the best tools of practice."

It appears to be working, he added: "The College Board analyzed 250,000 students from the high school graduating class of 2017 who took the new PSAT and then the new SAT. Students who took advantage of their PSAT results to launch their own free personalized improvement practice through Khan Academy advanced dramatically. Twenty hours of practice was associated with an average 115-point increase from the PSAT to the SAT—double the average gain among students who did not [do the practice]. Practice is an equal opportunity employer. Practice advances all students without respect to high school GPA, gender, race and ethnicity, or parental education. And it's free."

All of this, in turn, enables the College Board to create another form of IA—"intelligent advice"—tailored just for you and informed by AI. "With students' and families' permission, we share with advisers not only data on the student but also the patterns in the data the College Board can see, to make sure the adviser is fully informed," said Coleman. And to make sure there are advisers and coaches for those who need them most, the College Board partnered with the Boys & Girls Clubs of America to guarantee that as many students as possible take advantage of its free practice tools around the country. It also partnered with the College Advising Corps to match free trained advisers to high-achieving, low- or moderate-income students to help guide them to their best college choices—and it built links to prospective college scholarship opportunities. The platform also identifies kids who could be successful in AP courses their junior or senior years in high school but may have been too intimidated to sign up or didn't think they were good enough. This often applies to students of color, who often get shunted aside from these opportunities, which is why Stefanie Sanford likes to say, "People say tests are biased; well, tests are not nearly as biased as people are." Intelligent assistants are color-blind.

The Khan–College Board collaboration is really worth studying because it's a microcosm of how we can beat the bots—how we can make the transition to a different education-to-work-to-lifelong-learning social contract in the age of accelerations. There are three basic ingredients of the Khan–College Board revolution: (1) More will be on you, and you'd better take ownership of that fact and seek out intelligent assistants and assistance everywhere that you can; (2) precisely because more is on you, government and social organizations need to get serious about providing you not just any tools, but much better tools—informed by artificial intelligence tailored exactly for you and your needs and reinforced by a caring adult or coach, wherever possible; and (3) technology can take you only so far. Concentration also matters. Coleman likes to say that today the "technology of interruption has outpaced the technology of concentration." Students need to learn the discipline of sustained concentration more than ever and to immerse themselves in practice—without headphones on. No athlete, no scientist, no musician ever got better without focused practice, and there is no program you can download for that. It has to come from within.

If you come, they will build it—turning AI into IA is only going to get more efficient every year. "In the old days someone would publish a calculus textbook and get no data and no feedback on what is working for people and what is not," explained Sal Khan. So they spent the next five years just changing page numbers. Today, he said, Khan Academy can put up a set of calculus tutorials and see within hours which ones are the most effective in helping students come up with the right answers, iterate immediately, and start scaling the best tutorials globally within a few more hours. The ability to refine content and make it better at scale is staggering.

"Having high literacy rates was an accelerant for growth of the developed world, but now imagine we have an accelerant for the developing world"—where instead of 5 percent being able to participate and then contribute, you have 50 percent, Khan added. Young people who are motivated to learn can now go to Khan's platform and go as fast as they want, and some have started to go very, very fast.

Said Khan: "There is no ceiling anymore."

## *The Brilliant Janitor*

Intelligent assistants are not simply websites you can access. They are also portable tools that can turn AI into IA in remarkable new ways so that so many more people, no matter how educated or dexterous, can live above the average adaptability line—and even thrive there.

Consider what it is to be a janitor today at the Qualcomm campus in San Diego. Hint: thanks to intelligent assistants, it's become a knowledge worker job. Ashok Tipirneni, director of product management for Qualcomm's Smart Cities project, explained to me why: Qualcomm has created a business in showing companies how they can retrofit wireless sensors to every part of their buildings in order to generate a real-time, nonstop sort of EKG or MRI of what is going on deep inside every one of their buildings' systems. To create a demonstration model, Tipirneni started with six buildings at Qualcomm's Pacific Center Campus in San Diego, which included parking garages, office spaces, and food courts; the area was about a million square feet in total and used by about 3,200 people. They retrofitted small, self-powered, clip-on sensors to transmit all their data—from doors, trash cans, bathrooms, windows, lighting systems, heating systems, wires, chillers, and pumps—to a receiver on the campus. The receiver sends all the data up to the supernova to be stored, analyzed, and turned into intelligent advice for the building maintenance men. "We didn't have to break open a single wall," said Tipirneni.

The first result was significant savings. Labs started competing with one another over who could save the most. "We discovered that a lot of the energy use was coming out of PCs in labs, and just by putting the PCs in hibernate mode in six buildings when they were not being used is projected to save roughly a million dollars a year—it was shocking [to find] such an easy fix," said Tipirneni. "This data is giving us those insights—it's amazing."

But the fun part is that they started to stream all the data into icons on a tablet and then outfitted each of their maintenance men with one. The minute a leak or short happens or a valve is left open, it shows up on the tablet. And if something breaks the tablet will immediately display the repair manual. If something breaks or leaks that the maintenance team doesn't know how to fix, they take a picture of it with their tablet. "The system will know that this part in this building is connected to a pipe on the fourth floor and that floor is assigned to this technician, and it automatically sends a ticket to him to fix it," said Tipirneni. "The device will know exactly where the pipe is behind the drywall," so there is no need to guess where to make the hole. "You save time and money and use only what you need in the most efficient way. And then you use the time saved from treating symptoms to fix root causes."

Qualcomm is putting these sensor clips on all forty-eight buildings in San Diego. Suddenly the building maintenance guys "got converted to data engineers, which is exciting for them," added Tipirneni. They made sure the data was "distilled in a way that is easy for them to understand and be actionable. In the old days, when a facilities manager looked at a building, he would say: "If there is a leak someone will call me or I will see it." They were reactive. Now, says Tipirneni, "We trained them to look at signals and data that will point them to a leak before it happens and causes destruction. They did not know what data to look at, so our challenge was [to] make sensor data easy for them to make sense of, so we don't overwhelm them with too much data and just say 'You figure it out.' Our goal was, 'We will give you information you can use.'"

"The cognitive load is too much," he added, "and technology has to reduce that cognitive load on the user. Everyone will need and everyone will have a personal assistant."

The maintenance team now feel more like building technicians, not just janitors. "They feel it is a step up," said Tipirneni. "They got very excited about the interfaces."

And the best part, he added: "We had forty officials from four different cities in for a demo, and some of the maintenance people here presented it and they showed what they learned, and it struck a chord with the city officials. They were confident enough to talk about these things in a matter of a few months."

That is what an intelligent assistant can do.

## *Intelligent Algorithms*

I could tell you why intelligent algorithms are so valuable for the world of work in the age of accelerations, but I would rather just tell you the story of how LaShana Lewis, a computer server engineer with MasterCard, got her job. I got to know Lewis at a panel discussion on how to "re-wire the U.S. labor market," organized by Opportunity@Work.

Lewis, an African American woman, now age forty, was born to a single mom (who was herself just fifteen years old when she had LaShana) in East St. Louis, Illinois. "My mom was on welfare, and we lived in public housing. Everyone around us was on welfare. We did not have many resources at home; there were no computers in the schools, which were funded with property taxes." But Lewis discovered early in life that she "had a knack for fixing things." So whatever broke around the house—from toasters to sinks—she would repair herself. Once she hit high school, where they had computers, she dived into the computer science course; she ended up tutoring other students and catching the eye of the teacher, who told her: "You need to go to college and study computers." She got a scholarship to attend Michigan Tech, but even with the scholarship she did not have enough to support herself and dropped out after three and a half years—without a degree. She would have graduated in 1998.

"So I came back home and tried to get a job in computing, but I was blocked every time," said Lewis. "People asked if I had graduated, and I would not lie and said 'no,' so I got a job driving black kids back and forth to a supplemental tutoring program from the high school I went to in East St. Louis to the local community college. So I am driving the van, and one day the computer science tutor at this tutoring program quits. So they asked me to fill in, which I did. And at the end of the month, I asked if I could do the job full-time and they said, 'No, you don't have a degree.' So after that frustration, I went to a hiring firm and they got me into a help desk." She worked at help desks for ten years, helping knuckleheads like me reset their passwords and the like.

Her break came while she was working at the help desk at Webster University in St. Louis, when a colleague and faculty member saw just how talented she was. (She was constantly hanging around with the IT team, working as a backup technician.) One day, while Lewis was taking a refresher course on computing at Webster, her professor, who'd learned about a new intelligent assistant—LaunchCode.org—told Lewis to check it out. LaunchCode's aim is to help "you find the best resources online and in your community to prepare yourself for a job in tech." Its promise is: "Don't sweat your credentials, just show us what you can do. Apply online for a LaunchCode apprenticeship and we will help you grow your skills and passion for tech, while matching you with mentors and providing feedback on your progress. LaunchCode matches you with one of our 500 employer partners for a paid apprenticeship, typically twelve weeks long. Hone your skills on the job, learning from experienced developers. Nine out of ten apprentices are converted to full-time hires."

Lewis signed on with LaunchCode in June 2014 and was hired by MasterCard in St. Louis as an apprentice in September of that year and promoted to full-time assistant systems engineer by November, helping the credit card company manage its giant server network. In March 2016, she was promoted to systems engineer.

And as Lewis told me with a twinkle in her eye, "I still don't have my BA."

Roughly estimated, there are about thirty-five million LaShana Lewises in America today who started college but never finished. Imagine how much more productive we could be as a country if we could find ways to value and capture the learning those thirty-five million have. We simply cannot continue with this binary system of degree or no degree, where the key to inclusion is pedigree and not what you actually know and can actually do. The emergence of intelligent algorithms and networks such as LaunchCode, which can be used by employers as trusted validators to sow people into the system and not weed them out of it, holds the promise of unlocking a lot of wasted talent.

Says Lewis: "If you can do the job, you should get the job."

Fortunately, intelligent algorithms and intelligent networks are emerging and enabling a new social contract. There are actually a lot of people who have the skills certain employers are seeking, but may not have the traditional credentials to be appreciated. There are many people who would be happy to learn those skills but don't have the information on what they are or access to learning platforms, some of which are unconventional and not covered by traditional government loans. There are employers who have employees with the skills—or the aspiration to acquire the skills—for new jobs, but the employers don't know who they are or are not currently set up to offer them online training opportunities. And there are schools that are actually great at teaching those skills, but no one knows which schools do that the best.

As we develop more intelligent algorithms "to overcome these labor market failures," argued Byron Auguste, we can put so many more people to work—work better aligned to their talents that contributes more to our economy and society—no matter how many machines and robots are out there. These intelligent algorithms or networks are called "online talent platforms."

At the high end of the labor ladder, professionals already have a global intelligent algorithm to draw on: LinkedIn, the career professional social networking site. But its founders now want to extend that intelligent algorithm to the whole world of work by creating a global "economic graph." Here is how LinkedIn's CEO, Jeff Weiner, describes it on his company blog:

> Reid Hoffman and the other founders of LinkedIn initially created a platform to help people tap the value of their professional networks, and developed an infrastructure that could map those relationships up to three degrees. In doing so, they provided the foundation for what would eventually become the world's largest professional graph.
>
> Our current long-term vision at LinkedIn is to extend this professional graph into an economic graph by digitally manifesting every economic opportunity [i.e., job] in the world (full-

> time and temporary); the skills required to obtain those opportunities; the profiles for every company in the world offering those opportunities; the professional profiles for every one of the roughly 3.3 billion people in the global workforce; and subsequently overlay the professional knowledge of those individuals and companies onto the "graph" [so that individual professionals could share their expertise and experience with anyone].

Anyone will be able to access intelligent networks such as LinkedIn's global graph, see what skills are in demand or available, and even offer up online courses. You might teach knitting or editing or gardening or plumbing or engine repair. So many more people will be incentivized to offer their expertise to others, and the market for it will be vastly expanded.

Added Weiner:

> With the existence of an economic graph, we could look at where the jobs are in any given locality, identify the fastest growing jobs in that area, the skills required to obtain those jobs, the skills of the existing aggregate workforce there, and then quantify the size of the gap. Even more importantly, we could then provide a feed of that data to local vocational training facilities, junior colleges, etc., so they could develop a just-in-time curriculum that provides local job seekers the skills they need to obtain the jobs that are and will be, and not just the jobs that once were.
>
> Separately, we could provide current college students the ability to see the career paths of all of their school's alumni by company, geography, and functional role.

For instance, go to linkedin.com/edu. LinkedIn has studied its hundred-million-worker database to determine which schools seem to be launching more graduates into the top companies in various professional fields. You might be surprised: Accounting? Villanova and Notre Dame. Media? New York University and Hofstra. Software developers? Carnegie Mellon, Caltech, and Cornell. Whether you want to be a plumber or a surgeon, it is valuable to know which schools' alums keep rising at the leading companies.

LinkedIn is already busy building its graph, starting with several pilot cities, and if it succeeds in creating such an intelligent algorithm for the whole world one day, it will be a hugely valuable achievement. But how do we offer an intelligent tool like that for the half of the labor market not yet networked like LinkedIn professionals? That question is a reason LinkedIn's cofounder Reid Hoffman is one of the main backers of the intelligent algorithm Opportunity@Work, headed by Auguste and Karan Chopra, which is trying to fix the lower end of the labor market, from whence LaShana Lewis emerged, and where there are even bigger "talent arbitrage" opportunities to be found.

There are too many people like Lewis who have developed skills on their own but don't necessarily have the certificates, badges, or degrees that employers have grown accustomed to relying upon in hiring—way too accustomed in an era when people have so many more options to learn on their own.

Opportunity@Work is trying to solve this problem by working at the community level to create intelligent networks that help employers who are ready—even desperate—to hire anyone who can effectively do the tech jobs that they need filled. Many employers say that college degrees don't come with the skills they need—yet the screening tools they use for hiring mean that many people who have those skills are currently overlooked because they lack the diplomas or degrees or badges to prove it.

If someone has the skills—but not the academic pedigree or professional résumé—to be an IT systems administrator or web developer, Opportunity@Work tests them out on its TechHire.org platform, certifies their mastery of skills for various tech occupations, then connects them with the right employers or the right training to earn or learn more.

"We have to move to more hiring based on mastery, not history," argued Chopra. "We can steepen the slope of the learning curve, but if that learning and those skills are not recognized in the labor market, there is no incentive and no payoff." Too many companies today are investing in screening software to keep people out, based on pedigrees, rather than learning and matching software that could tap everyone's highest and best use.

How crazy is that? Here's an interesting data point from a 2015 labor survey by Burning Glass Technologies: 65 percent of new job postings for executive secretaries and executive assistants now call for a bachelor's degree, but "only 19% of those currently employed in these roles have a B.A." So four-fifths of secretaries today would be barred from being considered for two-thirds of the job postings in their own field because they don't have a degree to do the job they're already doing.

Message from employers: if you're a working secretary today without a BA and want to change jobs, another employer will consider you, but first you need to quit, go into debt for eighty thousand dollars to get a BA, and then interview for another opening for the exact job you are already doing. Welcome to the American job market today, where, Burning Glass notes, an "increasing number of job seekers face being shut out of middle-skill, middle-class occupations by employers' rising demand for a bachelor's degree" as a job-qualifying badge, even though it may be irrelevant to the job or your true capabilities.

What Opportunity@Work is trying to create through its networks, said Auguste, is nothing less than "a new demand signal for human capital." That would be a signal that says: "Anyone who can do these tasks to this standard in this context gets a try. We don't care how you learned it. We hire on mastery not pedigree—not everyone gets the specific job, but anyone has a shot." And if there are skills that you are missing, here are the local schools or learning platforms where you can fill them in on your own time.

Right now, no employer has the incentive to build out that platform, which is why you need groups such as Opportunity@Work or LinkedIn to create the intelligent networks that show everyone how it can work. The current system, in which there is one job winner and a thousand losers, is simply wasting too much human capital, and in the age of accelerations that is politically dangerous. Chopra and Auguste are certain that if they can get a critical mass of employers hiring on the basis of demonstrated skills, not history, and also link prospective employees with schools, coaches, or tutors, to help them respond to the skills most in demand, they can tip the labor market.

If you are a community college administrator, these intelligent networks are a great way to learn what employers are looking for and therefore what skills you should be teaching. You can then mix in innovations in intelligent financing, says Auguste: Imagine a micro-equity investment in a talented low-income student's tuition and living expenses for a fifteen-week "coding boot camp," which converts into debt only once she gets her first job as a software developer. We can open up job opportunities, solve our skills mismatch, and unlock immense value in our human capital if we move past the current antiquated frameworks of public and private student lending to more personalized, talent-based, pay-it-forward financing systems—where both educational institutions and employers have more skin in the game to ensure the payoff to students of securing an in-demand job.

"Our institutions spend so much time working on how to optimize returns on financial capital," said Auguste. "It is about time we started thinking more about how to optimize returns on human capital."

## *Come the Revolution*

Throughout this book I have stressed that technology moves up in steps—from platform to platform. But not all platforms are created equal. And it is my contention that the last two steps we've taken—the one that made connectivity *fast, free, easy for you, and ubiquitous,* around the year 2000, and the one that made complexity *fast, free, easy for you, and invisible* around 2007—constituted a fundamental inflection point in the power of men, machines, groups, and flows, an inflection point so profound that it has blown apart the basic workplace that we have known since the Industrial Revolution blew away the guild-based workplace. Thanks to the supernova, the workplace is being globalized, digitized, automatized, and roboticized at a speed, scope, and scale we've never seen before. It is hard to think of any career not being touched by this process, which is why it is posing such a fundamental challenge to how we think about educating people for work, organizing people at work, and helping people adjust to both new realities.

Most good middle-class jobs today—the ones that cannot be outsourced, automatized, roboticized, or digitized—are likely to be what I would call *stempathy* jobs. These are jobs that require and reward the ability to leverage technical and interpersonal skills—to blend calculus with human (or animal) psychology, to hold a conversation with Watson to make a cancer diagnosis and hold the hand of a patient to deliver it, to have a robot milk your cows but also to properly care for those cows in need of extra care with a gentle touch.

"In the 19th century most Americans spent their time working with animals and plants outdoors in the country," wrote the historian Walter

Russell Mead in a May 10, 2013, essay in *The American Interest* entitled "The Jobs Crisis: Bigger Than You Think":

> In the 20th century most Americans spent their time pushing paper in offices or bashing widgets in factories. In the 21st century most of us are going to work with people, providing services that enhance each other's lives . . .
>
> We are going to have to discover the inherent dignity of work that is people to people rather than people to things. We are going to have to realize that engaging with other people, understanding their hopes and their needs, and using our own skills, knowledge and talent to give them what they want at a price they can afford is honest work.

It is for this reason I believe that Dov Seidman is right when he argues that more work will move from "hands to heads to hearts." There will be an increasing amount of work around "all the things that the heart can do," he explained. "While machines can reliably interoperate, humans, uniquely, can build deep relationships of trust." We will still need manual labor, and people will continue working with machines to do extraordinary things. Seidman is simply arguing that the tech revolution will force humans to create more value with hearts and between hearts.

One of the fastest-growing U.S. restaurant franchises in recent years has been Paint Nite, which runs paint-while-drinking classes for adults. *Bloomberg Businessweek* explained in a July 1, 2015, story that Paint Nite "throws after-work parties for patrons who are largely lawyers, teachers and tech workers eager for a creative hobby." The artist-teachers who work five nights a week can make $50,000 a year connecting people to their hearts.

Economies get labeled according to the predominant way people create value, pointed out Seidman. So, the industrial economy, he noted, "was about hired hands. The knowledge economy was about hired heads. The technology revolution is thrusting us into 'the human economy,' which will expand value creation to include more hired hearts—all the attributes that can't be programmed into software, like passion, character, and collaborative spirit."

The technological revolution of the twenty-first century is as consequential as the scientific revolution, argued Seidman, and it is "forcing us to answer a most profound question—one we've never had to ask before: 'What does it mean to be human in the age of intelligent machines?'" If machines can compete with people in thinking, what makes us humans unique? And what will enable us to continue to create social and economic value? The answer, said Seidman, is the one thing machines will never have: "a heart."

The latest research backs this up. In an essay in *The New York Times* on October 18, 2015, entitled "Why What You Learned in Preschool Is Crucial at Work," Claire Cain Miller pointed out that "for all the jobs that machines can now do—whether performing surgery, driving cars or serving food—they still lack one distinctly human trait. They have no social skills. Yet skills like cooperation, empathy and flexibility have become increasingly vital in modern-day work."

Those occupations requiring strong social skills, she added,

> have grown much more than others since 1980, according to new research. And the only occupations that have shown consistent wage growth since 2000 require both cognitive and social skills . . .
>
> Yet to prepare students for the change in the way we work, the skills that schools teach may need to change. Social skills are rarely emphasized in traditional education.
>
> "Machines are automating a whole bunch of these things, so having the softer skills, knowing the human touch and how to complement technology, is critical, and our education system is not set up for that," said Michael Horn, co-founder of the Clayton Christensen Institute, where he studies education.

Miller consulted David Deming, an associate professor of education and economics at Harvard University and author of a new study on this subject. As Miller explained, Deming's research shows that in the tech industry "it's the jobs that combine technical and interpersonal skills that are booming, like being a computer scientist working on a group project." Miller quoted David Autor, an economist at MIT specializing in labor issues, as noting "if it's just technical skill, there's a reasonable chance it can be automated, and if it's just being empathetic or flexible, there's an infinite supply of people, so a job won't be well paid. It's the interaction of both that is virtuous."

## *What it All Means for Learning*

All of this has huge implications for education and learning. As business strategist Heather McGowan puts it, the accelerations in Moore's law and digital globalization have now reached a point where "work has become unbundled from jobs, and work and jobs have both become unbundled from companies, many of which are now really platforms."

What does that mean? "A job," she explains, "otherwise known as engaging individuals as employees, was once the most efficient means" to produce a lot of goods and services at scale. And for two centuries at least the "company" was the container in which we mostly did all these things called "jobs." "The company broke the workflow into subcomponents, or tasks, best achieved in command-and-control structures," added McGowan. "Our work environments, which grew out of the Industrial Revolution, were organized like assembly lines of factory production. So, too, were our learning systems that prepared people for this type of work—as specialized experts."

But as the age of accelerations took hold—and Moore's law started digitizing everything and therefore making it much easier to disaggregate and globalize every element of supply, design, and manufacture—the whole notion of a job began to get transformed.

"For work that is both complicated and can be digitized, its discrete subcomponents could now be performed anywhere around the globe by the lowest cost provider," explained McGowan, and soon platforms emerged both within companies and outside them—such as UpWork, TaskRabbit, Uber, Hourly Nerd, and the like—"to enable the production of atomized work."

That is what she meant when she declared that "work got unbundled from jobs." The production of work was no longer necessarily tied to the container of a job, it could be done anywhere—"and jobs and work got unbundled from a single company. They could be done by anyone, anywhere."

A 2016 study by economists Lawrence Katz of Harvard University and Alan Krueger at Princeton University found that "94% of net job growth in the past decade was in the alternative work category," said Krueger. "And over 60% was due to [the rise] of independent contractors, freelancers and contract company workers." That is why programs like Obamacare that disconnect health care from jobs have become so important.

"Just as jobs were once the most efficient means of managing worker production, a company was once the most efficient container for value creation," explained McGowan. But no more. Digitization means work can occur both inside and outside the company with equal efficiency, she added, which is why many of the most dynamic new companies today are just platforms where value is created ON the platform by individual workers rather than fixed containers in which value is created.

The most prominent example is Uber, said McGowan: "A traditional taxi company owns a fleet of cars that is both an asset and a liability. The company also owns the relationship with a set of workers who are either employees or fixed contractors. Uber does not own those assets and liabilities but rather effectuates the transaction between those in need of a ride (customers) and those willing to provide transportation with their own cars under their own efforts (driving)."

This is the context within which all discussions about education and learning have to take place—and, specifically, why lifelong learning is now the necessary but not sufficient foundation for lifetime employment.

"When work was predictable and the change rate was relatively constant, preparation for work merely required the codification and transfer of existing knowledge and predetermined skills to create a stable and deployable workforce," explained McGowan. "Now that the velocity of change has accelerated due to a combination of exponential growth in technology and globalization, learning can no longer be a set dose of education consumed in the first third of one's life." Studies show that as much as 50 percent of the content in an undergraduate degree may be obsolete within five years—for many students, that is before the degree is achieved, McGowan added. "Hard-won monolithic degrees earned early in life could be obsolete within a decade—well before the debt incurred to secure that degree is repaid and the investment realized."

As McGowan likes to say, in this age of accelerations "learning becomes more important than knowing." And when that is the case, she adds, "the new killer skill set is an agile mind-set that values learning over knowing. This mind-set positions the individual with an expectation of lifelong learning and adapting with a focus on the uniquely human skills of empathy, social and emotional intelligence, judgment, creativity, divergent thinking, and an entrepreneurial outlook for a long career that may include, according to recent studies, up to twenty jobs across as many as five industries."

This means individuals, governments, and companies will all have to adjust. "We once prepared workers for work by way of a transfer of skills and knowledge paradigm—the artisan and the apprentice," said McGowan. "Then, companies employed the workers to produce value, and the result for Ford, for example, was the company's ability to produce and sell a great car built by workers on an assembly line." Today, she noted, the goal of Tesla is not only to produce and sell a car, but to sell a car and constantly collect data off of it in order to learn more about driving behavior in order to sell better cars and possibly whole different products gleaned from the digitized data stream. "The Tesla Model S autopilot feature collects data from human interactions about driving and pathways in order to refine GPS data and learn better models of driving for improved safety of future autonomous Tesla vehicles, a concept collected into a system called 'fleet learning network,'" noted McGowan.

Amazon sells Alexa to consumers not only to make them smarter faster, but to make Amazon smarter and faster about them and sell them more books and clothes that they want as well as whole new products no one ever knew they needed or wanted.

"This same paradigm shift needs to take place with workers," she added. The most valued worker today is not the one that can just take on more and more complex tasks, but the worker who can also learn by doing that task and can identify ways to fork off from it to create new opportunities, markets, products, or services—for themselves or their firm.

That's why I always told my girls that when I graduated from college I got to *find* a job—a job that I have done now for almost forty years. But they will need to *invent* their jobs, and keep inventing them for each new cycle of change. Every new job opportunity has to be seen as an opportunity to learn, and every new learning opportunity has to be leveraged into a new job. And in an era when more jobs become jump balls that not only more people, but more machines and robots can compete for, you've got to be willing and able to jump. Similarly, added

McGowan, the best companies will be defined not just by the products or services they offer but by their ability to learn from every product they sell and from every consumer they encounter, to make more products and engage more consumers.

And that is why McGowan concludes that there are three questions you should never ask a kid or college student today, because the questions only calcify their mind-set and betray your own ignorance.

Question one: What are you going to be when you grow up? "This is an increasingly absurd question when most of the jobs they may do don't exist yet," she explains.

Question two: What is your major? "Higher Ed institutions increasingly require students to declare a major absent of context, exposure, and knowledge about themselves and the world in which they will live," explains McGowan. If the very nature of work and the very nature of the company is in as much flux as it seems to be today, says McGowan, "and a student gets locked into a major with rapidly expiring content, they become fixed on a notion or way of thinking that may be rapidly obsolete. You can bet their college debt will last longer than that first job or even first industry."

Question three: What do you do for a living? "This further calcifies their sense of self around a set role," argues McGowan. "As we see more industries implode and transform, much of that self-identity may go with them."

McGowan says never waste time asking a kid "what do you want to be when you grow up?" Instead ask them these questions: "How are you going to be when you grow up?"; "What is your passion?" (which is an evolving question); and "How will you make your passion productive?" That is, how will you translate your passion and purposes into new entrepreneurial opportunities, how will you not just find a job but keep inventing ones?

Again, it comes back to the idea that the greatest gift a teacher can give a student or a parent can give a child is a "mind-set": What is your attitude toward the global flows? How disposed are you to learning from them and adding to them? How equipped are you to filter the junk and fake news from the real knowledge and facts?

Who cares where you are on the right-left political spectrum today? What matters is where you are on the open-closed, love-to-learn, don't love-to-learn, spectrums.

## *It Takes a Family*

For all of these reasons, it should be obvious that the new workplaces being reshaped by the age of accelerations will demand multiple new social contracts so workers can get the best out of them and cushion the worst. One is between bosses and employees: bosses will have to learn to hire more people on the basis of what they can provably do, not just the pedigree they can ostentatiously produce, and to provide multiple avenues for lifelong learning within the company's framework. One is between you and yourself: if the bosses create the learning opportunities and help with the tuition, you will have to provide the grit and self-motivation to take advantage of both—to own your learning and your constant relearning. You cannot stress this last point enough: more will be on you—to figure out what to learn and to go out and learn it. The "digital divide" will soon disappear. Fairly soon, virtually everyone will have a screen and an Internet connection. In that world, argues futurist Marina Gorbis, the big divide will be "the motivational divide"—who has the self-motivation, grit, and persistence to take advantage of all the free or cheap online tools to create, collaborate, and keep learning over an entire lifetime when Mom and Dad will not be around to ask if you have finished your homework.

Again, those attitudes and values cannot be downloaded. They have to be uploaded in the old-fashioned way—in two-parent households supported by healthy, nurturing neighborhoods and communities. Intelligent assistance, assistants, and algorithms can do a lot. But they cannot make up, at scale, for the absence of stable families with parents who read to their children and instill the values of self-motivation, grit, and persistence in an age of accelerations, when more will be on you.

Just as small errors in leadership navigation can now get a whole country far offtrack when everything starts moving faster, so small errors in parenting navigation can now do the same for kids. All the old stuff that can only be built slowly one human to another matters more than ever.

Therefore, for all the reasons laid out above, there also has to be a new social contract between educators and students: companies no longer have the patience to wait around for universities to figure out their market, adapt their curricula, hire the right professors, and teach the students the new skills, especially when emerging online education platforms are doing all that now faster and from day one. If traditional postsecondary schools are going to remain relevant in a world where everyone will require lifelong learning, educators need to provide those opportunities at a viable speed, price point, and level of on-demand mobility. Finally, we will need a new social contract between governments and citizens: we need to create every possible regulatory and tax incentive for every company to provide, and every worker to get access to, intelligent assistance, intelligent assistants, intelligent networks, and intelligent financing for lifelong learning.

## *Robots Only Win if Let Them*

Right now we are in the grip of what seem to me artificially precise forecasts about the future—"47 percent" of jobs (not 48 percent!) will be lost by 2051. I say: Who knows? We should be modest in predicting the future, either way. What if 48 percent of jobs will be reinvented by 2052 in ways that will make them more satisfying? There are many plausible futures. Don't underestimate human ingenuity and the will to create new industries, new forms of work, and new ways for humans to translate their passions into profits.

Indeed, before you get all teary-eyed about the end of the worker Holocene that we're now experiencing, pause for a moment and consider the potential upsides of the new workplace. Marina Gorbis shared a memo she drafted for her institute on how this could actually work out better for many workers, if we put the right foundations in place:

> Imagine that you, as a worker, can decide when and how you want to earn income, using a platform that has information about your skills, capabilities, and previous tasks completed. You are seamlessly matched with the task that optimizes your income opportunity. Imagine that the same or another platform could direct you to learning opportunities that would maximize your earnings potential or support your desire to acquire new skills. Suppose that instead of having to come into the office, you can work at home or in a number of co-working spaces in your neighborhood, providing you with social con-

nections, community, and the necessary infrastructure to support your tasks. And imagine that in this world, the social safety net—all your benefits—are not tied to your employer but are portable. Every time you work for pay, independent of the platform or an organization, your benefits accrue to your personal security account. Pieces of this new ecosystem of work are already beginning to take shape, but the process is happening piecemeal, with many gaps and missteps.

In the summer of 2017, I wrote a column noting that roughly a decade ago two new "platform" companies burst out of California. The one that dominated the headlines was called Uber, which created a platform where with one touch of your phone you could summon a cab, direct the driver, pay the driver, and rate the driver. It grew like a weed, as all kinds of people became taxi drivers in their spare time. But Uber made clear that its ultimate goal was self-driving cars. The other was called Airbnb. It created a trust platform so efficient that people all over the world were ready to use it to rent out their spare bedrooms to total strangers. Airbnb is growing so fast that it's now adding the equivalent of one entire Hilton hotel chain's worth of rooms for rent each year.

But while Uber aspires to self-driving cars, Airbnb has a different goal: enabling what I call self-driving people. And that's why I won't be surprised if in five years Airbnb is not only still the world's biggest home rental service, but also one of the world's biggest jobs platforms. You read that right. Very quietly Airbnb has been expanding its trust platform beyond enabling people to rent their spare rooms to allowing them to translate their passions into professions, and thereby empowering more self-driving people.

It all started with people who were renting rooms saying to their customers: "Hey, hope you enjoy the room. By the way, I'm also a great cook; would you like me to prepare a dinner party for you?" Or, "I'm an amateur historian; would you like me to give you a tour of the city?" Now this trend has just taken off.

"We created a garden and planted one plant—and that was home-sharing," explained Brian Chesky over breakfast in San Francisco. "And now we're seeing what other things can grow in this garden."

To see what's growing, go to Airbnb's website and click not on "homes" but on "experiences." You'll find an endless smorgasbord of people turning their passion into profit and their inner artisan into second careers.

Take for instance the team of Luca & Lorenzo. They explain in endearing broken English: "We are 100 percent Italian food lovers; we were used to cook with our grandmothers since we were child. We continued to have this passion through the years, so it makes sense founded our company Lovexfood." For $152 a person, they will take seven people visiting Florence, Italy, on a trip to "make pasta from scratch in the woods outside the city" in an "old house . . . surrounded by a garden with aromatic plants. We are between the hills where is produced the famous Chianti wine."

For $35 a person, Lee Marvin will take five people in Havana on a tour of three-on-three neighborhood basketball games. "Christina" posted a message on his site on July 18, saying: "I signed my teenage son up for this & it was one of the best activities of the trip. It was supposed to end at 8 p.m. or so. Well, my son felt so welcomed that he & Lee Marvin's gang hung out for several hours after they played basketball. They learned about each other's lives, told jokes, talked sports and really bonded. Talk about a great emersion into the Cuban culture." Also, not a bad way for a Cuban to earn $175 a night, minus Airbnb's commission.

Chesky believes that the potential for Airbnb experiences could be bigger than home-sharing. I agree.

"The biggest asset in people's lives is not their home, but their time and potential—and we can unlock that," he explained. "We have these homes that are not used, and we have these talents that are not used. Instead of asking what new infrastructure we need to build, why don't we look at what passions we can unlock? We can unlock so much economic activity, and this will unlock millions of entrepreneurs."

When he retires, said Chesky, age thirty-five, "I'd like to say that Airbnb created 100 million new entrepreneurs in the world." I wouldn't bet against him. Because the world is full of artisans and people with passions waiting to be unlocked. In America, though, there is a surplus of fear and a poverty of imagination in the national jobs discussion today—because "all we are focusing on are the things that are going away," said Chesky. "We need to focus on what's coming. Do we really think we're living in the first era in history where nothing will ever again be created by humans for humans, only by machines? Of course not. It's that we're not talking about all of these human stories."

Are platforms like Airbnb's the only answer for the American middle-class jobs challenge? Of course not. There is no one answer. That's the point. We have to do fifty things right to re-create that broad middle class of the '50s and '60s, and platforms like Airbnb's are just one of them. So much of what companies did in the past, concluded Chesky, "was unlocking natural resources to build the stuff we wanted." Today's new platforms are unlocking human potential to "be the people we wanted."

In short, weep not for that nine-to-five work era of old. It's gone and it is not coming back. But once we get through this transition, and it will be rough, I am convinced there's a high probability that a better and fairer workplace is waiting on the other side, if we can learn to combine the best of what is new—artificial intelligence—with the best of what never changes and never will change: self-motivation, caring adults and mentors, and practice in your area of interest or aspiration.

Just before the school year started in 2014, Gallup released a massive poll it had conducted of college graduates who had been in the workplace for at least five years. The poll tried to answer this question: What are the things that happen at a college or technical school that, more than anything else, produce "engaged" employees on a fulfilling career track?

"We think it's a big deal" where we go to college, Brandon Busteed, the executive director of Gallup's education division, explained to me for a column I wrote about the poll. "But we found no difference in terms of type of institution you went to—public, private, selective or not—in long-term outcomes. *How* you got your college education mattered most."

And two experiences stood out from the poll of more than one million American workers, students, educators, and employers: *Successful students had one or more teachers who were mentors and took a real interest in their aspirations, and they had an internship related to what they were learning in school.* The most engaged employees, said Busteed, consistently attributed their success in the workplace to having had a professor or professors "who cared about them as a person," or having had "a mentor who encouraged their goals and dreams," or having had "an internship where they applied what they were learning." Those workers, he found, "were twice as likely to be engaged with their work and thriving in their overall well-being."

There's a message in that bottle.

NINE

# *Control vs. Kaos*

***The violent chaos in Yemen isn't orderly enough to merit being called a civil war.***

***—Simon Henderson, "The Rising Menace from Disintegrating Yemen,"* The Wall Street Journal, *March 23, 2015***

From 1965 to 1970, American television audiences were entertained by a popular sitcom called *Get Smart*. The show was a spoof on James Bond, and starred Don Adams as agent Maxwell Smart, who went by the code name "Agent 86," with Barbara Feldon playing his sidekick, "Agent 99." Written by Buck Henry and Mel Brooks, *Get Smart* famously introduced the shoe phone to American audiences, but the show also introduced something else: its own version of geopolitics and a bipolar world.

Do you remember the name of the intelligence agency Maxwell Smart worked for? It was called "Control." And do you remember the name of Control's global enemy? It was called "Kaos"—"an international organization of evil."

The creators of *Get Smart* were way ahead of their time. After all, it increasingly appears that the most important division in the "post–post–Cold War world," which we're now in, is between regions of "Control" and regions of "Kaos"—or, as I prefer to describe them, "the World of Order" and "the World of Disorder."

That was not what a lot of Americans and Europeans were expecting after the Cold War. The Cold War was a struggle between two competing systems of order, dominated by two competing superpowers, who could, relatively speaking, keep their allies ideologically in line, physically intact, and militarily in check. The relevant geographic and ideological dividing lines were East–West, communist–capitalist, totalitarian–democratic.

In the post–Cold War world, from 1989 until the early 2000s, the dominant struggle—and it was not really much of a struggle at all—was between an American hegemon and everybody else. Our economic and political system had "won." The communist system had lost, and for the most part we thought that the only problem going forward would be the pace at which everyone adopted our democratic-capitalist formula for success—and then all would be right with the world.

And because America and its allies had such a surplus of military and economic power, they chose to use some of it to oppose holdouts against this democratizing trend, like Saddam Hussein in Iraq, the military rulers of Haiti, and Slobodan Milošević in Serbia/Bosnia—and to pressure China with a human rights campaign and Russia with NATO and European Union expansion campaigns. It seemed as if it were only a matter of time before the whole world would come our way.

As the Johns Hopkins foreign policy professor Michael Mandelbaum argued in his book *Mission Failure: America and the World in the Post–Cold War Era*, during this window of overwhelming American dominance, "the main focus of American foreign policy shifted from war to governance, from what other governments did beyond their borders to what they did and how they were organized within them."

Referring to U.S. operations in Somalia, Haiti, Bosnia, Kosovo, Iraq, and Afghanistan and to Chinese human rights policy, Russian democratization, NATO expansion, and the Israeli-Palestinian peace process, Mandelbaum wrote: "The United States after the Cold War . . . became the equivalent of a very wealthy person, the multibillionaire among nations. It left the realm of necessity that it had inhabited during the Cold War and entered the world of choice. It chose to spend some of its vast reserves of power on the geopolitical equivalent of luxury items: the remaking of other countries."

But that era came a cropper when the interventions in Iraq and Afghanistan became mired in failure and the Great Recession of 2008 dampened American growth. This all combined to sap American power and self-confidence—the self-confidence that it knew the right things to do to stabilize the world and how to do them, and that it *could* do them. All of that was reflected in the foreign policy of President Barack Obama, which was characterized by pinched aspirations, a humility about whether America knew best, a skepticism toward foreigners, particularly from the Middle East, who claimed that they shared our values and beckoned us to come partner with them, and the dispatching of troops abroad with an eyedropper, almost counting them one by one. I say all of this without meaning to criticize; there were good reasons for Obama's circumspection when it came to the Middle East. Elsewhere, such as in eastern Europe and Asia, Obama actually reinforced America's military presence to balance Russia and China, and his use of the American military to stem the outbreak of Ebola in West Africa was decisive in preventing a global pandemic. So the notion that America under Obama just withdrew from the world is nonsense. But there was a pulling back in the Middle East, and it had two major consequences: it abetted the rise of the Islamic State in Iraq and Syria (ISIS), and it contributed to the massive outflow of refugees from that region into Europe. That outflow in turn helped to create the anti-immigration backlash that fueled the British withdrawal from the European Union and the rise of populist/nationalist politics inside almost every EU member state.

It's important to remember that America is such an important player on the world stage that even small shifts in how we project power can have decisive impacts. And it's this combination of shrinking American power in one part of the world plus the reshaping of the world more broadly by the accelerations in the Market, Mother Nature, and Moore's law that defines the era we are in today, which I call the post–post–Cold War world. It is a world characterized by some very old and some very new forms of geopolitical competition all swirling together at the same time. That is, the traditional great-power competition, primarily among the United States, Russia, and China, is back again (if it ever really went away) as strong as ever, with the three major powers again jockeying over spheres of influence, along golden-oldie fault lines such as the NATO–Russia frontier or the South China Sea. This competition is propelled by history, geography, and the traditional imperatives of great-power geopolitics, and is reinforced today by the rise of nationalism in Russia and China. Its contours will be determined by the balance of power between these three big nation-states. This story has been well documented, and is not my main focus.

What I am most interested in is what is new in this post–post–Cold War world: how the simultaneous accelerations of the Market, Mother Nature, and Moore's law are reshaping international relations and forcing America in particular and the world generally to reimagine how we stabilize geopolitics. As much as at any time since the start of the Cold War, we are, again, *Present at the Creation*, which was how Secretary of State Dean Acheson titled his memoir about his tenure at the State Department in that very plastic period after World War II (1949–1953), a period that saw the emergence of the Soviet Union as a global superpower, the spread of nuclear weapons, the fading of empires, and the emergence of a multiplicity of new states.

The age of accelerations in geopolitics is an equally plastic period, but it is not yet clear that we have the ability or imagination to build the alliances and global institutions to stabilize it the way the post–World War II statesmen did—yet that is our calling.

I see several new challenges emerging. The first come with the world's rising interdependence; in particular, interdependence has created some unusual geopolitical inversions that now influence every decision America makes in foreign policy. For example: During the Cold War your allies helped to protect you from your enemies. In the post–post–Cold War world, where we are now so interdependent, your allies—such as Greece—can now kill you faster than your enemies. If Greece cannot pay back its sovereign and private debts, or the European Union begins to fracture because of the British exit, it could set off a chain of falling dominoes that would undermine the European Union and NATO—just as profoundly and quickly as anything Russia or China could do. That would have huge strategic consequences for the United States, since the EU is the other great center of democratic capitalism in the world and is America's primary partner in promoting those values globally and in stabilizing the world generally.

A parallel inversion governs America's relations with Russia and China in an age of interdependence. It is not clear today what threatens America more, their strength or their weakness. If either were to collapse into the World of Disorder it would be a disaster. Russia spans nine time zones and still has thousands of nuclear warheads that need to be controlled and hundreds of nuclear bomb designers. We need a reasonably functioning Russian state to keep a lock on its nuclear weapons, Mafia bosses, drug traffickers, and cybercriminals. And we need a stable Russia to serve as a counterbalance to China, to be a global energy supplier to Europe, and to take care of its aging citizenry. If China, for its part, were to collapse into disorder, it would negatively impact everything from the cost of the shoes on your feet and the shirt on your back to the mortgage on your house to the value of the currency in your wallet. China may be America's rival, but, in today's interdependent world, its collapse would be far more threatening to America than its rise. Probably the worst thing a rising China might do is bully all its neighbors into toeing its line, take over more islands in the South China Sea, or demand more economic concessions from foreign investors. But a falling China could melt down the U.S. stock market and trigger a global recession, if not worse.

While this high degree of interdependence poses one set of new challenges, the rising risk of state failure in a number of countries poses another. These risks can be seen around the world. Julian Lindley-French, vice president of the Atlantic Treaty Association and a visiting research fellow at the National Defense University in Washington, D.C., warns of what he calls "weakism" or "disintegrationism"—which is disintegration down to the level of gangs and tribes and the emergence of groups such as the Islamic State and Boko Haram that fill power vacuums. The very real disintegration of weak states in Africa and the Middle East is now reaching a scale that is creating large emerging zones of disorder, or Kaos, to borrow from *Get Smart*, and these are spilling out so many refugees and economic migrants that the stability of the World of Order is starting to be threatened—witness the splintering of the European Union.

In the Cold War, the biggest challenge for American foreign policy was almost always managing strength—our own strength, that of our allies, such as the European Union and Japan, and that of our main rivals, Russia and China. Today, the American president spends much more time managing and navigating weakness: the weakness of our allies in the EU and Japan, the weakness of an angry, humiliated, and economically frail Russia, the weakness of states that have disintegrated, and the economic weakness of America after 9/11 and the 2008 crash.

Managing weakness is an enormous headache. If you are America, and you don't intervene to prop up disintegrating states, you get spreading disorder. And if you put your foot down and intervene, you can find that your foot has gone right through the floorboards, which can be excruciatingly painful to remove. You also get a big bill. (See: Afghanistan, Somalia, and Iraq.)

The weakism and disintegrationism that Lindley-French speaks of coincide with and help shape another new challenge that we face today: Moore's law and the Market are also birthing a whole new category of international actors that I call super-empowered breakers. We discussed earlier how the power of makers was being amplified by the supernova. That same energy source is, however, enabling jihadists, rogue states such as North Korea, angry lone wolves, and cybercriminals to compete with superpowers and super-empowered makers across a vastly expanded attack surface—including your home computer, which cyberattackers can now lock up until you agree to pay a ransom.

Put all of these old and new challenges together and you would understand why we, in America at least, had it so relatively easy during the Cold War, when we could latch on to a single unifying policy—containment of the Soviet Union—and find that it answered almost every question we had when it came to foreign policy. The challenge of the post–post–Cold War world, as it is being reshaped by the age of accelerations, is so much more complex. It requires deterring traditional big power rivals, as in days of yore, and simultaneously doing what we can to shrink the World of Disorder and stem the disintegration of weak states, with their spillover human migrations that threaten the cohesion of the European Union in particular; at the same time, we must contain and degrade the super-empowered breakers—all this in a much more interdependent world.

That is why reimagining geopolitics in the age of accelerations is so vital today—but bring your humility. As Henry Kissinger testified before the Senate Armed Services Committee on January 29, 2015: "The United States has not faced such diverse crises since the end of the Second World War." The problem of peace, he added, "was historically posed by the accumulation of power—the emergence of a potentially dominant country threatening the security of its neighbors. In our period, peace is often threatened by the disintegration of power—the collapse of authority into 'non-governed spaces' spreading violence beyond their borders and their region." This is particularly acute in the Middle East, Kissinger noted, where "multiple upheavals are unfolding simultaneously. There is a struggle for power within states; a contest between states; a conflict between ethnic and sectarian groups; and an assault on the international state system. One result is that significant geographic spaces have become ungovernable, or at least ungoverned."

The standard American foreign policy playbooks were not written for this world; our traditional tools were not designed for this world; global institutions have not yet adapted to this world; and our domestic debates are not really attuned to the challenges of this world. What does it mean to be a "liberal" or a "conservative" in foreign policy terms in this post–post–Cold War world?

We have no idea because none of us has been here before. That was one of the main conclusions of the "Global Trends 2035" report put out in January 2017 by the U.S. National Intelligence Council, a product of the entire U.S. intelligence community. "For better and worse," it concluded, "the emerging global landscape is drawing to a close an era of American dominance following the Cold War. So, too, perhaps is the rules-based international order that emerged after World War II. It will be much harder to cooperate internationally and govern in ways publics expect. Veto players will threaten to block collaboration at every turn, while information 'echo chambers' [and fake news] will reinforce countless competing realities, undermining shared understandings of world events."

So, yes, we are indeed present again at the creation of something new in the geopolitical arena, and much responsibility will fall to America to figure it out and offer the policy innovations, and generosity, to manage it. What follows is my take on how we got here and how to at least begin thinking about navigating forward.

But one piece of advice: if any president calls and offers you the job of secretary of state, tell him or her you'd love the plane, but you really had your heart set on secretary of agriculture.

## *The Holocene of Geopolitics*

It is easy to forget today just how much the global order that settled on our planet after World War II and lasted through the post–Cold War era was—in retrospect—the geopolitical equivalent of the Holocene climate era. That is, just as the Holocene presented the perfect Garden of Eden climate for Mother Earth, and the perfect economic climate for middle-class workers, it also presented the perfect climate for newly independent states. And there were a lot of them.

In the wake of World War I and the fall of several empires, scores of new independent nations were created. The Austro-Hungarian Empire gave way to Austria, Hungary, Czechoslovakia, and Yugoslavia. Russia ceded Finland, Estonia, Latvia, and Lithuania. And Russia and the Austro-Hungarian Empire also birthed a new Poland and Romania. The Ottoman Empire gave way to a raft of newly independent or colonized states, including Lebanon, Egypt, Syria, Iraq, Jordan, Cyprus, and Albania. And in Africa the dismantled German Empire was carved into states such as Namibia and Tanzania. Then, in the wake of World War II, a wave of decolonization was unleashed, giving birth to an independent India, Pakistan, Libya, Sudan, Tunisia, Ethiopia, Morocco, Mali, Senegal, the Republic of the Congo, the Somali Republic, Niger, Chad, Cameroon, Nigeria, Algeria, Rwanda, Eritrea, Zambia, Indonesia, Vietnam, Laos, Cambodia, Thailand, Malaysia, Singapore, and South Korea—and more. And then after the collapse of the Soviet Union in the early 1990s, all of its peripheral satellite states were set free, including Kazakhstan, Kyrgyzstan, Tajikistan, Turkmenistan, Uzbekistan, Armenia, Azerbaijan, and Moldova, not to mention the various parts of Yugoslavia—Slovenia, Croatia, Bosnia and Herzegovina, Serbia, Montenegro, and Macedonia. Lithuania, Latvia, Georgia, Ukraine, and Estonia also became independent.

While very few of these new states had the economic, natural, or human resources to develop into strong industrial democracies, or even autocracies, their weaknesses were masked for many years—during the Cold War and immediately after—by a variety of factors that made being an "average" or "below average" state quite sustainable.

For starters, the global geopolitical environment around them was, relative to a century that saw two world wars, quite stable. Neither system was led by a Hitler or a jihadist. The two superpowers even maintained a "hotline"—a special communication system connecting the White House and the Kremlin—so each could clear up any misunderstanding with the other to prevent any direct hot wars with nuclear weapons. Strategically, both sides deployed enough nuclear weapons to guarantee not only a first-strike capability but also a retaliatory second-strike capability if the other side fired first, creating a system of "mutual assured destruction," or MAD, which all but guaranteed that neither side would ever use any of their atomic weapons.

More important, though, the intense competition between America and the Soviet Union to collect allies on their respective sides of the chessboard provided a steady flow of resources to create and reinforce order in so many of these new states, which enabled many of them to get by with just C+ leadership—or, to put it in human terms, to get by without exercising regularly, lowering their cholesterol, building muscle, studying hard, or increasing their heart rate. Why should they, when the two superpowers would ply them with cash to build roads, technical assistance to run their governments, and weapons to build internal security services to control their borders and their people? Moscow and Washington also sent billions of dollars and rubles in foreign aid to average countries and leaders to help them balance their budgets, run their schools, and build their stadiums. They also offered their youth scholarships to enroll at the Peoples' Friendship University of Russia or the University of Texas in America.

Because the stability of every square in the global chessboard mattered to Washington and Moscow, the Soviet Union was ready to rebuild the defeated army of Syria after it lost three wars to Israel in 1967, 1973, and 1982—and the United States was ready to support corrupt governments from Latin America to the Philippines, year in and year out. And when aid didn't work, they intervened directly to prop up allies—the Russians in Eastern Europe and Afghanistan and the Americans in Latin America and South Vietnam. The Americans wanted to make sure that as their exhausted European allies lost their colonies or gave them independence, Russian-backed local communists would not take over; meanwhile, the Kremlin would spend almost any amount of money to hold Eastern Europe under its thumb or flip a country in Central America from the American camp to the Soviet one.

At that same time, it was not so hard to influence another country. Because the populations of new nations were relatively small and uneducated, and relatively few people were able to compare their circumstances with the circumstances of people elsewhere, foreign aid went a long way.

Iran's population in 1980, for instance, was forty million, as opposed to more than eighty million today—and climate change had not reached the disruptive extremes we're seeing now, so growing seasons were more reliable. At the time, China had shut itself in and was not a threat to the low-wage workers of every country in the world. And of course there were no robots that could milk a cow or sew textiles.

Meanwhile, economic and demographic trends were also making it easy for America to support a lot of average countries. As Erik Brynjolfsson and Andrew McAfee point out in their book, the four key measures of an economy's health (per capita GDP, labor productivity, the number of jobs, and median household income) all rose together for most of the Cold War years. "For more than three decades after World War II, all four went up steadily and in almost perfect lockstep," Brynjolfsson noted in a June 2015 interview with the *Harvard Business Review*. "Job growth and wage growth, in other words, kept pace with gains in output and productivity. American workers not only created more wealth but also captured a proportional share of the gains."

In hindsight, the period from World War II up to the fall of the Berlin Wall was "an incredible period of economic moderation," argued James Manyika, one of the directors of the McKinsey Global Institute. And economic moderation drove political moderation and stability. It made inclusion and immigration easier to tolerate. Most countries were also still benefiting from improved health care and decreased child mortality, producing a demographic dividend of bulging youth populations and relatively few older people to take care of. This made more generous pensions easier to handle in many countries. And most countries had not eaten through their natural capital. All in all, it was relatively easy to be an "average" democracy or autocracy during the Cold War and even into the post–Cold War period. It was a geopolitical Holocene.

Well, say goodbye to all that as well.

## *Average Is Over for Countries*

Virtually all of those things that made being an average weak state in the Cold War era and beyond relatively easy have now disappeared. Just go down the list: China or Vietnam can now vacuum up so many of the low-wage labor jobs in the world, particularly in first-rung-of-the-ladder industries, such as textiles. Robots can now milk cows. Oil prices have fallen globally, meaning both the petro-states and those indirectly propped up by them are weakened. At the same time, slower growth in China has lately shrunk its voracious appetite for African, Australian, and Latin American commodities. China accounted for more than a third of global growth in recent years, and its growth engine multiplied the growth of many of the countries that exported raw materials to Beijing. That has slowed. China's total debt has grown from roughly 150 percent of its GDP in 2007 to around 240 percent today—a massive increase in one decade that is dampening its growth and its imports and shrinking China's wallet for foreign aid and investment in African and Latin American commodity-exporting countries.

In May 2011, I spent some time in Egypt covering the post–Hosni Mubarak turmoil. After being gone for about two weeks from my wife, I headed home. I had some time to kill at the Cairo airport, so I rummaged through the "Egyptian Treasures" shop, hoping to find a few souvenirs to bring back. I didn't care much for the King Tut paperweights or the pyramids ashtrays, but I was intrigued by a stuffed camel, which, if you squeezed its hump, emitted a camel honk. When I turned it over to see where it was manufactured, I read: "Made in China." Same with the pyramids ashtrays. So Egypt, a country where nearly half the population lived on two dollars a day and at least 12 percent of the population, and far more young people, were unemployed, found itself suddenly competing in a world where a country half a world away could make its national icons into an ashtray or a honking-humped camel, ship them transcontinentally, and still make a profit more efficiently than Egyptians could. Meanwhile, domestic unrest kept tourists from coming over to ride real camels.

As Warren Buffett says, "You only find out who is swimming naked when the tide goes out." All of this withdrawal of support from the big powers and these changes in the global economy were exposing who actually had built a domestic economy and who was just riding on the agricultural commodity and oil booms. Turns out, a lot of countries were buck naked. And some, such as Venezuela, which spent as it went and saved nothing for a rainy day, are now falling apart. But that's not all. Climate change is now hammering many developing countries much harder, particularly in the Middle East and Africa, undermining their agricultural production. And in Africa and parts of the Arab world, as we've already shown, continued high population growth rates are multiplying every stress—all while the Internet, cell phones, and social media are making it much easier for the disgruntled to organize to take governments down and much harder to organize stable alternatives.

And that tide could recede a lot more. Antoine van Agtmael, the investor who coined the term "emerging markets," argues that we are at the start of a paradigm shift around manufacturing that could actually bring a lot of jobs back to America and Europe from the developing world. "The last twenty-five years was all about who could make things cheapest, and the next twenty-five years will be about who can make things smartest," says van Agtmael. The combination of cheap energy in America and more flexible, open innovation—where universities and start-ups share brainpower with companies to spin off discoveries; where manufacturers use a new generation of robots and 3-D printers that allow more production to go local; and where new products integrate wirelessly connected sensors with new materials to become smarter and faster than ever—is making America, says van Agtmael, "the next great emerging market." Good for us, but maybe not for the emerging markets of old.

Put it all together and you see why it was so much easier to be an average developing country in the Cold War and post–Cold War eras than it is today—and why some states are starting to fall into the World of Disorder. Today, this world includes parts of Somalia, Nigeria, South Sudan, Senegal, Iraq, Syria, Egyptian Sinai, Libya, Yemen, Afghanistan, western Pakistan, Chad, Mali, Niger, Eritrea, the Congo, and various swaths of Central America, including parts of El Salvador, Honduras, and Guatemala, and the pirate-infested waters of the Indian Ocean. It also includes the warlord-run zones that Russia carved out of neighboring states on its periphery—in the eastern Ukraine, Abkhazia, Chechnya, South Ossetia, and Transnistria. What all of these places have in common is that central authority either has collapsed or can barely extend its writ beyond the capital. In some cases these states were destabilized by the United States and its allies decapitating their governments—Iraq and Libya—and not effectively building successor authorities. Others have disintegrated on their own from the stresses of civil war, environmental degradation, and extreme poverty, and they are now hemorrhaging refugees in all directions.

Maybe it is just a coincidence, but many, though not all, of the failing countries have borders that are almost entirely straight lines. Those lines and borders with ninety-degree angles were largely produced by imperial and colonial powers—corresponding to their particular interests at the colonial stage of history and not to real ethnic, religious, racial, tribal, or even geographic logic on the ground, let alone the voluntary association of people bound together in nationhood by social contracts.

These states are least able to handle the age of accelerations. They are like caravan homes in a trailer park, built on slabs of cement, with no real foundations or basements. People often wonder, "Why do tornadoes always hit trailer parks?" They don't. It's just that trailer parks are enormously frail and vulnerable when they do get hit. That is what's happening today to so many of these average countries. The three accelerations are plowing through many of these flimsy, contrived, artificial states like a tornado through a trailer park.

But this problem is not only afflicting the states with straight-line borders. It is impacting weak states of all shapes and sizes. I have spent time reporting from the World of Disorder in the past few years, looking at the states hardest hit by the age of accelerations. Here is a quick sample survey from Madagascar to Syria to Senegal to Niger that highlights how the end of the Cold War world and the rise of a world shaped by accelerations in the Market, Mother Nature, and Moore's law have stressed these already frail states to the limit—and beyond.

## *Madagascar*

Madagascar, the island nation off the eastern coast of Africa, is one of the ten poorest countries in the world. I visited in the summer of 2014. It is the poster child for how "average is over" thanks to the three accelerations. Where do I start? The population of Madagascar has exploded in the last two decades, with a growth rate of 2.9 percent, among the highest in Africa. The island's population increased by more than three million people just from 2008 to 2013—to twenty-three million, almost double what it was in 1990. This is an island. It is not getting any bigger. The combination of dwindling foreign aid after the end of the Cold War and damage from increasingly severe cyclones has ravaged the country's roads, power, and water infrastructure. I took a two-hour jeep ride into the interior on a major artery that was so badly eroded it involved driving from pothole to pothole. I can say it was easily the worst road I have ever traveled on in my life on Planet Earth. More than 90 percent of Madagascar's population lives on less than two dollars a day, and, therefore, not surprisingly, some six hundred thousand children who should be in school are not.

Madagascar received foreign aid from everywhere during different stages of the Cold War. The United States paid to have a NASA satellite tracking station there for a while; the French gave foreign aid to its former colony and sent arms for the Malagasy Armed Forces, including MiG-21 fighter jets; the Cubans sent teachers, and the Chinese sent road builders and even built a sugar factory. And, finally, and you cannot make this up, the country's gleaming white Presidential Palace—Madagascar's version of the U.S. White House—was designed and built in the 1970s by the North Koreans, who also trained the Madagascar president's security detail and provided assistance with agriculture and irrigation.

Today, much of that aid is no longer being extended, and parts of the island are being washed away. The soil for agriculture in Madagascar is iron rich, nutrient poor, and often very soft. Ninety percent of Madagascar's forests have been chopped down for slash-and-burn agriculture, timber, firewood, and charcoal over the last century. Indeed, most hillsides have no trees anymore to hold the soil when it rains. Flying along the northwestern coast, you can't miss the scale of the problem. You see a giant red plume of red soil bleeding into the Betsiboka River, bleeding into Bombetoka Bay, bleeding into the Indian Ocean. The mess is so big that astronauts have taken pictures of it from space. It is as if the whole country were bleeding.

This is a tragedy for everyone: "98 percent of Madagascar's land mammals, 92 percent of its reptiles, 68 percent of its plants and 41 percent of its breeding bird species exist nowhere else on Earth," according to the World Wildlife Fund. Madagascar is also home to "two-thirds of the world's chameleons and 50 species of lemur, which are unique to the island." Unfortunately, too many have been hunted. Thanks to the globalization of illicit flows, illegal wildlife trading has left Madagascar exposed to Chinese merchants, who work with corrupt officials to illegally export everything from valuable rosewood timber to rare tortoises.

For a while, globalization did bring some textile manufacturers to Madagascar to create jobs. They set up factories and provided low-skilled employment, but then pulled up stakes and moved to Vietnam and elsewhere when the local politics became too unstable. These manufacturers had options, and as soon as they were spooked they left. The empty factories tell the tale. And in the post–post–Cold War world what Madagascar once thought was average is now way below average. Mandatory education in Madagascar is only through age fifteen, and it's in the local Malagasy language, making it rather hard to compete for higher-wage work with, say, Estonia, which is now teaching computer coding in first grade.

It is hard to see how Madagascar reverses these trends. Said Russ Mittermeier, the renowned primatologist from Conservation International, who has worked in Madagascar to help preserve its environment since 1984: "The more you erode, the more people you have with less soil under their feet to grow things." And the more insecure people feel, the more kids they have as insurance.

## *Syria*

Syria is the geopolitical superstorm of the age of accelerations. It's what happens when every bad trend converges in one place—extreme weather, extreme globalization, extreme population growth, extreme Moore's law, and a newly extreme unwillingness of the United States and many other smaller powers to decisively intervene because all they win is a bill.

To fully understand it, though, you have to start with Mother Nature. In 2014, I traveled to northern Syria to write columns and do a documentary on the impact of the drought—called the *jafaf* in Arabic—on the civil war there for the TV series *Years of Living Dangerously,* which was then on Showtime. "The drought did not cause Syria's civil war," the Syrian economist Samir Aita explained to me, but the government's failure to deal with it was a critical stressor fueling the uprising.

This is the story, he explained: After Bashar al-Assad took power from his late father in 2000, he opened up the regulated agricultural sector in Syria to large-scale farmers, many of them government cronies, to buy up land and drill as much water as they wanted, eventually severely diminishing the water table. This began driving small farmers off the land into towns, where they had to scrounge for work. Because of the population explosion that really got cranking in the 1980s and 1990s due to shrinking mortality rates, those leaving the countryside came with huge families and settled in towns around cities such as Aleppo. Some of those small towns swelled from two thousand people to four hundred thousand in a decade or so. The government failed to provide proper schools, jobs, or services for this youth bulge.

Then Mother Nature showed up. Between 2006 and 2011, some 60 percent of Syria's landmass was ravaged by the worst recorded drought in its modern history. With the water table already too low and river irrigation shrunken, this drought wiped out the livelihoods of between eight hundred thousand and a million Syrian farmers and herders. And this happened at a time when the population of Syria had doubled twice in sixty years. As a result, half the population in Syria between the Tigris and Euphrates rivers left the land for urban areas, beginning in the early 2000s, said Aita. And with Assad doing nothing to help the drought refugees, a lot of very simple farmers and their kids got politicized.

The idea of state government "was invented in this part of the world, in ancient Mesopotamia, precisely to manage irrigation and crop growing," said Aita, "and Assad failed in that basic task." Young people and farmers starved for jobs—and land starved for water—were a prescription for revolution.

That was the specific message of the drought refugees, like Faten, whom I met in May 2013, in her simple flat in Sanliurfa, a Turkish city near the Syrian border. Faten, then thirty-eight, a Sunni Muslim, had fled there with her son Mohammed, nineteen, a member of the Free Syrian Army, who'd been badly wounded in a firefight a few months earlier. Raised in the northeastern Syrian farming village of Mohasen, Faten, who asked me not to use her last name, told me her story: She and her husband "used to own farmland . . . We tended annual crops. We had wheat, barley, and everyday food—vegetables, cucumbers, anything we could plant instead of buying in the market. Thank God there were rains, and the harvests were very good before. And then suddenly, the drought happened."

What did it look like? "To see the land made us very sad," she said. "The land became like a desert, like salt." Everything turned yellow.

Did Assad's government help? "They didn't do anything," she said. "We asked for help, but they didn't care. They didn't care about this subject. Never, never. We had to solve our problems ourselves."

So what did you do? "When the drought happened, we could handle it for two years, and then we said, 'It's enough.' So we decided to move to the city. I got a government job as a nurse, and my husband opened a shop. It was hard. The majority of people left the village and went to the city to find jobs, anything to make a living to eat."

The drought was particularly hard on young men who wanted to study or marry but could no longer afford either, Faten added. Families married off daughters at earlier ages because they couldn't support them. Faten, her head conservatively covered in a black scarf, said the drought and the government's total lack of response radicalized her, her neighbors, and her sons, who joined the opposition fighters. So when the first spark of revolutionary protest was ignited in the small southern Syrian town of Dara'a, in March 2011, Faten and other drought refugees couldn't wait to sign on. "Since the first cry of '*Allahu akbar*,' we all joined the revolution. Right away." Was this about the drought? "Of course," she said, "the drought and unemployment were important in pushing people toward revolution." (Indeed, she was in Turkey to get medical care for her son Mohammed, who sat quietly during our interview, alternately looking at battle pictures on his cell phone and watching a satellite TV broadcast from a rebel station inside Syria.)

Abu Khalil, forty-eight, was one of those who didn't just protest. A former cotton farmer who had to become a smuggler to make ends meet for his sixteen children after the drought wiped out their farm, he became the Free Syrian Army commander in the Tel Abyad area. We met at a crushed Syrian Army checkpoint in Tel Abyad when I crossed into Syria's Rakah province—ground zero for the drought. After being introduced by our Syrian go-between, Abu Khalil, who was built like a tough little boxer, introduced me to his fighting unit. He did not introduce them by rank but by blood, pointing to each of the armed men around him and saying: "My nephew, my cousin, my brother, my cousin, my nephew, my son, my cousin . . ." Free Syrian Army units are often family affairs. In a country where the government for decades wanted no one to trust anyone else, it's no surprise.

"We could accept the drought because it was from Allah," said Abu Khalil, "but we could not accept that the government would do nothing." The bottom line, said Abu Khalil, was that this "was a revolution of the hungry." Before we parted, he pulled me aside to say that all that his men needed were antitank and antiaircraft weapons and they could finish Assad off. "Couldn't Obama just let the Mafia send them to us?" he asked. "Don't worry, we won't use them against Israel."

Some diplomats saw all of this coming. On January 21, 2014, I wrote a column in *The New York Times* quoting a November 8, 2008, cable from the U.S. Embassy in Damascus to the State Department that had been unearthed by WikiLeaks. This was in the middle of the Syrian drought. The embassy was telling the State Department that Syria's U.N. food and agriculture representative, Abdullah bin Yehia, was seeking drought assistance from the U.N. and wanted the United States to contribute.

Here are a few key passages:

> The U.N. Office for Coordination of Humanitarian Affairs launched an appeal on September 29 requesting roughly $20.23 million to assist an estimated one million people impacted by what the U.N. describes as the country's worst drought in four decades . . .
>
> Yehia proposes to use money from the appeal to provide seed and technical assistance to 15,000 small-holding farmers in northeast Syria in an effort to preserve the social and economic fabric of this rural, agricultural community. If UNFAO efforts fail, Yehia predicts mass migration from the northeast, which could act as a multiplier on social and economic pressures already at play and undermine stability . . .
>
> Yehia does not believe that the [government of Bashar al-Assad] will allow any Syrian citizen to starve . . . However, Yehia told us that the Syrian minister of agriculture . . . stated

> publicly that economic and social fallout from the drought was "beyond our capacity as a country to deal with." What the U.N. is trying to combat through this appeal, Yehia says, is the potential for "social destruction" that would accompany erosion of the agricultural industry in rural Syria. This social destruction would lead to political instability.

It is impossible to disconnect the Arab Spring from the climate disruptions of the years just before, 2009–2010. For instance, Russia, the world's fourth-largest wheat exporter, suffered its worst drought in a hundred years in that period. Dubbed the "Black Sea Drought," it included a heat wave that set fires that burned down huge acreage of Russian forests. The drought parched farm fields and shrank the country's breadbasket so much that the Russian government banned wheat exports for a year.

At the same time, wrote Christian Parenti, author of *Tropic of Chaos: Climate Change and the New Geography of Violence*, in an essay on CBS .com on July 20, 2011, massive flooding occurred in Australia, another significant wheat exporter. This coincided with excessive rains in the American Midwest and Canada, damaging more corn and wheat production, while "freakishly massive flooding in Pakistan, which put some 20% of that country under water, also spooked markets and spurred on the speculators."

The result: the FAO Food Price Index hit its all-time high in February 2011—just when the Arab Awakenings hit—throwing some forty-four million people into poverty, according to the U.N. Those climate-driven prices began dramatically boosting bread prices in Egypt, sparking that country's upheaval. And the more upheaval there was in the Middle East, the more oil headed toward $125 a barrel, making everything worse as the costs of fertilizer and operating tractors increased.

In June 2013, I was in Cairo; I got up one morning at five to watch the operations of a bakery selling government-subsidized bread in the dirt-poor Imbaba neighborhood. In the background, through an open window, I heard children in a Koranic school cheerfully repeating verses for their teacher. As soon as the baker opened his shutters, a scrum of men, women, and children jostled to get their bags of pita, the staple of their diet. They had to get there early, they knew, because the baker sold only so many subsidized pita loaves; he sold the rest of his government-subsidized flour on the black market to private bakers who charged five times the official price. He had no choice, he told me, because his fuel costs were spiking. I actually watched the subsidized-flour bags, marked with government stamps, being carried on shoulders by young men right out the side door. "This is the hardest job in Egypt," the bakery owner told me. Everyone is always mad at him, especially those who lined up early and still left with no bread.

It wasn't for nothing that the main chant of the protesters who brought down President Mubarak in 2011 was "Bread, freedom, dignity"—and the bread came first. Such is politics in the age of accelerations.

## *Senegal and Niger*

I first met Monique Barbut, who heads the U.N. Convention to Combat Desertification, at the U.N. Paris climate meeting in late 2015. Her calling card was a presentation with three maps of Africa, each with an oblong outline around a bunch of dots clustered in the middle of the continent. Map number one: the most vulnerable regions of desertification in Africa in 2008. Map number two: conflicts and food riots in Africa, 2007–2008. Map number three: terrorist attacks in Africa in 2012. All three overlapped in the same sub-Saharan heart of Africa. "Desertification acts as the trigger," explained Barbut, "and climate change acts as an amplifier of the political challenges we are witnessing today—economic migrants, interethnic conflicts, and extremism."

Barbut's point was that this expanding World of Disorder problem is not just a Middle East war story. It's an Africa climate, desertification, and population story. It breaks your heart to see the news footage of overcrowded, rickety boats full of migrants overturning in rough seas on the Mediterranean, as people scramble to get out of the World of Disorder into the World of Order, but what is often lost sight of, notes Barbut, is that only about one-third of those refugees are coming out of Syria, Iraq, and Afghanistan. The other two-thirds are coming primarily from a cluster of very arid African states: Senegal, Niger, Nigeria, Gambia, and Eritrea. The best place to start to understand the spreading disorder in parts of Africa is by going to the headwaters of the human migration flows and then following the migrants northeast through Niger to Libya, where they try to sail to Europe. You can see all three accelerations at work.

Let's start in Ndiamaguene, a village in the far northwest of Senegal. If I were giving you directions, I'd tell you that it's the last stop after the last stop—it's the village after the highway ends, after the paved road ends, after the gravel road ends, and after the desert track ends. Turn left at the last baobab tree. It's worth the trek, though, if you're looking for where and why these migration flows start.

I visited in April 2016 to write columns about the connection between climate change and human migration and to do another documentary with the team making *Years of Living Dangerously*, now for National Geographic Channel. The day we arrived, April 14, 2016, it was 113 degrees—far above the historical average for the day, a crazy level of extreme weather. But there was an even bigger abnormality in Ndiamaguene, a farming village of mud-brick homes and thatch-roof huts. The village chief gathered virtually everyone in his community to receive us, and they formed a welcoming circle of women in colorful prints and cheerful boys and girls with incandescent smiles, home from school for lunch. But the second I sat down with them, I realized that something was wrong with this picture.

There were almost no young or middle-aged men in this village of three hundred. They were all gone.

It wasn't disease. They had all hit the road. The village's climate-hammered farmlands could no longer sustain them, and with so many kids—44 percent of Senegal's population is less than fourteen years old—there were just too many mouths to feed from the declining yields. So the men had scattered to the four winds in search of any job that might pay them enough to live on and send some money back to their wives or parents. This trend is repeating itself all across West Africa. Tell these young African men that their odds of getting to Europe are tiny and they will tell you, as one told me, that when you don't have enough money to buy even an aspirin for your sick mother, you don't calculate the odds. You just go.

"We are mostly farmers, and we depend on farming, but it is not working now," the village chief, Ndiougua Ndiaye, explained to me in

Wolof, through a translator. After a series of on/off droughts in the 1970s and 1980s, the weather patterns stabilized a bit, "until about ten years ago," the chief added. Then, the weather got really weird. The rainy season used to always begin in June and run to October. Now the first rains might not start until August, then they stop for a while, leaving fields to dry out, and then they begin again. But they come back as torrential downpours that create floods. "So whatever you plant, the crops get spoiled," the chief said. "You reap no profits."

The chief, who gave his age as seventy but didn't know for sure, could remember one thing for certain: when he was young he could walk out to his fields anytime during the planting season "and your feet would sink into" the moist earth. "The soil was slippery and oily and it would stick to your legs and feet and you would have to scrape it off." Now, he said, picking up a fistful of hot sand, the soil "is like a powder—it is not living anymore."

Had he ever heard of something called "climate change"? I asked. "We heard about it on the radio and we have seen it with our own eyes," said Ndiaye. "The winds from east to west have changed, and the winds from the west are now warmer. Winter doesn't last long anymore. And this year, there wasn't even any winter. We live in constant summer."

*We live in constant summer.* The chief's rough impressions were not wrong. Senegal's national weather bureau says that from 1950 to 2015, the average temperature in the country rose 2 degrees Celsius, much faster than anticipated. Since 1950, the average annual rainfall has declined by about fifty millimeters (about two inches). So the men of Ndiamaguene have no choice but to migrate to bigger towns or out of the country. The lucky few find ways to get smuggled into Spain or Germany, via Libya. Libya, it seems, was like a cork in Africa, and when the United States and NATO toppled the Libyan dictator—but did not put troops on the ground to help secure a new order—they essentially uncorked Africa, creating a massive funnel to the Mediterranean coast.

The less lucky find work in Dakar or Libya or Algeria or Mauritania, and the least lucky get marooned somewhere along the way—caught in the humiliating twilight of having left home and gained nothing and having nothing to return to. This is creating more and more tempting recruiting targets for jihadist groups such as Boko Haram, which can offer a few hundred dollars a month—a king's ransom when you are living on two dollars a day.

The chief introduced me to Mayoro Ndiaeye, the father of a boy who had left to find work. "My son left for Libya one year ago, and since then we have no news—no telephone, nothing," he explained. "He left a wife and two children. He was a tile fixer. After he made some money [in the nearby town] he went to Mauritania and then to Niger and then up to Libya. But we have not heard from him since."

The father started to tear up. These people live so close to the edge. One reason they have so many children is that the offspring are a safety net for aging parents. But the boys are all leaving and the edge is getting even closer. Which means they are losing the only thing they were rich in: a deep sense of community. Here, you grow up with your family, parents look after children and children then look after parents, and everyone eats and lives together.

But now with the land no longer producing enough, they are losing their community, said the chief. "Everyone has a [male] family member who has had to leave . . . When I was young, my brothers and I would go together to cultivate the field for our father. Our mothers would wait for us to bring the production home so they could take care of the rest. And the whole family would be here to enjoy the harvest. If this situation continues, there will come a time when we won't be able to stay here, because we won't make a living. We will be compelled to follow our children to other places."

All the data points in that direction. Ousmane Ndiaye, head of the climate unit for Senegal's National Civil Aviation and Meteorology Agency, trained at Columbia in climate science. In his drab office at the Dakar airport, Ndiaye clicked through his climate graphs for me on his Dell desktop, telling a horrifying story.

"Last week the weather was five degrees Celsius above the normal average temperature, which is a very extreme temperature for this time of year," he explained. Click to graph two. "From 1950 to 2015 average temperature in Senegal has gone up two degrees Celsius," said Ndiaye, adding that the whole 2016 Paris U.N. climate conference was about how to avoid a two-degree rise in the global average temperature since the Industrial Revolution . . . and Senegal is already there. Click. The U.N. Intergovernmental Panel on Climate Change "in 2010 gave four scenarios for Senegal, and the worst was unbelievable—and now," he said, "the observation says we're following that path even faster than we imagined, and it leads to four degrees Celsius rise in average temperature by 2100. People are still doubting climate change, and we are living it." Click.

"You live here and you see on TV people having a good life, and democracy [in Europe]," he added, "and here you are in a poor life, people have to do something . . . They don't have the tools to survive here. The human being is just a more intelligent animal, and if [he or she] is pushed to the extreme, the animal instinct will come out to survive."

To complete the picture of the refugee flow, you have to move west and north to Agadez, Niger, at the southern edge of the Sahara. Starting in 2015 a regular ritual is repeated there every Monday evening and only Monday evenings: thousands of young men, crammed into the back of Toyota pickup trucks, gather in a large caravan to make the long trek from the (mild) World of Disorder—Niger—through the (wild) World of Disorder—Libya—in hopes of catching some of kind of boat into the World of Order—Europe. The caravan's assembly is quite a scene to witness. Although it is early evening, it's still 105 degrees Fahrenheit outside. Two of our cameramen were overcome by heat, lugging round their equipment. This is desert, just on the edge of Agadez, and so there is little more than a crescent moon to illuminate the night.

Then, all of a sudden, the desert comes alive.

Using the WhatsApp messaging service on their smartphones, the local smugglers, who are tied in with networks of human traffickers extending all across West Africa, start coordinating the surreptitious loading of migrants from safe houses and basements across the city. These almost entirely young men have been gathering in Agadez all week—from Senegal, Sierra Leone, Nigeria, Ivory Coast, Liberia, Chad, Guinea, Cameroon, and Mali, as well as towns in Niger. With fifteen or so men crammed together into the back of each Toyota pickup, their arms and legs spilling over the sides, the vehicles pop out of alleyways and follow scout cars that have zoomed ahead to make sure that no pesky police officers or border guards are lurking who have not been paid off. It's like watching a symphony, but you have no idea where the conductor is. Eventually, all the vehicles converge at a gathering point north of the city, forming a giant caravan of one hundred to two hundred vehicles, depending on the Monday. They need this strength in numbers to ward off desert bandits.

I was standing at the Agadez highway control station watching this convoy. As the Toyotas whisked by me, kicking up dust, they painted the desert road with stunning moonlit silhouettes of these young men, silently standing in the back of each vehicle. They will have to stand for more than twenty-four hours as they head for Libya and the coast. The thought that their promised land is war-ravaged Libya tells you how desperate are the conditions they're leaving. Between nine thousand and ten thousand men are making this journey every month.

Agadez used to live off adventure tourism and trade. With its ornate mud-baked structures, it is a Unesco World Heritage Site, because of its "numerous earthen dwellings and a well-preserved group of palatial and religious buildings including a 27 m high minaret made entirely of mud brick, the highest such structure in the world," according to Unesco.org. Now all the tourist vehicles are being repurposed for the human trafficking of people out of the World of Disorder into the World of Order. Or, as one human smuggler told us: "Before, we were in tourism. It's the tourism industry that we did here in Agadez. And tourism, it no longer exists. Now, we have our vehicles. That's how we make our living. We transport. We live off of this."

A few of those being smuggled out agreed to stop and talk—nervously. One group of very young men from elsewhere in Niger told me they're joining the rush to pan for gold in Djado in the far north of Niger. More typical were five young men who had their faces covered in, yes, ski masks and spoke in Senegalese-accented French. They told a familiar tale: no work in the village, went to the town, no work in the town, heading north.

Here and elsewhere, desertification acts as the trigger; climate change and population growth act as amplifiers; interethnic and tribal conflicts are the political by-product, and WhatsApp provides both an alluring picture of where things might be better—Europe—and a cheap tool for hopping a migration caravan to get there. "In the old days," says Barbut, "we could just give them a Live Aid concert in Europe or America and then forget about them. But that won't work anymore. They won't settle for that. And the problem is now too big."

No walls will permanently hold them back. I interviewed twenty men from at least ten African countries at the International Organization for Migration aid center in Agadez—all had gone to Libya, tried and failed to get to Europe, and returned, but were penniless and unable to get back to their home villages. I asked them, "How many of you and your friends would leave Africa and go to Europe if you could get in legally?"

"*Tout le monde*," they shouted, and they all raised their hands. I don't know much French, but I think that means "everybody."

What is most striking about this explosion of both refugees and economic migrants that we are witnessing on the world stage today is that it is largely the result of nation-states melting down, not interstate wars. Indeed, noted David Miliband, president of the International Rescue Committee, which oversees relief operations in more than thirty war-affected countries, more people in the world today are "fleeing a conflict" at a time when wars between nations "are at a record low." That is because we now have nearly thirty civil wars under way in weak states that are "unable to meet the basic needs of citizens or contain civil war," a sign of states cracking from inside under the pressure of the age of accelerations.

The United States has not been immune to this flood. Since 2014, some 150,000 unaccompanied children under age twenty-one from Guatemala, El Salvador, and Honduras (and Mexico) have been sent by their parents on a journey through Mexico to the United States to escape the gang violence that has gripped these states as they have descended into the world of disorder. Most are able to qualify for political asylum because they are fleeing chaos. As soon as they get inside the U.S. border, they turn themselves over to U.S. immigration authorities and file for asylum and to be connected with relatives already in America. But how many more can the United States take? "They're fleeing from threats and violence in their home countries," noted Vox.com, "where things have gotten so bad that many families believe that they have no choice but to send their children on the long, dangerous journey north." Honduras, Guatemala, and El Salvador are among the most environmentally degraded and deforested regions in Central America. They cut their forests; we got their kids.

It is not only Europe and America that have become the promised land for economic and climate migrants from the World of Disorder. So too has the Promised Land. In recent years, Israel has been flooded with some sixty thousand illegal immigrants, mostly from Eritrea and Sudan. Stroll the blocks around the Central Bus Station in Tel Aviv, where many have found shelter, and you'll see African men on cell phones on every street. They sailed, walked, or drove to Israel's borders and either slipped in on their own or were smuggled in by bedouins across Egypt's Sinai Desert. They were attracted not by Zionism or Judaism, but just by the hope of order and work.

That is why the two fences Israel has built since 2014 were not against Arab armies but migrants from the world of disorder. One was a 240-kilometer barrier along the Egyptian border, through Israel's Negev desert, and the second was a 30-kilometer fence along Israel's border with Jordan to keep out Syrian refugees. On September 6, 2015, Prime Minister Benjamin Netanyahu said at his weekly cabinet meeting that Israel will not allow itself to be "submerged by a wave of illegal migrants and terrorist activists . . . Israel is a small country, very small, without demographic or geographic depth. That is why we must control our borders."

Today you don't even have to be a historically "promised land" to be worried about being swamped by refugees from the world of disorder. On February 7, 2016, BBC.com reported that Tunisia had "completed the first part of a 200km (125-mile) barrier along its border with Libya, designed to deter terrorism. The barrier is made of sand banks and water trenches. Tunisia's defense minister said the second phase of the project would involve installing electronic equipment with the help of Germany and the U.S. Security forces said the defenses—which aim to make the border impassable by vehicles—had already helped to reduce smuggling."

On June 20, 2016, the United Nations Refugee Agency (UNHCR), which tracks forced displacement worldwide based on data from governments, partner agencies, and UNHCR's own reporting, issued a report stating that a total of 65.3 million people were displaced at the end of 2015, compared with 59.5 million just twelve months earlier. At the end of 2013, that number had stood at 51.2 million, and a decade ago at 37.5 million. Moreover, the report said the situation was likely to worsen further. Globally, 1 in every 122 humans is now either a refugee, internally displaced, or seeking asylum. If this were the population of a country, the report said, it would be the world's twenty-fourth biggest.

## *The Inequality of Freedom*

The accelerations in the Market, Mother Nature, and Moore's law are stressing frail states not only from outside but also from below. That is, both technology and globalization today are empowering "political makers," who want to remake autocratic societies into more consensual ones, and "political breakers," who want bring down governments in order to impose some religious or ideological tyranny, even though they may lack any ability to govern effectively.

Let's look at both—starting with the political makers. The historian Walter Russell Mead once pointed out that after the 1990s revolution that collapsed the Soviet Union, Russians liked to say: "It's easier to turn an aquarium into fish soup than to turn fish soup into an aquarium."

It is never easy, under the best of conditions, for inhabitants of a country to reconstitute it as a going concern after it has collapsed, but it may be even more difficult in the age of accelerations. The lifelong learning opportunities you need to provide to your population, the infrastructure you need to take advantage of the global flows, and the pace of innovation you need to maintain a growing economy have all become harder to achieve. And if your country has spent the post–post–Cold War destroying itself—in an age when no superpowers will dive in to rebuild it for free or even for a fee anymore—catching up is going to be very, very difficult. And then there is one more—surprise—factor: the Internet. There is mounting evidence that social networks make it much easier to go from imposed order to revolution than to go from revolution to some kind of new sustainable, consensual order.

Influenced by Isaiah Berlin's concept of "positive" and "negative" liberty, Dov Seidman argues that all over the world we now see people creating unprecedented levels of "freedom from—freedom from dictators, but also freedom from micromanaging bosses, from networks forcing us to watch commercials, and freedom from the neighborhood stores, freedom from the local banker, freedom from hotel chains."

But when it comes to politics, the freedom people cherish most, he argues, is "freedom to"—the freedom to live the way they want because their freedom is anchored in consensual elections, a constitution, the rule of law, and a parliament. There are growing swaths of the world today where people have secured their freedom from, but failed yet to build the freedom to. And that explains a lot of the spreading and stubborn disorder. Seidman calls the gap in those countries, such as Libya or Syria or Yemen, or Egypt after the fall of President Hosni Mubarak, that have secured their freedom from but not their freedom to, "the inequality of freedom." And it may be the most relevant inequality in the world today.

"'Freedom from' happens quickly, violently, and dramatically," notes Seidman. "'Freedom to' takes time. After the Jews got their freedom from Pharaoh in Egypt, they had to wander in the desert for forty years before they developed the laws and moral codes that gave them their freedom to."

It turns out that social networks, cheap cell phones, and messaging apps are really good at both enabling and impeding collective action. They enable people to get connected horizontally much more easily and efficiently, but they also enable individuals at the bottom to pull down those at the top more easily and efficiently—whether they are allies or enemies. Military strategists will tell you that the network is the most empowered organizational form in this period of technological change; classical hierarchies do not optimize in the flat world, but the network does. Networks undermine command-and-control systems—no matter who is on top—while strengthening the voices of whoever is on the bottom to talk back. Social media is good for collective sharing, but not always so great for collective building; good for collective destruction, but maybe not so good for collective construction; fantastic for generating a flash mob, but not so good at generating a flash consensus on a party platform or a constitution.

And this challenge of governing is only going to get worse, as Moore's law and global flows keep accelerating. The National Intelligence Council's Global Trends 2035 report points out that "Greater public access to information about leaders and institutions—combined with stunning elite failures such as the 2008 financial crisis and Petrobras corruption scandal—has undermined public trust in established sources of authority and is driving populist movements worldwide. Moreover, information technology's amplification of individual voices and of distrust of elites has in some countries eroded the influence of political parties, labor unions, and civic groups, potentially leading to a crisis of representation among democracies." It is also leading to a crisis for revolutions. Even revolutionaries are now finding that it's harder to translate protesting into governing. Facebook and Twitter are two-edged swords.

One need only listen to some of the key players in some of the "Square Revolutions" of the last decade to learn how they have learned the hard way about the limits of the Internet as a political tool. On a visit to Hong Kong in 2014, I interviewed Alex Yong-Kang Chow, twenty-four, a senior studying literature at Hong Kong University and at the time a leader of the Hong Kong Federation of Students, which spearheaded the pro-democracy "Occupy Central" civil disobedience movement, which began in Hong Kong on September 28, 2014. It aimed to curb Beijing's influence over Hong Kong's more democratic politics and to partially shut down the center of that city-state. It was not a total failure, but hardly an unmitigated success, either.

"What was missing in the [Occupy Central] movement was a mechanism to allow different points of view to debate and settle differences," Chow told me. "If those disputes could not be settled within the movement, afterward there would be a lot of discord and bitterness. Every time when these ideas were brought out, people would veto them. There was no way to settle disputes—no single organization could gain enough trust from all the participants. And the Hong Kong people lack the political culture of resolving disputes through debates."

What about Facebook and other social networks? I asked him.

"Technology is useful in communicating," he responded. "People would divide into different teams [and occupy different parts of Central Hong Kong]. Some would observe the police and alert others about their movements; others would monitor online discussions and update people on the front lines. This gave us a rapid way of circulating information and for people to react instantly and in a rapid way. Activists would go online to Facebook and observe new information from Facebook . . . [These technologies] are very useful tools to propel the progressive movement or to counter government propaganda."

But was there a downside?

"The government was also observing and decoding the messages when we were using these apps and social media—people sent by the Chinese government," said Chow. "Smartphones were being monitored."

In the end, Chow asked himself the most analog question of all: "How can an organization gain trust and legitimacy and connect with the people? The Hong Kong Federation of Students was accountable to students. But it also had to be accountable to the one million people of Hong Kong who were mobilized in the umbrella movement. So how does a single student organization balance the need to channel the aspirations of one million people and serve the students at the same time?" It needs, he answered, "trust and connections," and those take time to build face-to-face. "That is what was missing to have the strategy be sustainable. With trust and connections [you can have] a great alliance to counter your opponents. Without trust and connections, it is very hard to sway the authorities and easy for the government to topple you."

What distinguished Tunisia from all the other Facebook-driven Arab Awakenings, and made it so far the most successful, were some very analog attributes, most notably the deep roots of Tunisian civil society—trade unions, lawyers' associations, women's groups, business associations, human rights organizations. It was collective, face-to-face efforts to bridge the differences between Islamists and secularists after the fall of the Tunisian dictatorship that won several of these organizations the 2015 Nobel Peace Prize.

Elsewhere, the difficulty of achieving genuine political order has led to growing numbers of "un-free" people in the world. Income inequality is destabilizing, "but so is freedom inequality," said Seidman. When "freedom from" outstrips "freedom to," amplified actors in the grip of destructive ideas "will cause more harm and destruction, unless they become inspired and enlisted in constructive human endeavors," he argued. "They will be like inmates on the loose."

No one has given better testimony to the difference between securing freedom from and freedom to than Wael Ghonim, aka "the Google guy," who helped launch the revolution against the Egyptian president Hosni Mubarak in 2011. I was in Cairo at the time, and the day before Mubarak resigned I followed Ghonim on the Friday noon broadcast from Cairo on Al Arabiya satellite TV. He had just gotten released from jail and was full of anger against the regime and passion for the democratic revolution and the role that social media had played stoking it. But that revolution in the end got derailed by the failure of the progressive forces to unite, the desire by the Muslim Brotherhood to divert it into a religious movement, and the Egyptian Army's ability to exploit the weakness of all these civil groups in order to maintain its grip on both the Egyptian deep state and its economy.

In December 2015, Ghonim, who has since moved to Silicon Valley, posted a TED talk that I wrote about in a column. In the talk, he asked what went wrong—squarely addressing this question: Is the Internet better for creating freedom from than freedom to? This is the essence of what he concluded: "I once said, 'If you want to liberate a society, all you need is the Internet.' I was wrong. I said those words back in 2011, when a Facebook page I anonymously created helped spark the Egyptian revolution. The Arab Spring revealed social media's greatest potential, but it also exposed its greatest shortcomings. The same tool that united us to topple dictators eventually tore us apart."

In the early 2000s, Arabs were flocking to the Web, Ghonim explained: "Thirsty for knowledge, for opportunities, for connecting with the rest of the people around the globe, we escaped our frustrating political realities and lived a virtual, alternative life." This included him personally. Then, in June 2010, he noted, the "Internet changed my life forever. While browsing Facebook, I saw a photo . . . of a tortured, dead body of a young Egyptian guy. His name was Khaled Said. Khaled was a twenty-nine-year-old Alexandrian who was killed by police. I saw myself in his picture . . . I anonymously created a Facebook page and called it 'We Are All Khaled Said.' In just three days, the page had over a hundred thousand people, fellow Egyptians who shared the same concern."

Soon Ghonim and his friends used Facebook to crowdsource ideas, and "the page became the most followed page in the Arab world," he said. "Social media was crucial for this campaign. It helped a decentralized movement arise. It made people realize that they were not alone. And it made it impossible for the regime to stop it." Ghonim was eventually tracked down in Cairo by Egyptian security services, beaten, and then held incommunicado for eleven days. But three days after he was freed, the millions of protesters his Facebook posts helped to galvanize brought down Mubarak's regime.

Alas, the euphoria soon faded, said Ghonim, because "we failed to build consensus, and the political struggle led to intense polarization." Social media, he noted, "only amplified" the polarization "by facilitating the spread of misinformation, rumors, echo chambers, and hate speech. The environment was purely toxic. My online world became a battleground filled with trolls, lies, hate speech." Supporters of the army and the Islamists used social media to smear each other, while the democratic center, which Ghonim and so many others occupied, got marginalized. Their revolution was stolen by the Muslim Brotherhood and, when it failed, by the army, which then arrested many of the secular youths who first powered the revolution. The army now has its own Facebook page to defend itself.

Having had time to reflect, said Ghonim, "it became clear to me that while it's true that polarization is primarily driven by our human behavior, social media shapes this behavior and magnifies its impact. Say you want to say something that is not based on a fact, pick a fight, or ignore someone that you don't like. These are all natural human impulses, but because of technology, acting on these impulses is only one click away."

Ghonim sees five critical challenges facing today's social media in the political arena:

> First, we don't know how to deal with rumors. Rumors that confirm people's biases are now believed and spread among millions of people. Second, we create our own echo chambers. We tend to only communicate with people that we agree with, and thanks to social media, we can mute, unfollow, and block everybody else. Third, online discussions quickly descend into angry mobs. All of us probably know that. It's as if we forget that the people behind screens are actually real people and not just avatars. And fourth, it became really hard to change our opinions. Because of the speed and brevity of social media, we are forced to jump to conclusions and write sharp opinions in one hundred forty characters about complex world affairs. And once we do that, it lives forever on the Internet, and we are less motivated to change these views, even when new evidence arises. Fifth—and in my point of view, this is the most critical—today, our social media experiences are designed in a way that favors broadcasting over engagements, posts over discussions, shallow comments over deep conversations. It's as if we agreed that we are here to talk at each other instead of talking with each other.
>
> There's a lot of debate today on how to combat online harassment and fight trolls. This is so important. No one could

> argue against that. But we need to also think about how to design social media experiences that promote civility and reward thoughtfulness. I know for a fact if I write a post that is more sensational, more one-sided, sometimes angry and aggressive, I get to have more people see that post. I will get more attention. But what if we put more focus on quality? . . . We also need to think about effective crowdsourcing mechanisms, to fact-check widely spread online information, and reward people who take part in that. In essence, we need to rethink today's social media ecosystem and redesign its experiences to reward thoughtfulness, civility and mutual understanding.
>
> Five years ago, I said, "If you want to liberate society, all you need is the Internet." Today, I believe if we want to liberate society, we first need to liberate the Internet.

The stories of Ghonim and Chow are vivid reminders, observes the veteran international pollster Craig Charney, that while the Internet "improves the ability to connect, it is no substitute for political organizations, culture, or leadership—and spontaneous movements tend to be weakest in all of these." Many Arab Awakening efforts ultimately failed because they could not build an organization and politics that could translate their progressive ideas into a governing majority. Writing in the *Financial Times*, on February 28, 2014, Mark Mazower, a professor of history at Columbia and the author of *Governing the World: The History of an Idea, 1815 to the Present*, pointed out:

> The fundamental Leninist insight still holds: nothing can be done without organisation. If Solidarity was able to transform itself into a long-term force in Polish politics, it was because its leaders understood the need to organise themselves, and because its roots in union activism gave it an inherent structure to begin with . . .
>
> Removing tyrants sometimes does indeed lead to freedom. At other times it merely leads to new kinds of tyranny. Happy the revolution where the revolutionaries are both freedom-loving and effectively organized for the long haul of political struggle.

Sometimes you have to go through the analog steps of knocking on doors, printing out leaflets, and persuading neighbors face-to-face, one at a time, to build the institutional muscles and civic habits that are needed most the morning after the revolution.

All of those old and slow steps are what actually create a pipeline for leadership. Social media is great at quickly mobilizing a lot of people around a lowest common denominator objective. But just as you can't speed up the gestation period for a baby—you can't speed up the gestation period for leadership either. Charles de Gaulle, Dwight Eisenhower, or George Washington had to struggle and fail and pick themselves up and build coalitions brick by brick to become leaders and then public figures. The byproduct was their celebrity. In today's networked world, it is so easy for anyone to become famous for fifteen minutes and become a celebrity. Some then try to use their celebrity to get elected and then they hire professionals to backfill them with the knowledge and skills to lead. It's all backwards.

An anonymous Greek proverb says that "Society grows great when old men plant trees whose shade they know they shall never sit in." Celebrity-leaders don't do shade. They look for instant gratification, which is why we are likely to see more and more societies securing their freedom from—but failing to build their freedom to.

## *The Breakers*

In November 2004, I went to Iraq, accompanying the visiting chairman of the Joint Chiefs of Staff, General Richard Myers. Of all the images I saw on that trip, none stayed with me longer than a display that the Twenty-Fourth Marine Expeditionary Unit, in the Sunni Triangle, near Ramadi, prepared for the visiting chairman. It was a table covered with defused roadside bombs made from cell phones wired to explosives. You just call the phone's number when a U.S. vehicle goes by and the whole thing explodes. The table was full of every color and variety of cell phone bomb you could imagine.

I thought to myself: "If there is a duty-free electronics store at the gates of hell, this is what the display counter looks like."

The three accelerations have reshaped geopolitics not only by making us so much more interdependent and by blowing up weak states and stressing strong states, but also by super-empowering individuals to create even more disorder.

The supernova "serves as a kind of amplifier of human behavior," observed Richard K. Miller, president of the Franklin W. Olin College of Engineering. "In each successive generation, a smaller and smaller number of people is enabled to affect the lives of larger and larger numbers of other people through the application of technology. The effects may be intentional or unintentional, and they may be beneficial or they may not. The relentless development of new technology raises the stakes on social, economic, and political consequences in each generation."

We have referred to the super-empowered individuals and groups who use these powers constructively as "the makers." But, as noted earlier, the same technologies also spawn super-empowered angry men and women—"the breakers." When it is a great time for makers, it is, unfortunately, also a great time for breakers. If you want to break something now at scale, this is your era. In the old days, "technological advances of importance were not part of a system that [immediately enabled] their global distribution, so they did not immediately show up in the hands of malevolent types at the same rate and leverage that you see today," explained Craig Mundie. "When these technologies were only available to states, you could talk about nonproliferation as an enforceable goal." No more. Today many of these tools, or instructions for how to build them, are easily downloadable from the cloud by anyone with a Visa card. So breakers can tap this energy source to amplify their power of one—and to connect, communicate, and collaborate with those of like mind—just as easily as any maker.

To put it another way, what we're seeing with this proliferation of super-empowered breakers is "a new kind of globalization that combines the technologies of the future with the enmities of the past," Mark Leonard, who heads the European Council on Foreign Relations, observed in a February 28, 2017, essay on Project Syndicate. It is a frightening combination.

Because if today's breakers are much more empowered, they are also less easily deterred. There is no mutual assured destruction—MAD—doctrine keeping Al Qaeda or ISIS from going to extremes. Just the opposite: for the jihadist suicide bombers, mutual assured destruction is like an invitation to a party and a date with ninety-nine virgins. As the Harvard University strategist Graham Allison summed it up: "Historically, there has always been a gap between people's individual anger and what they could do with their anger. But thanks to modern technology, and the willingness of people to commit suicide, really angry individuals can now kill millions of people if they can get the right materials." And that is becoming steadily easier with the globalization of flows and the rise of 3-D printing, by which you can build almost anything in your basement, if it can fit.

In writing this book I could not decide which frightening examples to include of how much and how easily super-empowered angry people can now spread disorder. Here are the stories that made the cut:

- "Somewhere between more than half to two-thirds of Americans killed or wounded in combat in the Iraq and Afghanistan wars have been victims of IEDs planted in the ground, in vehicles or buildings, or worn as suicide vests, or loaded into suicide vehicles, according to data from the Pentagon's Joint IED Defeat Organization or JIEDDO," *USA Today* reported on December 19, 2013. "That's more than 3,100 dead and 33,000 wounded. Among the worst of the casualties are nearly 1,800 U.S. troops who have lost limbs in Iraq and Afghanistan, the vast majority from blasts, according to Army data . . . The bombs radically affected how the American military could move around the war zone, creating a heavy reliance on helicopters and other aircraft in order to avoid roads, says Army Lt. Gen. John Johnson, JIEDDO director. 'They've caused us a lot of pain . . . a lot of effort and a lot of treasure,' Johnson says . . . The IED has given rise to a multibillion-dollar industry in vehicle and body armor, robots, ground-penetrating radar, surveillance, electrical jamming, counterintelligence, computer analysis and computerized prostheses. The Government Accountability Office says it's impossible to estimate the total U.S. cost of fighting the bombs over two wars. But the Pentagon has spent at least $75 billion on armored vehicles and tools for defeating the weapons." You can make an IED for one hundred dollars.

- On January 26, 2015, *The New York Times* reported that a "White House radar system designed to detect flying objects like planes, missiles and large drones failed to pick up a small drone that crashed into a tree on the South Lawn early Monday morning" and that "the crash raised questions about whether the Secret Service could bring down a similar object if it endangered President Obama." It turned out that a government employee who was reportedly drunk at the time had been operating the device. The *Times* further reported that "a Secret Service officer who was posted on the south grounds of the White House 'heard and observed' the drone, the agency said, but the officer and others stationed at the residence were unable to bring it down before it passed over the White House fence and struck a tree. The drone was too small and flying too low to be detected by radar, officials said, adding that because of its size, it could easily have been confused for a large bird." Though the president and his wife were in India at the time on a government visit, their two daughters, Sasha and Malia, were at home.

- On January 27, 2015, in an interview with CNN's Fareed Zakaria, after that two-foot-wide quadcopter drone crashed on the White House grounds, President Obama made the following observation: "The drone that landed in the White House you buy in RadioShack."

Thanks to big data and the supernova, we can now find the needle in the haystack with incredible ease. At the same time, the superempowered breakers can now inject that needle into the rest of us with incredible force and accuracy. The future will be a test of who finds who first. Consider this February 18, 2016, story from New Scientist.com:

> Extortion is bigger business than ever, and now it doesn't have to rely on people depositing bags stuffed with cash. Earlier this month, cybercriminals attacked a hospital in Los Angeles, then demanded payment in bitcoin to let the hospital regain access

to their computers. It's the most high-profile case yet of cyber-extortion using software known as ransomware.

The attack on Hollywood Presbyterian Medical Center effectively knocked it offline. As a result, patients had to be diverted to other hospitals, medical records were kept using pen and paper, and staff resorted to communicating by fax.

The attackers demanded 9,000 bitcoins—around $3.6 million. After a two-week stand-off, the hospital yesterday paid out $17,000 . . .

"Ransomware has really exploded in the last couple of years," says Steve Santorelli, a former U.K. police detective who now works for Team Cymru, a threat intelligence firm based in Florida. One ransomware package, CryptoLocker 3.0, is thought to have earned attackers $325 million in 2015 alone.

"These guys are crazy sophisticated," says Jake Williams, the founder of cybersecurity firm Rendition Infosec . . .

Ross Anderson, a security researcher at the University of Cambridge, says bitcoin has helped cybercriminals to access payments without being caught. "In the old days, collecting ransom was really hard. The police would just put a radio tracker in the carpet bag full of £20 notes, and they would always get the guy. Now it's possible to collect ransoms by bitcoin. Lots of people are doing it."

Last story: In a February 9, 2016, worldwide threat assessment report to the Senate Armed Services Committee, James Clapper, the U.S. director of national intelligence, added gene editing—for the first time—to a list of threats posed by "weapons of mass destruction and proliferation." As the *MIT Technology Review* noted on that day, "Gene editing refers to several novel ways to alter the DNA inside living cells. The most popular method, CRISPR, has been revolutionizing scientific research, leading to novel animals and crops, and is likely to power a new generation of gene treatments for serious diseases. It is gene editing's relative ease of use that worries the U.S. intelligence community, according to the assessment." Clapper's report said: "Given the broad distribution, low cost, and accelerated pace of development of this dual-use technology, its deliberate or unintentional misuse might lead to far-reaching economic and national security implications."

## *No Known Home Address*

It is not just the fact that individual breakers can now do more damage more cheaply and more easily that is so unnerving. It's that they no longer need a traditional organization to arm or direct them—an organization that can be tracked and destroyed by police or armies.

In recent years we have seen the steady rise of the "loner terrorist." We've seen individuals, couples, or very small groups, sometimes brothers and cousins, often psychologically disturbed, who get radicalized in a very short period of time after tapping into jihadist or other flows online. They then go out and perpetrate grand acts of violence against innocent civilians, many of them only claiming allegiance to an Islamist or other cause ex post facto.

On July 14, 2016, such a man drove his truck into the Bastille Day crowd in Nice, France, killing eighty-six people and wounding hundreds. The entire phenomenon is condensed into a few paragraphs of a *Daily Telegraph* report:

> Tunisian-born Mohamed Lahouaiej Bouhlel—described as a "weird loner" who "became depressed" when his wife left him—was a French passport holder who lived in the Riviera city and was regularly in trouble with the law.
>
> Bouhlel was reportedly not on a terrorist watch list and investigators are seeking to establish his motives—and are also looking for possible accomplices.
>
> A psychiatrist who saw him in 2004 told *L'Express* that Bouhlel had come to him because of behavioral problems and that he diagnosed him as suffering from "the beginnings of psychosis."
>
> The French interior minister, Bernard Cazeneuve, said the attacker "appears to have become radicalised very quickly" as one neighbor of his estranged wife added: "Mohamed only started visiting a mosque in April . . ."

> Bouhlel's phone is said to be full of messages, videos and photographs, including ones of men and women he had recently slept with . . .
>
> He visited gyms and salsa bars regularly, and would also visit websites "showing pictures of executions," said BFM TV.
>
> "The busy sex life of a man who had recently discovered a religious faith is shown by the data on the device," BFM added.
>
> The divorced father-of-three also used his phone to prepare his attack on civilians, including hundreds of children enjoying a Bastille Day fireworks display.
>
> He also took a selfie of himself inside the hired truck just before heading off on his killing and maiming spree, [e-mailing] it to family members in his native Tunisia.

It is as if accelerated global flows through social networks are heating up certain people who live on the margins of societies and inspiring and encouraging them to engage in hero-in-their-own-minds acts of violence. They want to go out with a bang that the whole world gets to see—even though they are not formal members of any organization.

The strategist George Friedman, chairman of the research firm Geopolitical Futures, explains why these self-motivated lone wolves and small groups may be the future of terrorism—and why they are so hard to deter. In the decade after 9/11, "at its heart, the United States' strategy was to identify terrorist groups and destroy them," Friedman wrote on GeopoliticalFutures.com on July 26, 2016. "The assumption was that terrorism required an organization. Progress in this strategy meant identifying an organization or a cell planning terror operations and disrupting or destroying it . . . Operationally, the strategy worked. Terrorists were identified and killed. As the organizations were degraded and broken, terrorism declined—but then surged."

The reason it surged anew may be because breakers can so easily come together now, just like a maker's start-up, and act on their own. As a result, groups like ISIS may depend less on command and control and more on being the inspirer, the organization that heats up the molecules through social networks and then just sits back and enjoys the show.

As Friedman put it, "The essential problem has been a persistent misunderstanding of radical Islamism. It is a movement, not an organization." Organizations can be penetrated, broken and their leadership structure and headquarters annihilated. That is much harder with a diffuse movement. It is why the Pentagon keeps announcing that it killed this or that "senior ISIS leader," but the movement only continues.

"For 15 years, the operational focus for the U.S. has been the destruction of terrorist organizations," Friedman added. "The reason for this is that destroying a particular group creates the illusion of progress. However, as one group is destroyed, another group arises in its name. For example, al-Qaida is being replaced by the Islamic State. The real strength of Islamist terrorism is the movement that the organization draws itself from and that feeds it. So long as the movement is intact, any success at destroying an organization is, at best, temporary and, in reality, an illusion."

It should be clear by now that our conventional, special operations–based approach to defeating this phenomenon is not succeeding. The only thing that might work, argued Friedman, is "to bring pressure on Muslim states to make war on the jihadists and on other strands of Islam to do so as well. The pressure must be intense and the rewards substantial. The likelihood of it working is low. But the only way to eliminate this movement is for Muslims to do it." To deter this kind of breaker our first line of defense has to be their families, psychiatrists, schoolteachers, and neighbors, who can detect changes in personal behavior far faster than any intelligence agency. It takes a village to deter a breaker of this kind.

"Hybrid" warfare, the ability of super-empowered individuals and small groups, as well as weak nation-states, to leverage the tools of the digital age to sap the strength of stronger nation-states and engage them in a kind of "gray zone" warfare on multiple fronts is becoming increasingly problematic, and the stronger states have yet to come up with a sustainable deterrent. Moises Naim, the author of *The End of Power,* elaborated on this point in a February 13, 2017, essay on Atlantic.com. We now know, noted Naim, that "Al-Qaeda spent an estimated $500,000 on the attacks, which killed almost 3,000 people and cost hundreds of billions of dollars in material losses. The reactions that followed were even larger and more consequential than the attacks themselves: The United States launched what is to date its longest war ever (in Afghanistan), and its third-longest (in Iraq), at the estimated combined cost of $3 trillion to $5 trillion."

While al-Qaeda taught a new generation of Americans about how kinetic asymmetric warfare can be the perfect weapon for super-empowered individuals, noted Naim, WikiLeaks and the Kremlin have taught America and other democracies that cyber warfare can be the perfect asymmetric weapon of weak superpowers. While al-Qaeda relied "on physical violence to kill people, destroy buildings, and disable critical infrastructure," he explained, the Russians relied on the "internet and other cyber tools, which can cause not only physical damage but also weaken the institutions that are critical for the functioning of a democratic government . . . Russian interference in the 2016 U.S. presidential election, involving what U.S. intelligence believes were Kremlin-directed hacks and leaks of emails damaging to the candidacy of Hillary Clinton . . . represented a political cyber-Pearl Harbor. And that cyber confrontation was asymmetrical, not because America was at a technological disadvantage (the U.S. is among the world's leaders in the technologies needed to wage cyberwars), but because Russia was able to exploit the weak points of America as a democracy. What made America uniquely susceptible to the attack from an authoritarian Russia is emblematic of what makes other democracies particularly vulnerable, relative to their authoritarian counterparts, to political cyberattack. For one thing, the 2016 election attack targeted the democratic process itself. In the words of the intelligence community's January 2017 report on the incident, the hacks and leaks worked to 'undermine public faith in the U.S. democratic process, denigrate Secretary Clinton, and harm her electability and potential presidency.' They aimed to take advantage of the free flow of information in a democratic society, the affect of that information on public opinion, and the electoral mechanisms through which public opinion determines a country's leadership."

The asymmetry of the Russian attack did not only stem from how little it cost to do so much damage—probably less than the cost of a single MiG-29 fighter jet used in conventional warfare—but also from the fact that a cyberattack by Russia on Western democracies is an ideal weapon because the Russian autocracy is so much less vulnerable to a counterattack.

If "a hacker leaked damaging information about Vladimir Putin, there are various obstacles in the way of its having an electoral effect," explained Naim. "Restrictions on the media in Russia could prevent the information from circulating widely. Even if it did manage to attract publicity and sway public opinion, what then? Putin has tight control over the country's electoral apparatus, meaning that a voting citizenry inclined to punish him for leaked evidence of misdeeds has no real mechanism to do so . . . And if democratic politicians are more vulnerable to the effects of leaks, democracies are also more likely to produce leakers to begin with. The legal protections individuals enjoy in the democratic states make it hard to deter this type of behavior. But the cost of leaking in an autocratic society like Russia, where political opponents of Putin have been known to wind up dead, could be far higher, obviously posing a major disincentive."

Add it all up and you realize that in the last thirty years the weapons that did the most damage to the United States were weapons of the weak—IEDs and suicide bombers undermined us in Iraq and Afghanistan and cyber-hacking from Russia polluted America's 2016 democratic election and, in the process, helped to elect a president, Donald Trump, who threw NATO and the whole Western alliance into political turmoil, not to mention America itself.

## *The New Balance of Power*

For all of these reasons we need to rethink what constitutes "the balance of power." During the Cold War, if you wanted to assess the global balance of power you would likely look at the annual survey *The Military Balance*, published by the London-based International Institute for Strategic Studies, and self-described as the most "trusted military data on 171 countries: size of armed forces, defense budgets, equipment." That book would tell you the relative strengths of their armies, navies, and air forces (their hard power), and their "soft power": the relative strengths of their economies, their societal appeal, and the degree of entrepreneurship in their culture. And if you added up all those numbers, you would have a rough measure of the balance of power between different nation-states.

Not anymore. Assessing today's balance of power requires a much wider lens. "In the old days, when you talked about the balance of power, you were really talking about conventional forces, nuclear forces, and the framework of arms control to regulate them," John Chipman, the director of the International Institute of Strategic Studies, told me. "There was an easy consensus on every measure of power and how you count them. It was purely a math problem." But today, conventional military power, while still important, is only one factor. If you want to measure the balance of power now, if you want to explain geopolitics, let alone manage it, you need to consider the power of one, the power of machines, and the power of flows and how all of that is breaking down weak states and empowering breakers—all in a more interdependent world.

In trying to manage such a world, "you can't just pull out the old playbook," observed U.S. Defense Secretary Ashton Carter. "Destructive power of greater and greater magnitude now can be delivered by smaller and smaller hands . . . You're kidding yourself if you think you're in a world where all you have to think about are states."

## *Learning to ADD*

For all these reasons, geopolitics has to be reimagined in the age of accelerations, just like everything else. To be sure, you can still score a lot of points these days on the op-ed pages or on the campaign trail by blithely pretending that the United States can do what it always did—as President Kennedy put it in his inaugural, "pay any price, bear any burden, meet any hardship, support any friend, oppose any foe to assure the survival and the success of liberty." But that is not going to happen. The post–post–Cold War world, alas, has been a cold shower for America's (not to mention my own) can-do optimism. We have learned the hard way in Iraq and Afghanistan that liberty taking root depends much less on what we do than on what *they* do, and if they are not ready to pay that price, bear that burden, meet that hardship, support one another, and collectively oppose the foes of liberty, we cannot do it for them. We've learned that while we can transplant hearts, and we can transplant boots on the ground, we cannot transplant political culture—and particularly an ethic of pluralism—where there is no topsoil of trust.

And, finally, we have learned that when the major threat of instability is the "weakism" and "disintegrationism" of existing states, and the rise of super-empowered individuals and small bands of breakers, our traditional power-tool kit is insufficient to meet these threats—on our own. We cannot put every Humpty Dumpty state back together again—

on our own. And we cannot find every needle, every super-empowered angry person, in a haystack before it sticks us—on our own.

In short, we have to face two fundamental facts about geopolitics today:

Fact #1: *The necessary is impossible.*

Fact #2: *The impossible is necessary.*

That is, while we cannot repair the wide World of Disorder on our own, we also cannot just ignore it. It metastasizes in an interdependent world. If we don't visit the World of Disorder in the age of accelerations, it will visit us. This is especially true when you know that the age of accelerations is going to continue to hammer frail states and produce migration flows, particularly from Africa and the Middle East toward Europe, as well as more super-empowered breakers.

So what to do?

In an earlier historical epoch, we could usually count on some giant imperialist power sweeping into these regions of disorder—like northern Nigeria, Libya, Yemen, Somalia, or Syria—and imposing order from the outside and crushing "the breakers." In the Cold War, Russia effectively occupied all of Eastern Europe, suppressing not only its freedoms but also its ethnic conflicts. For five centuries the Ottomans managed most of the Middle East in the same way. But today we live in a postimperial and a postcolonial world. *No great power wants to occupy anybody.* As we've seen, the major powers have all learned the hard way that when you occupy another country all that you win is a bill. It is much easier to import a country's labor and natural resources—or their brainpower online—than it is to take them over.

Also, in an earlier historical epoch, like World War II or the Cold War, we could more easily galvanize an alliance of like-minded democracies to combat this threat to global stability. Today, "weakism" and "disintegrationism" do not have the galvanizing power of Nazism or the Red Menace. They also don't lend themselves to the traditional tools of warfare—tanks, planes, and troops—and they don't hold out the satisfying prospect of a "V-E Day"—a final victory—and a ticker-tape parade with "When Johnny Comes Marching Home." Nation-rebuilding or disorder-containing and breaker-deterring are much more diffuse and longer-term projects and much less morally satisfying.

Moreover, while we don't have the resources to solve the problem of disorder by intervening *over there*, we also cannot solve the problem of disorder from *over here*—in the West. The sudden and massive influx of refugees from Africa and the Middle East has overwhelmed the absorptive capacity of the European Union and triggered a populist-nationalist backlash, while also prompting the EU to start limiting its policy of free movement of people between countries. The June 2016 British vote to withdraw from the EU was driven in no small degree by anti-immigration sentiment.

And we still cannot ignore the challenges posed to international order by our rival superpowers Russia and China, which, because they are authoritarian states, are not as vulnerable to disorder or breakers as are open societies in the West.

Add it all up, ladies and gentlemen, and what you have is a perfect example of a "wicked problem"—many stakeholders, but no agreement on the problem definition or on the solution. And doing nothing will become increasingly unsustainable.

So I repeat, what to do?

If I were reimagining geopolitics from an American/Western perspective in such a world, I would begin with the most honest statement I can offer: I don't know what is *sufficient* to restore order to the World of Disorder—one should be very humble in the face of such a wicked problem—but I am fairly certain of what is *necessary*.

It's a policy that can be called ADD, because those are its initials: amplify, deter, and degrade.

## *Knowledge Is Power*

Let's go through the logic of each element and see why they can add up to a national security strategy for a country like America today—starting with "amplify." It is a truism, but one worth repeating, that disorder and the rise of super-empowered breakers on the scale that we are seeing in the Middle East and Africa is a product of failed states unable to keep up with the age of accelerations and enable their young people to realize their full potential. But these trends are exacerbated by climate change, population growth, and environmental degradation, which are undermining the agricultural foundations that sustain vast African and Middle

Eastern populations on rural lands. The combination of failing states and failing agriculture is producing millions of young people, particularly young men, who have never held a job, never held power, and never held a girl's hand.

That terrible combination of humiliating pathologies is then preyed upon by jihadist-Islamist ideologues (with money), who promise these young people redemption or ninety-nine virgins in heaven if they double down on backwardness—if they go back to a seventh-century Islamist puritanical lifestyle. As George Friedman pointed out above, we cannot reverse these trends on our own; the will has to come from within these societies. But we can raise the odds that they will do it themselves by raising the number of people with the will to do so. What America and the West can do—and have not done nearly enough of—is to invest in and amplify the islands of decency and the engines of capacity-building in countries in, or bordering on, the World of Disorder. When we invest in the tools that enable young people to realize their full potential, we are countering the spread of humiliation, which is the single biggest motivator for people to go out and break things.

In May 2012—a year after the Arab awakening erupted—the United States made two financial commitments to the Arab world that each began with the numbers 1 and 3. The U.S. gave Egypt's military regime $1.3 billion worth of tanks and fighter jets. It also gave Lebanese public school students a $13.5 million merit-based college scholarship program, putting 117 Lebanese kids through local American-style colleges that promote tolerance, gender and social equality, and critical thinking. Having visited both countries at that time, I noted in a column that the $13.5 million in full scholarships bought the Lebanese more capacity and America more friendship and stability than the $1.3 billion in tanks and fighter jets ever would. So how about we stop being stupid? How can sending planes and tanks to a country—Egypt—where half the women and a quarter of the men can't read possibly end well?

The American embassy in Beirut introduced me to four of the 2012 Lebanese scholarship students—they attend either the Lebanese American University or Haigazian University, both of which offer modern U.S.-style bachelor's degrees. As I noted in my column, Israa Yassin, an eighteen-year-old from the village of Qab Elias who was studying computer science, told me: "This whole program is helping to make the youth capable of transforming this country into what it should be and can be. We are good, and we have the capabilities and we can do a lot, but we don't get the chance. My brother just finished high school, and he could not afford [university]. His future is really stopped. The U.S. is giving us a chance to make a difference . . . We will not be underestimated anymore. It is really sad when you see a whole generation in Lebanese villages—hundreds of guys doing nothing—no work, not going to college." Wissal Chaaban, eighteen, from Tripoli, who also was attending the Lebanese American University and studying marketing, told me that the program is in America's interest because it sends young people to colleges that "encourage openness, to accept the other, no matter how different, even if he was from another religion."

A few days after talking to these students, I went to Amman, Jordan, where I interviewed some public school teachers at Jordan's Queen Rania Teacher Academy, which was working with a team from Columbia University to upgrade teaching skills. I talked to them about the contrast between the $13.5 million in U.S. scholarships and the $1.3 billion in military aid, and Jumana Jabr, an English teacher in an Amman public school, summed it up better than I ever could: One is "for making people," she said, "and the other is for killing people." If America wants to spend money on training soldiers, she added, well, "teachers are also soldiers, so why don't you spend the money training us? We're the ones training the soldiers you're spending the $1.3 billion on."

In June 2014, I was invited to give the commencement address at the American University of Iraq, in Sulaimaniya, Kurdistan. As I wrote in my column at the time, I am a sucker for commencements, but this one filled me with many different emotions. For starters, the scene was stunning in the highlands of Kurdistan. As Dina Dara—the student speaker and valedictorian of the 2014 graduating class—took the stage, the sun was just setting, turning Azmar Mountain in the background into a reddish-brown curtain. The class was about 70 percent Kurds, with the rest coming from every corner, religion, and tribe of Iraq. Parents bursting with pride, cell phone cameras in one hand and bouquets in the other, had driven up from Basra and Baghdad dressed in their finest to see their kids get their American-style college degrees. Three Kurdish TV stations carried the ceremony live.

"It has been quite a journey," Dara, who had been accepted to graduate school at Tufts, told her classmates. (Since the university opened, in 2007, all the valedictorians had been Iraqi women.) "We went through a whole different experience living in the dorms. This evening . . . we are armed with two things: first, the highly valued American education that makes us as competent and qualified as the rest of the students in the world. And, second, the empowerment of a liberal arts education." "[As we] exercise critical thinking techniques that have been the core of our education here, and as we try to move beyond the traditional conventions, beyond what others suggest, we may struggle. But isn't this how nations are built?"

Karwan Gaznay, twenty-four, a Kurd, told me he grew up reading books about Saddam: "Now we have this American education. I did not know who Thomas Jefferson was. I did not know who James Madison was. So when the government is doing something wrong, now we can say: 'This is wrong. I have been educated.' . . . I ran for student president, and Arab guys voted for me. We are living as a family in the university. I am not pessimistic about Iraq. We can work together if we want to."

The best long-term investment the American government could make to help stabilize the World of Disorder and widen the islands of decency there would be to help fund and strengthen schools and universities throughout the Middle East, Africa, and Latin America that promote American-style liberal arts and technical education. Unfortunately, there are so many huge defense industry lobbies that promote funding the tools for killing people and so few advocates for funding the schools for building people. That has to change. Education alone is not a cure-all, but drones alone are a cure-nothing. Islands of decency can spread. Drones are one and done.

## *It Takes a Chicken*

Even as we amplify educational opportunities, we also have to amplify the opportunities for the poorest of the poor, particularly in Africa, to remain in their home villages on their land. If it wants to stop the spread of disorder, the developed world needs to do this at a scale we have never attempted before. Two of the smartest people I know on this subject are

Bill Gates and the U.N. Convention to Combat Desertification's Monique Barbut. It is worth listening to what they both have to say—and what they have to say is basically the same thing: you have to stabilize the basic foundations of life in disordered societies, particularly Africa. That can mean starting with something as simple as a chicken coop.

Gates put it to me this way: "For good stuff to happen, it requires a lot of things to go well—you need many pieces to get stability right." None of it is going to happen overnight, but we need to work with the forces of order that do still exist in the World of Disorder to start building a different trajectory, beginning with all the basics: basic education, basic infrastructure—roads, ports, electricity, telecom, mobile banking—basic agriculture, and basic governance. The goal, said Gates, is to get these frail states to a level of stability where enough women and girls are getting educated and empowered for population growth to stabilize, where farmers can feed their families, and where you "start to get a reverse brain drain" as young people feel that they have a chance to connect to and contribute and benefit from today's global flows by staying at home and not emigrating.

Believe it or not, he argued, a good place to start is with chickens—a solution he demonstrated for me and other interested visitors by erecting a huge model chicken coop on the sixty-eighth floor of 4 World Trade Center. "If you were living on $2 a day, what would you do to improve your life?" Gates asked on his blog. "That's a real question for the nearly 1 billion people living in extreme poverty today. There's no single right answer, of course, and poverty looks different in different places. But through my work with the foundation, I've met many people in poor countries who raise chickens, and I have learned a lot about the ins and outs of owning these birds . . . It's pretty clear to me that just about anyone who's living in extreme poverty is better off if they have chickens. In fact, if I were in their shoes, that's what I would do—I would raise chickens."

Here's why, he explained:

> They are easy and inexpensive to take care of. Many breeds can eat whatever they find on the ground (although it's better if you can feed them, because they'll grow faster). Hens need some kind of shelter where they can nest, and as your flock grows, you

might want some wood and wire to make a coop. Finally, chickens need a few vaccines. The one that prevents the deadly Newcastle disease costs less than 20 cents.

They're a good investment. Suppose a new farmer starts with five hens. One of her neighbors owns a rooster to fertilize the hens' eggs. After three months, she can have a flock of 40 chicks. Eventually, with a sale price of $5 per chicken—which is typical in West Africa—she can earn more than $1,000 a year, versus the extreme-poverty line of about $700 a year.

[Chickens] help keep children healthy [by keeping them fed]. Malnutrition kills more than 3.1 million children a year.

And, maybe most important of all, he added,

They empower women. Because chickens are small and typically stay close to home, many cultures regard them as a woman's animal, in contrast to larger livestock like goats or cows. Women who sell chickens are likely to reinvest the profits in their families . . .

Dr. Batamaka Somé, an anthropologist from Burkina Faso who has worked with our foundation, has spent much of his career studying the economic impact of raising chickens in his home country [and attests to their value] . . .

Our foundation is betting on chickens . . . Our goal: to eventually help 30 percent of the rural families in sub-Saharan Africa raise improved breeds of vaccinated chickens, up from just 5 percent now.

When I was growing up, chickens weren't something you studied, they were something you made silly jokes about. It has been eye-opening for me to learn what a difference they can make in the fight against poverty. It sounds funny, but I mean it when I say that I am excited about chickens.

Barbut shares Gates's view that you have to get the basics right that stabilize the bottom of the pyramid so people are not forced to either "flee or fight."

You have to build solutions "at the source," Barbut said to me. "You know, we live in a world where we believe that technology is going to bring the solution to everyone, and it's very difficult to make people say, 'Please, maybe not all the world is yet ready for that. You need to deal with the small farming agriculture piece first. Today in the world, you have five hundred million farms which are less than three hectares, and those five hundred million farms [are providing] the direct living of 2.5 billion people. It means one-third of the planet lives on those small entities." If they are wiped out by climate change and desertification, as is starting to happen all over West Africa and the Sahel region now, "you are going to have major crises . . . Eighty percent of the population of Niger lives off of the land. If you lose your little land, then you have lost everything."

In the past, she notes, when you had a drought, people would migrate for the season, until the drought went away. Then they came back and tried again. "But what we are seeing—and we think that it is very much linked to the climate change—the droughts are becoming harder and harder," said Barbut. "Now it's every three to four years . . . [So] instead of seasonal migration, you get definitive migration, because people have lost their land . . . It becomes totally unproductive forever, at least if you don't do big measures to restore it. And this is a phenomenon that we see increasing very much." If this trend continues, millions of people in the southern part of Africa and the Horn "are going to lose their means of living. But what does that mean? It means also that those farmers are not going to be able to feed the population they were feeding, and so it's going to have repercussions on the food prices." It also means millions of Africans will either flee to southern parts of Africa and destabilize those regions or try to cross the Mediterranean to get into Europe.

Barbut has her own idea for an affordable modern-day Marshall Plan for Africa. "To restore a hectare of degraded land, it costs between one hundred and three hundred dollars," she said, while a day in the refugee camp for one refugee in Italy costs the host government forty-two dollars. "So, please, we are not talking about a huge amount of money," she noted. Her proposal: in the thirteen countries from Mali to Djibouti, fund a "Green Corps" of five thousand people—one per village in each country—give them basic training and seedlings for planting trees that can retain water and soil, and pay them each two hundred dollars a month to take care of their plantings. This is an idea that actually originated with African leaders. It's called "the Great Green Wall": a ribbon of land restoration projects stretching across the entire southern edge of the Sahara, to hold the desert back—and help anchor people in the communities where they actually want to live. It makes a lot more sense than building expensive leaky walls around Europe that will never hold if millions of Africans have to migrate.

"Today people are putting walls all over the place," said Barbut, "and me also, I dream of a wall—a wall that we have called 'the Great Green Wall.' We have to stop the deserts coming down [from the Sahara]. We are going to need to replant enough vegetation so that we stop the desert advancing and we restore the fertility of our land and the storage of our water. It will bring hundreds of millions back to work. It will feed the people, and you could store up . . . $CO_2$ emissions. So it will help on the climate change."

In addition to these no-tech solutions for amplifying decency and capacity, there's one high-tech concept worth investing in: nothing would help create more local economic growth than bringing high-speed wireless broadband connectivity to every village in Africa. Every study on this subject indicates that connecting the poor to the world of flows—of education, commerce, information, and good governance—drives economic growth and enables people to generate income while staying in their homes.

It's not a pipe dream. It would take so little investment to have huge impacts on these communities. As Jeff Raikes, cofounder of the Raikes Foundation, former CEO of the Bill & Melinda Gates Foundation, and a board member of the Alliance for a Green Revolution in Africa, wrote in a December 28, 2016, essay on Project Syndicate: "Agriculture comprises almost two-thirds of Sub-Saharan Africa's workforce, and in 2003 the African Union called for countries to increase their investment in the sector to an ambitious 10% of all government spending. Only 13 countries answered that call, but their investments—in research and development, services that help farmers take advantage of new research findings, credit and financing initiatives, commodity exchanges, and other marketing efforts—have already paid dividends. Those 13 countries have experienced marked improvements in agricultural production, per capita GDP, and nutrition . . . Only about 6% of rural households in

Sub-Saharan Africa receive loans from financial institutions. Moreover, almost two-thirds of African farmland soil is missing key nutrients, and many farmers lack the technical knowledge and resources to restore their land's fertility, leaving them unable to take full advantage of new technologies."

Chicken coops, gardens, and Webs—it's either some combination of those or: *Would the last one out please turn off the lights . . .*

## *Deter and Degrade*

Though the Cold War has long since ended, deterrence remains a crucial tool in a world in which superpower rivalry has not gone away. Russia still really would like to break up the NATO alliance—just as NATO still really does see the most important part of its mission today as containing any possible Russian aggression. China really would like to see the United States retreat from the South China Sea and shrink its power profile in Asia generally; the United States really does believe that its role in maintaining the openness of global sea lanes requires making certain that China doesn't alone write the rules of the road for the South China Sea, let alone the Pacific. And both Russia and China still have nuclear weapons targeting America—and the rogue state of North Korea clearly aspires to have the same. The power of all of them has to be balanced by a strong American nuclear deterrent. Without that, every country on the border of Russia and China would seek nuclear weapons to protect itself.

But that's not all. Deterring today's Russia, in particular, is a complex challenge that requires more than building missiles. On July 28, 2016, the *Washington Post* columnist Anne Applebaum, an expert on Eastern Europe, noted that President Vladimir Putin has evolved a "hybrid foreign policy, a strategy that mixes normal diplomacy, military force, economic corruption and a high-tech information war." Indeed, on any given day, the United States has found itself dealing with everything from cyberattacks by Russian intelligence hackers on the computer systems of the U.S. Democratic Party, to disinformation about what Russian troops, dressed in civilian clothes, are doing in Eastern Ukraine, to Russian attempts to take down the Facebook pages of widows of its soldiers killed in Ukraine when they mourn their husbands' deaths, to hot money flows into Western politics or media from Russian oligarchs connected to the Kremlin. In short, Russia is taking full advantage of the age of accelerating flows to confront the United States along a much wider attack surface. While it lives in the World of Order, the Russian government under Putin doesn't mind fomenting a little disorder—indeed, when you are a petro-state, a little disorder is welcome because it keeps the world on edge and therefore oil prices high.

China is a much more status quo power. It needs a healthy U.S. economy to trade with and a stable global environment to export into. That is why the Chinese are more focused on simply dominating their immediate neighborhood.

But while America has to deter these two other superpowers with one hand, it also needs to enlist their support with the other hand to help contain both the spreading World of Disorder and the super-empowered breakers. This is where things start to get tricky: on any given day Russia is a direct adversary in one part of the world, a partner in another, and a mischief-maker in another.

In Syria, the Obama administration has constantly wrestled with a fiendishly difficult question: Should America and its allies work to take out the murderous Syrian president Bashar al-Assad first—in which case they would lose the support of Iran and Russia and likely introduce even more near-term disorder into Syria? Or should it take out ISIS first—with the tacit support of Iran and Russia—and allow Assad to stay in power, containing total disorder but also crushing the more secular, democratic Syrian opposition? As of the writing of this book, America has not resolved that dilemma.

In other parts of the world, the United States needs China's help—for instance, to contain North Korea's nuclear missile program and prevent it from proliferating nuclear materials in the World of Disorder. One could imagine China agreeing to help—but only if the United States cuts Beijing more slack in the South China Sea.

As for the breakers, be they individuals or groups like ISIS and Al Qaeda, they cannot be deterred. They can, though, be contained and degraded in their various theaters of operation, by using air power, special forces, drones, and local forces. In the end, however, they can only be sustainably destroyed by their host communities' delegitimizing their narrative and ultimately killing or jailing their leaders. Outsiders can help degrade them, but ultimately only the village can destroy them.

Yes, it makes for a rather messy strategic environment. Which only reinforces why, as Waylon Jennings might have put it in song, "Momma, don't let your daughters grow up to be secretaries of state." You need to juggle drones and walls where you must; invest in chickens, gardens, and schools where you can; amplify islands of decency wherever you find them; deter competing superpowers—whenever you're not also enlisting their help; learn to live with the fact that a foreign policy of amplify, deter, and degrade will more often than not require us to side with the *least bad* over the *worst*; and finally, appreciate that widening decency is the necessary precursor to electoral democracy—and in many places is more important.

## *Captain Phillips*

None of these ideas are the stuff of great geopolitical doctrines; but the age of accelerations will be a graveyard for fancy big ideas. When the necessary is impossible but the impossible is necessary, when no power wants to own the World of Disorder but, increasingly, no power can ignore it, it is going to take these hybrid combinations of drones and walls, aircraft carriers and Peace Corps volunteers, plus chickens, gardens, and Webs, to begin to create stability in the age of accelerations.

Since we began this chapter with a TV sitcom that foreshadowed the future, let's end with a movie that highlights the present—and with any luck *doesn't* foreshadow the future. It's the film *Captain Phillips*, which was based on the very real 2009 hijacking of the unarmed U.S. container ship *Maersk Alabama* by a gang of Somali pirates in a speedboat. The film centers around the struggle between the *Alabama*'s commanding officer, Captain Richard Phillips, played by Tom Hanks, and the Somali pirate captain, Muse—played by Barkhad Abdi, a Somali actor who was living as a refugee in Minnesota—who takes Phillips and his ship hostage. The Somali pirates seize the ship while it is transiting the Indian Ocean off East Africa. While interrogating the Boston-bred Phillips and learning of his background, Muse nicknames him "Irish."

In a critical scene, Phillips tries to reason with the Somali hijacker, but in doing so only displays his ignorance of the depth of despair haunting the World of Disorder today. At one point he says to the pirate: "There's got to be something other than being a fisherman or kidnapping people."

To which Muse replies, "Maybe in America, Irish. Maybe in America."

Muse's musing is deeply poignant, and our goal should be to both rewrite it and appreciate it. We need to rewrite the notion that the only way for some people in parts of the World of Disorder to sustain themselves is by either fishing or kidnapping—that will no longer suffice. That would lead to a nightmarish world. A policy of amplify, deter, and degrade is intended to facilitate an alternative.

At the same time, Americans need to appreciate just how much their country is the last, best hope for a lot of people and an irreplaceable source of order. Just one tiny recent example: In 2014, when the Ebola virus broke out in West Africa, it was the American military that sent in three thousand troops and $3 billion to wipe it out. There was no Russian or Chinese aid mission that jumped into the breach. Yes, I am glad there is a United Nations, a World Bank, and global flows that are knitting the world together through Facebook and Google. But at the end of the day they all depend on a healthy American economy, a strong American military able to project power and deter autocracies, and an unwavering willingness to defend the values of pluralism and democracy against those who would threaten them from abroad or within. With the European Union, the other great center of democracy and free markets, weakening in recent years, America's pivotal role in upholding those values globally becomes even more important.

Lately, many Americans have lost sight of both their country's achievements and its vital role in stabilizing the global commons. An immigrant friend of mine from Zimbabwe, Lesley Goldwasser, once remarked to me, "You Americans kick this country around like it's a football. But it's not a football. It's a Fabergé egg. You can break it." She is right. We can break it. And in an era when liberty, free markets, pluralism, and the rule of law—all the pillars of a stable society—will be challenged by breakers, bullies, and disorder, we do so at our own peril.

TEN

# *Mother Nature as Political Mentor*

Charles Darwin is often quoted as saying that it is not the strongest species that survives but the most adaptable. But according to QuoteInvestigator.com (QI), he did not write that in his classic *On the Origin of Species*—and there's no evidence that he even said it somewhere else. QI's research suggested that the quote emerged over time from a speech delivered by a Louisiana State University business professor, Leon C. Megginson, at the convention of the Southwestern Social Science Association in 1963.

Megginson reportedly said:

> Yes, change is the basic law of nature. But the changes wrought by the passage of time affect individuals and institutions in different ways. According to Darwin's *Origin of Species*, it is not the most intellectual of the species that survives; it is not the strongest that survives; but the species that survives is the one that is able best to adapt and adjust to the changing environment in which it finds itself. Applying this theoretical concept to us as individuals, we can state that the civilization that is able to survive is the one that is able to adapt to the changing physical, social, political, moral, and spiritual environment in which it finds itself.

Thank you, Professor Megginson!

That is so well said—whether Darwin said any of it or not. To paraphrase, it's not the strongest quote that survives, it's the most adaptable! And this one is so relevant for our times. In the first decade and a half of the twenty-first century, we went through a major technological inflection point—connectivity became fast, free, easy for you, and ubiquitous, while complexity became fast, free, easy for you, and invisible. And this has unleashed flows of energy that, in combination with climate change, have, as we've already explored, reshaped the workplace and geopolitics and prompted us to reimagine how we approach both. But that reimagining cannot succeed in isolation. It also requires us to reimagine our domestic politics—both in order to deliver the kinds of specific policy fixes we need in the workplace and in geopolitics, and also more generally to create a society with the kind of resilience we'll need to thrive when the Market, Mother Nature, and Moore's law are all accelerating. This is going to require some very different approaches to politics generally, and that political realignment appears to already be under way.

In the previous chapter I argued that in the age of accelerations some weak states would explode. What seems to be happening to strong states is that their politics implodes—that is, their borders hold but their political parties begin to crack up, because in their present forms they cannot adequately and coherently respond to the simultaneous and interrelated challenges posed by the accelerations in technology, globalization, and the environment.

To put it in the context of the previous two chapters, just as average is over for every worker and for every weak developing state, so average is also over for every strong developed democracy as well—particularly America, the European Union, and Japan. Three key pillars, three key assumptions or expectations that kept their societies relatively stable and democratic—oscillating between sturdy center-right and center-left political boundaries—have come under serious stress.

The first, and most important, pillar is the expectation that each new generation can enter the middle class and enjoy better living standards than their parents did—by just doing average work with average skills. As we noted in chapter eight on AI and IA, in order to sustain a middle-class lifestyle everyone has to work harder and relearn faster.

The second pillar is the broad commitment to pluralism and to welcoming a steady stream of immigrants, both low-skilled high-energy and high-skilled. That pillar, too, has been stressed by the age of accelerations, due to the rise of the world of disorder and the millions of migrants and refugees trying to escape to the world of order. The industrialized democracies of the Northern Hemisphere simply cannot accommodate the numbers of people now beating on their doors, and in Europe, where many of those immigrants are Muslims, the integration challenge has created even more stress.

The last pillar was the expectation that the economic fates of urban and rural communities would be broadly similar. While salaries in upstate New York and New York City might not be the same, on a relative basis you could live a decent middle-class life in both. But that expectation, too, has faded in the age of accelerations.

As social scientist Jonathan Rauch pointed out in a March 2017 essay in *The Atlantic,* "the United States in 2016 offered a particularly vivid example" of the new sharp contrast between urban and rural communities, as "Hillary Clinton carried only 472 counties, out of more than 3,000, but those 472 were predominantly urban and accounted for nearly two-thirds of the country's total economic output."

In other words, to enjoy an average lifestyle in the Cold War era, it did not matter whether you lived in a rural or urban setting. The American Dream was available in both regions. But in an age of accelerations there are fewer one-company towns, as more manufacturing work has moved abroad, and in those that exist the company has been automated, so there are fewer jobs. So, unless you have a really dynamic local leadership, or are close to a university (and there are a lot more of these rural communities than you might think, but not enough), increasingly the only way to hold on to the American Dream is by living in a globally connected, multicultural, lifelong-learning-rich, urban context. That is why we've seen such a pronounced split between the fates of many urban and rural communities, which affected both the Brexit vote and the Trump election.

That is why I contend that for political parties in democratic systems to remain relevant today they have to develop responses to the big three accelerations and how they are shaking the key pillars that have upheld stable, pluralistic Western industrial democracies for nearly a century. But up to now, traditional center-right and center-left parties in America and Europe have struggled to do so.

Their failure has created a gaping political/emotional hole that has produced or propelled a collection of populist-nationalist parties—who advocate going back to an idealized past, to in effect try to slow down the age of accelerations. It also creates an opening for new political movements, like Emmanuel Macron's in France, that are responding to the three accelerations in constructive ways the traditional parties were unable to do, and winning support from voters for doing so.

After 2007, in particular, citizens in America and Europe felt they were being hurtled along into the future so much faster than ever, but the governments and institutions they had long relied upon were not being reformed and updated accordingly.

On May 16, 2016, *The New York Times* carried a story about a divisive Austrian election, featuring two quotes that spoke for so many voters across the industrialized world. One was from Georg Hoffmann-Ostenhof, a columnist for the liberal weekly magazine *Profil*. "We are in a situation where people don't understand the world anymore, because it is changing so fast. And then came the migrants, and people were told that politicians had lost control of the borders. That just heightened the overall sense that control was gone." The other was from Wolfgang Petritsch, a veteran diplomat and former chief aide to Austria's erstwhile center-left chancellor Bruno Kreisky: "Social democracy was always driven by ideas," he said. "But the ideas have gone missing."

This chapter aspires to remedy that shortage of ideas by addressing the political innovation that needs to accompany the age of accelerations if we want to maintain a world with many stable, pluralistic democracies.

So here's where I begin: you can't have a relevant political strategy if you don't start with a brutally honest assessment of what world you are living in. We are living in a world experiencing three "climate" changes at once: a change in the climate of technology, in the climate of globalization, and in the actual climate and environment, thanks to their simultaneous accelerations.

What is it that you want for citizens and society when the climate changes? You want two things for certain: resilience and propulsion. You want citizens and societies that can absorb shocks, because disruptions will come faster and more often when you have three climates change at once. And you want propulsion. You don't want your citizens curled up in a ball under their beds. You want them to feel empowered and enabled to get the most out of these accelerations and cushion the worst—not to put up walls but to be able to live without them.

That takes a whole new kind of political agenda, and this chapter is my contribution to that rethinking. I set myself the task of starting with a blank sheet of paper and asking not what it means to be a "conservative" or a "liberal" today (frankly, who cares?), but rather how we maximize the resilience and self-propulsion of every citizen and community in America—that is, their ability to both absorb shocks and keep progressing in this age of accelerations. It is a different approach to politics—a necessary one, I believe—and it yields a political agenda unlike anything on offer in America today.

## *Mother Nature's Killer Apps*

Before taking out my blank sheet of paper, though, I did one vitally important thing: I looked for a mentor. I asked myself, who is the "person" with the most experience absorbing climate changes and retaining resilience and continuing to flourish? The answer came easily: I know a woman who has been doing that for about 3.8 billion years. Her name is Mother Nature.

I can think of no better political mentor today than she. As Johan Rockström noted, Mother Nature is not a living being, but she is a biogeophysical, rationally functioning, complex system of oceans, atmosphere, forests, rivers, soils, plants, and animals that has evolved on Planet Earth since the first hints of life emerged. She has survived the worst of times and thrived in the best of them for nearly four billion years by learning to absorb endless shocks, climate changes, surprises, and even an asteroid or two. That alone makes Mother Nature an important mentor. But she is even more relevant today because we human beings have now built—with our own hands, brains, muscles, computers, and machines—our own complex global system of networks. These networks have become so interconnected, hyperconnected, and interdependent in their complexity that, more than ever, they've come to resemble the complexity of the natural world and how its interdependent ecosystems operate.

"If we are evolving to be more like nature, we better damn well get good at it," observed the physicist and environmentalist Amory Lovins.

I agree. So let's first try to understand the basic strategies that Mother Nature employs to build resilient ecosystems that can absorb shocks and still move forward, and then try to translate that into policies that a party would advocate to help Americans better navigate this age of accelerations.

I am hardly the first to highlight the virtue of using nature as a metaphor. Janine Benyus, considered the mother of the biomimicry movement, likes to speak about nature as a "model," "measure," and "mentor." It is that model and mentor role that I am most interested in today. To be sure, everything that Mother Nature does is done unconsciously, and has evolved over millennia, but that doesn't mean we can't learn from and mimic her by choice. So, if Mother Nature could describe her killer apps for building resiliency to thrive in periods of climate change, what would she say?

She would surely start by telling us that she is incredibly adaptive over time through a variety of mechanisms, beginning with evolution through natural selection and the use of constant feedback loops. Mother Nature, to put it in modern management terms, is a lifelong learner. As Martin Reeves, the BCG business consultant, explained to me, feedback loops are the system Mother Nature uses to detect changes in the environment, to identify which plants or animals respond to those changes with the most resilience and propulsion and then to propagate their most desirable traits (i.e., genes) in the next generation of plants and animals.

It all happens quite naturally through trial and error. It is true, adds Lovins, that 99 percent of experiments that Mother Nature has tried didn't work and "got recalled by the Manufacturer." But the 1 percent that survived did so because they learned to adapt to a certain niche in the natural world and were able, therefore, to thrive and procreate and project their DNA into the future. Mother Nature also adapts "through social specialization," or learned behavior. These adaptations evolve over millennia, Lovins explained to me: "Some ants go out and look for food and some stay home and take care of the young, and that enables those who look for food to cover bigger areas. Specialized ant colonies have foragers and nest-keepers. This, too, is an adaptation, a learned behavior. It is not in their DNA. You cannot sequence such differentiated behaviors, but you can observe and mimic them, and doing so over time can become so powerful and advantageous that the organisms that do it dominate everyone else in their niche, just as we do as mammals." One cannot stress enough, Mother Nature believes in lifelong learning; species that don't keep learning and adapting disappear.

Oddly enough, one of the best ways to observe evolutionary adaptation via DNA is by visiting the desert. I say "oddly" because the desert would seem like the worst place to go on a safari. But in the hands of a good guide—whom my wife and I had when we visited the Serra Cafema Camp in the far northwest region of Namibia, overlooking the Kunene River on the border with Angola—not only do you discover that the desert is rich with biodiversity, but, because of how the smallest beetle stands out in the desert, you can see close up the genius of Mother Nature's adaptation-design skills. You can see close up the tiny percentage of insects and plants that learned to survive in the harsh desert by evolving unusual ways to capture and conserve water.

*Wired* magazine ran a story on November 26, 2012, about an American start-up that was

> developing a self-filling water bottle that sucks moisture from the atmosphere to create condensation, in the same way the humble Namib desert beetle does.
>
> The beetle, endemic to Africa's Namib desert—where there is just 1.3 cm of rainfall a year—has inspired a few proof-of-concepts in the academic community, but this is the first time a self-filling water bottle has been proposed. The beetle survives by collecting condensation from the ocean breeze on the hardened shell of its wings. The shell is covered in tiny bumps that are water attracting (hydrophilic) at their tips and water-repelling (hydrophobic) at their sides. The beetle extends and aims the wings at incoming sea breezes to catch humid air; tiny droplets 15 to 20 microns in diameter eventually accumulate on its back and run straight down towards its mouth.
>
> NBD Nano, made up of two biologists, an organic chemist and a mechanical engineer, is building on past studies that constructed structurally superior synthetic copies of the shell.

Another way Mother Nature produces resilience is by being relentlessly entrepreneurial—always looking for new niches to exploit and fill and always experimenting to see which plants and animals best coevolve. "If there is an open space in nature, some plant or animal will find a way to adapt to it and make a living there, in a way that no other species is; some other plant or animal will eat those species and produce waste that some other plant or animal will be eager to eat as food or fertilizer," remarks Lovins. "Nature is always innovating, creating new mutations as new opportunities arise."

And those mutations are tested in the context of the whole system to see if they are a good idea—if they fit into the system and make the whole more resilient. If, instead, they produce inadvertent toxins that harm the system, Mother Nature will innovate a correction. Mother Nature is the opposite of dogmatic—she is constantly agile, heterodox, hybrid, entrepreneurial, and experimental in her thinking. "Nature is restless, always exploring, inventing, trying, and failing," adds Tom Lovejoy, university professor in environmental science at George Mason University. "Each ecosystem, and each organism, is an answer to a set of problems."

In that sense another of Mother Nature's killer apps is her ability to thrive on diversity—both nurturing it and rewarding it in all species of plants and animals. Mother Nature understands that the best way to evolve and advance the best ideas is to have a large pool of them and through constant feedback loops—see which ones can adapt to which niches and also serve the whole. And so she is very pluralistic: she understands that nothing enhances the resilience of an ecosystem, or healthy interdependencies, more than a richly diverse cornucopia of plant and animal species, each adapted to the other and to a specific environmental niche.

High biodiversity means every niche is filled and playing its part to keep the whole in balance. "Think of the slow loris," says Lovins. "It's a small nocturnal primate that oozes along the branch of a tree very smoothly and softly—it looks like someone doing slow tai chi—to eat the leaves off the most slender twigs on the farthest tips of the most slender branches," and then converts those leaves into energy. Another, heavier loris specializes in leaves at the thicker part of the branch that can bear its weight. And other lorises eat still different things. Nature evolves organisms for every niche—and as long as there are physical niches to be filled, she will fill them with species that are ever better adapted and evolved for that niche, and the dynamic flux between all the units results in greater resilience, balance, and growth.

The University of Minnesota biologist G. David Tilman, one of the world's leading experts on biodiversity, wrote an article in the May 11, 2000, issue of *Nature* entitled "Causes, Consequences and Ethics of Biodiversity," which reviewed the main scientific field research on this subject. He stated:

> On average, greater diversity leads to greater productivity in plant communities, greater nutrient retention in ecosystems and greater ecosystem stability. For instance, grassland field experiments both in North America and across eight different European sites, ranging from Greece in the south and east to Portugal and Ireland in the west and Sweden in the north, have shown that each halving of the number of plant species within a plot leads to a 10–20% loss of productivity. An average plot containing one plant species is less than half as productive as an average plot containing 24–32 species. Lower plant diversity also leads to greater rates of loss of limiting soil nutrients through leaching, which ultimately should decrease soil fertility, further lowering plant productivity.

Another way Mother Nature builds resilience is by being very federal in how she self-organizes. She nests her communities—which are analogous to states, counties, and towns—within a flexible framework that makes the whole more than the sum of its parts. That is, she's built on trillions and trillions of small-scale networks, starting with microorganisms and building into bigger and bigger ecosystems. But each one is a little community, naturally adapting and evolving in order to survive and thrive.

"From microbe to top predator, ecosystems are a community, and they operate like one," added Lovejoy. And when you have trillions of small-scale networks woven together into ecosystems, the overall system is very hard to break. It's resilient. As Michael Stone puts it in the Center for Ecoliteracy's ecological principles handbook: "All living things in an ecosystem are interconnected through networks of relationships. They depend on this web of life to survive. For example: In a garden, a network of pollinators promotes genetic diversity; plants, in turn, provide nectar and pollen to the pollinators. Nature is made up of systems that are nested within systems. Each individual system is an integrated whole and—at the same time—part of larger systems." Life, he added, "did not take over the planet by combat but by networking"—one ecosystem to the next.

Mother Nature, in her own way, appreciates the power of ownership—and the virtues of belonging to a place. To be sure, natural systems have no owners, no self-interested managers per se, the way many human systems do. There is no Lion King in nature. Humans created the concept of one species managing the entire system for the collective interest—the idea of "dominion." That said, though, species coevolve with the places and niches best suited for them; each healthy ecosystem has a unique ecological balance of plants, animals, microorganisms, and the underlying processes and "plumbing" that connect them. That ever-evolving combination is what makes each ecosystem unique. And the unique set of species of plants and animals that evolve there is said to be *of the place, not just in it.* They are at home there, they are rooted, they fit, they belong, because they are in balance—and that balance produces enormous resilience. In that sense they "own" that place. When every niche is being filled by a plant or animal adapted to that niche, it's harder for any single invasive species to break in and disrupt the whole system—one alien or destructive element can't pull the whole thing down.

Still, that ecosystem and its balance have to be reproduced and defended every day; species rise and fall, and compete with one another, every second. Which is another of Mother Nature's killer apps—she never confuses stability with stasis. She understands that stability is produced by endless acts of dynamism. She would tell us that there is nothing static about stability. In nature a system that looks stable and seems to be in equilibrium is not static. A system that looks static and is static is a system that's about to die. Mother Nature knows that to remain stable you have to be open to constant change, and no plant or animal can take its position in the system for granted—just as a durable economy, says the University of Maryland's Herman Daly, is macro-stable but micro-variable.

"The most resilient ecosystems and countries," noted Glenn Prickett, chief external affairs officer for the Nature Conservancy, "are those that are able to absorb many alien influences and incorporate them into their system, while maintaining its overall stability." Think of the United States or India or Singapore.

Still another of Mother Nature's killer apps in producing resilient ecosystems is that she is very sustainable—through a highly complex circular system of food, eat, poop, seed, plant, grow, eat, food, poop, seed, plant, grow . . . Nothing is wasted. Everything cycles, a world without end.

Mother Nature also believes in bankruptcy. She believes individual plants and animals must be allowed to fail for the whole ecosystem to succeed. She has no mercy for her mistakes, for the weak, or for those who can't adapt to get their seeds, their DNA, into the next generation. Allowing the weak to die off unlocks more resources and energy for the strong. What markets do with bankruptcy laws, Mother Nature does with forest fires. "Nature kills off her failures to make room for her successes," wrote Edward Clodd, the English banker and anthropologist, in his 1897 book *Pioneers of Evolution from Thales to Huxley.* "The unadapted become extinct" and "only the adapted survive." From the ashes rises new life.

Mother Nature believes in the vital importance of topsoil—the top layer of soil in which all plants and trees sink their roots and derive their primary nutrients to grow into the world. Think about our planet. It is really just one big rock covered by an incredibly thin layer composed of the subsoil and topsoil. "The most basic thing that sustains any ecosystem is topsoil," notes the energy engineer Hal Harvey, founder of Energy Innovation. "And the first thing you learn about topsoil is that in most places it is really thin and can easily get washed away. It is just this sliver-like black layer coating the earth," covering a thousand miles below of lifeless, inhospitable rock. Topsoil on average is usually no more than six to ten inches deep. "And yet the ecosystem that emerges from topsoil is so rich, so plentiful, it's able to sustain this huge diversity of plant and animal life," observes Harvey. Conversely, as Jared Diamond and earlier historians chronicled, almost all failed civilizations collapsed because they didn't steward their topsoil.

Mother Nature believes in the virtue of patience. She knows that nothing strong comes from rushing. She's fine with being late. She is resilient precisely because she builds her ecosystems slowly, patiently. She knows that you can't rush the four seasons and condense them into two. Just as you can't speed up the gestation period for a baby elephant or a baby ant, you cannot force the resilient baobab tree to live for three thousand years by rushing its growth.

Finally, Mother Nature, because she practices all of the above strategies for building resilience, understands the virtues of what Dov Seidman calls "healthy interdependencies" versus "unhealthy interdependencies." In systems with healthy interdependencies, explains Seidman, "all the component parts rise together. In an interdependent system that is unhealthy, they all fall together."

What does healthy interdependence look like? It looks like all of Mother Nature's killer apps working together at once—adaptability, diversity, entrepreneurship, ownership, sustainability, bankruptcy, federalism, patience, and topsoil. In political terms, the United States and Canada have a healthy interdependency—they have risen together; Russia and Ukraine today have an unhealthy interdependency—they have fallen together.

I asked Russ Mittermeier of Conservation International to give me his most vivid example of healthy interdependencies in nature enabling an entire ecosystem to rise together. He offered up the ecosystem around spider monkeys and woolly monkeys in the tropical rain forests of Central and South America.

These primates survive, he explained, largely by eating fruits that grow on hardwood trees. Mother Nature, through evolution, learned to brightly color the shells to make them easy to find and attractive to frugivores. The monkeys crack them open and inside they find the seed, which is covered in aril, a sweet, sugar-rich layer that nature generated as monkey bait and bird bait. The monkeys don't have the time or dexterity to just suck off the aril, so they pop the whole seed in their mouth, enjoy it, and digest the sweet part and then pass the rest through their guts. (Some seeds actually won't germinate unless they have passed through the gut of such animals, whose bacteria secrete enzymes that crack the seed coat.) A few hours later they excrete out the seeds, nicely covered with their feces, which serves as fertilizer when the seed hits the rain forest floor. Those seeds eventually grow into what? More dense hardwood trees, so the monkeys are, in effect, creating gardens for their favorite food. But hardwood trees are also one of nature's most efficient tools for vacuuming up carbon out of the air and sequestering it. "Large birds, such as toucans, curassows, and guans, and even forest-dwelling tortoises play a similar role to the monkeys in eating and dispersing the seeds of the hardwood trees," Mittermeier explained.

But this resilient interdependent ecosystem can easily become unhealthy. Many of these same species that keep the tropical forest in a healthy interdependency—the spider monkeys and woolly monkeys, tortoises and toucans—"are often the most heavily hunted animals, to the point that they have been exterminated from many otherwise intact forests," noted Mittermeier. So then what happens? Kill too many spider monkeys, tortoises, and toucans and before you know it you've lost your seed dispersers and you end up with fewer hardwood trees, so you have a less dense forest and less carbon sequestration. And then before you know it you have more global warming and, in a few decades, you end up with a few additional inches (or, in time, many feet) of sea level outside your beach house. In nature, everything is connected to everything else—in either a healthy interdependence or an unhealthy one.

While there is much we humans can learn from Mother Nature, "one should never idealize nature," argued Mittermeier. "Nature is brutal. It is a system of conflict, stresses, and adaptation, where different species of plants and animals are beating the hell out of each other 24/7/365 in a dynamic struggle to reproduce themselves. The very engine of nature is each plant and animal's drive for reproductive success and its ability to adapt in ways that best enable them to produce offspring or seeds that go into the next generation"—all while other species try to eat them or oust them so they can procreate instead.

When you have a highly diverse system of plants and animals all striving to reproduce their genes at once, it may not be healthy or resilient for any one species or seed that gets eaten each day. Even so, the overall symphony when in balance can be very healthy and resilient—healthy in the sense that its parts thrive together and healthy in the sense that the whole is more resilient to any sudden changes in climate or development that get thrown its way. And this resilience comes from the melding of competition with collaboration—different organisms don't just feed one another; they also cocreate conditions in which all can thrive together.

## *Culture and Politics*

So let's stop for a moment and review what all of this discussion about Mother Nature has to do with our own societies. The answer can be found in Megginson's dictum: "The civilization that is able to survive is the one that is able to adapt to the changing physical, social, political, moral, and spiritual environment in which it finds itself." It's my contention that in this age of accelerations, the countries, cultures, and political systems that will be the most adaptive will be those that consciously choose to mimic Mother Nature's killer apps for producing resilience and propulsion. But the key words here are "consciously choose." Mother Nature evolved her adaptive skills over billions of years unconsciously, with utter moral indifference. We humans cannot be so brutal, or morally indifferent, as we look to build our resilience—nor do we have millennia to figure out how to perfect these tools. We have to translate Mother Nature's killer apps into human politics deliberately, consciously, wherever possible consensually—and as quickly as possible.

And to start with I would focus on five of these killer apps that have immediate application to governing today: (1) the ability to adapt when confronted by strangers with superior economic and military might without being hobbled by humiliation; (2) the ability to embrace diversity; (3) the ability to assume ownership over the future and one's own problems; (4) the ability to get the balance right between the federal and the local—that is, to understand that a healthy society, like a healthy tropical forest, is a network of healthy ecosystems on top of ecosystems, each thriving on its own but nourished by the whole; and, maybe most important, (5) the ability to approach politics and problem-solving in the age of accelerations with a mind-set that is entrepreneurial, hybrid, and heterodox and nondogmatic—mixing and coevolving any ideas or ideologies that will create resilience and propulsion, no matter whose "side" they come from.

Of course, the speed at which any society embraces these strategies will always be a product of the interplay between politics, culture, and leadership. Culture shapes a society's political responses, and its leadership and politics, in turn, shape culture. What exactly is culture? I like this concise definition offered by BusinessDictionary.com: culture is the "pattern of responses discovered, developed, or invented during the group's history of handling problems which arise from interactions among its members, and between them and their environment. These responses are considered the correct way to perceive, feel, think, and act, and are passed on to the new members through immersion and teaching. Culture determines what is acceptable or unacceptable, important or unimportant, right or wrong, workable or unworkable."

One of the worst mistakes you can make as a reporter is to underestimate the power of culture in how societies respond to big changes. Another is to conclude that culture is immutable and can never change. Cultures can change, and they often do—sometimes under the raw pressure of events and the need to survive, and sometimes thanks to political choices engineered by leaders. The late senator Daniel Patrick Moynihan famously observed: "The central conservative truth is that it is culture, not politics, that determines the success of a society. The central liberal truth is that politics can change a culture and save it from itself."

That is why I also like the definition of leadership offered by a Harvard University expert on the subject, Ronald Heifetz, who says the role of a leader is "to help people face reality and to mobilize them to make change" as their environment changes to ensure the security and prosperity of their community. Since the age of accelerations involves a change in the physical, technological, and social environment for so many people, leadership today is about nurturing the right cultural attitudes and specific policy choices that best enable the mimicking of Mother Nature's killer apps.

The power of a visionary leader to help a society and culture navigate its way through big moments requiring adaptation is beautifully depicted in one of my all-time favorite movie scenes. The film *Invictus* tells the story of how Nelson Mandela, in his first term as president of South Africa, enlists the country's famed rugby team, the Springboks, on a mission to win the 1995 Rugby World Cup and, through that, to start the healing of that apartheid-torn land. The almost all-white Springboks had been a symbol of white domination, and blacks routinely rooted against them. When the post-apartheid, black-led South African sports committee moved to change the team's name and colors, President Mandela stopped them. He explained that part of making whites feel at home in a black-led South Africa was not uprooting all their cherished symbols.

"That is selfish thinking," Mandela, played by Morgan Freeman, says in the movie. "It does not serve the nation." Then, speaking of

South Africa's whites, Mandela adds, "We have to surprise them with compassion, with restraint and generosity."

I love that line: We have to surprise them. There is no better way to change a culture than having a leader who surprises supporters and opponents by rising above his history, his constituencies, and his pollsters, and just doing the right things for his country. Through his enlightened leadership, Mandela did a lot to change the culture of South Africa. He created a little more trust and healthier interdependencies between blacks and whites and, in doing so, made that country more resilient.

With Mandela's example in mind, let's revisit Mother Nature's five most important killer apps and consider why they are so relevant today.

## *Being Adaptive When Confronted with the Stranger; or, The Need to Change*

One of the key differentiators when it comes to the openness of a culture or political system to adaptation is how it responds to contact with strangers. Is your culture easily humiliated by how much it's been left behind and therefore likely to dig in its heels, or is it more inclined to swallow some pride and try to learn from the stranger? In an age when contact between strangers is happening more than ever, this is a critical issue. Why some leaders and cultures are more adaptive than others when faced with big changes in their environment is one of the great mysteries of life and history, but it is impossible to ignore the differences. All I know is that since becoming a reporter in 1978, I have spent a lot of my career covering the difference between peoples, societies, leaders, and cultures focused on learning from "the other"—to catch up after falling behind—and those who feel humiliated by "the other," by their contact with strangers, and lash out rather than engage in the hard work of adaptation. This theme has so permeated my reporting that I have been tempted to change my business card to read: "Thomas L. Friedman, *New York Times* Global Humiliation/Dignity Correspondent."

There is a simple but well-known golf story that carries a deep truth about how cultural dispositions shape attitudes toward adaptation. In the September 2012 issue of *Golf Digest*, Mark Long and Nick Seitz wrote a story called "Caddie Chatter," in which Long related the following story told by Tom Watson's longtime caddie Bruce Edwards. Edwards had caddied for Watson for many years, then briefly for Greg Norman, and then went back to Watson. Edwards described how differently Watson and Norman would react if they each hit a perfect drive down the middle of the fairway but it ended up in a divot: "Years ago I asked Bruce Edwards how it was being back with Tom Watson after a couple of years with Greg Norman. Back then Greg was the man, but they'd gone a couple of years without a win together. Bruce said, 'Let's say you're three under for the day but you drive it in a divot at 16. Norman would look at me and say, 'Bruce, can you believe my bad luck?' Tom would look at the ball, look at the divot and say, 'Bruce, watch this!'"

There are people who are constantly cursing their luck, and there are people who will play the ball as best they can from wherever it lies and see it as a challenge. They know that the one thing they can control is not the bounce of the ball but their own attitude toward hitting it. In that context self-confidence and optimism are powers unto themselves. There are cultures that, when faced with adversity or a major external challenge, tend to collectively say, "I am behind, what is wrong with me? Let me learn from the best to fix it." And they learn to adapt to change. And there are those that say, "I am behind, what did *you* do to me? It is your fault."

Adaptability without humiliation, for instance, certainly describes nineteenth-century Japan, a country that had done its best to have no contact with strangers and to seal out the rest of the world. Its economy and politics were dominated by feudal agriculture and a Confucian hierarchical social structure, and they were steadily declining. Merchants were the lowest social class, and trading with foreigners was actually forbidden except for limited contact with China and the Dutch. But then Japan had an unexpected encounter with a stranger—Commodore Matthew Perry—who burst in on July 8, 1853, demanding that Japan's ports be open to America for trade and insisting on better treatment for shipwrecked sailors. His demands were rebuffed, but Perry came back a year later with a bigger fleet and more firepower. He explained to the Japanese the virtues of trading with other countries, and eventually they signed the Treaty of Kanagawa on March 31, 1854, opening the Japanese market to foreign trade and ending two hundred years of near isolation. The encounter shocked the Japanese political elites, forcing them to realize just how far behind the United States and other Western nations Japan had fallen in military technology.

This realization set in motion an internal revolution that toppled the Tokugawa Shogunate, which had ruled Tokyo in the name of the emperor since 1603, and brought Emperor Meiji, and a coalition of reformers, in his place. They chose adaptation by learning from those who had defeated them. They launched a political, economic, and social transformation of Japan, based on the notion that if they wanted to be as strong as the West they had to break from their current cultural norms and make a wholesale adoption of Western science, technology, engineering, education, art, literature, and even clothing and architecture. It turned out to be more difficult than they thought, but the net result was that by the late nineteenth century Japan had built itself into a major industrial power with the heft to not only reverse the unequal economic treaties imposed on it by Western powers but actually defeat one of those powers—Russia—in a war in 1905. The Meiji Restoration made Japan not only more resilient but also more powerful.

Alas, not every culture is able to deal with contact with strangers by swallowing its pride the way the Japanese did and vacuuming up everything they can learn from the stranger as fast as they can.

The Chinese actually had a phrase, "the century of humiliation," that they used to describe the years stretching from the 1840s, when China first tasted British imperialism, to its invasion by Japan and further debacles. As *The Economist* noted in an August 23, 2014, story about China: "For centuries China lay at the center of things, the sun around which other Asian kingdoms turned. First Western ravages in the middle of the 19th century and then China's defeat by Japan at the end of it put paid to Chinese centrality." But after opening to the world in the 1970s, China used its history to energize its future. Particularly under the leadership of Deng Xiaoping, it acknowledged it was in a divot, and reached out to the world to learn everything it could to adapt and catch up and reestablish its greatness.

By contrast, Russia let its humiliation get the best of it after the collapse of the Soviet Union, which President Putin once described as the "greatest tragedy of the twentieth century." Lawrence E. Harrison, writing in a collection he coedited, *Culture Matters in Russia—and Everywhere*, noted:

> [The] collapse of communism has left Russia humiliated—it has lost its great power status and is on the sidelines watching as former ally and competitor China moves toward that status. Russia's export profile looks like that of a Third World country, with the lion's share of exports dependent on natural resource endowment, above all petroleum and natural gas. The country that beat the United States into space has been unable to produce an automobile of export quality—not to mention its comparable shortfalls in information technology.
>
> During this period of national humiliation, one can well understand why the Russian leadership expressed intense concern over the relatively poor performance of Russian athletes in the 2010 Winter Olympics in Vancouver and the 2012 Summer Olympics in London.

But even to this day, Putin continues to look for dignity for Russia in all the wrong places—such as harassing Ukraine or diving into the Syrian civil war—rather than truly tapping and unleashing the greatness and talents of his own people.

Some Arab and Muslim nations and terrorist groups have clearly fallen into a "Who did this to us?" mind-set. Asra Q. Nomani is a former *Wall Street Journal* reporter and an Indian-born Muslim who worked with Daniel Pearl before Pearl was murdered in Pakistan. On June 20, 2012, she testified to the U.S. House of Representatives Committee on Homeland Security, on the subject "The American Muslim Response to Hearings on Radicalization Within Their Community":

> In 2005, Joe Navarro, a former FBI special agent, coined the concept of terrorists as "wound collectors" in a book, *Hunting Terrorism: A Look at the Psychopathology of Terror*, which incorporated years of experience analyzing terrorists worldwide from Spain to today's Islamic movements. He wrote that "terrorists are perennial wound collectors," bringing up "events from decades and even centuries past." He noted: "Their recollection of these events is as meaningful and painful today as when they originally took place. For them there is no statute of limitations on suffering. Wound collection to a great extent is

> driven by their fears and their paranoia, which coalesces nicely with their uncompromising ideology. Wound collecting serves a purpose, to support and vindicate, keeping all past events fresh, thus magnifying their significance into the present, a rabid rationalization for fears and anxieties within."
>
> To me, this phenomenon extends to the larger Muslim community, where there are wounds expressed in living room debates that earn many Muslims status as "couch jihadis," as one U.S. law enforcement official referred to them in conversation with me. I grew up eavesdropping on these "couch jihadis" in the men's sections of our dinner parties. Indeed, Mr. Navarro told me, "Collecting wounds becomes cultural" for communities worldwide. Clearly, knowing a community's wounds is important to understanding its history, Mr. Navarro said, but he noted, "The beauty of extremism is that it doesn't allow forgiveness."

I have covered a lot of wound collectors in the Middle East, but, again, it is not universal. The same Arab Muslim world that produced Nasser and bin Laden, so consumed with lashing out to overcome their humiliation, also produced dignity in Tunisia under Habib Bourguiba and Dubai under Sheikh Mohamed bin Rashid al Maktoum, who chose instead to dig deep, embrace change, learn from the other, and build out. The very same Latin America that produced the dictator Hugo Chávez in Venezuela produced the dynamic democratic president Ernesto Zedillo in Mexico. The very same Russia that produced Putin produced Mikhail Gorbachev, with his relatively more liberal vision for his country. The very same Southeast Asia that produced the genocidal Pol Pot in Cambodia produced the builder Lee Kuan Yew in Singapore.

## *Embracing Diversity*

As for embracing diversity, it is more vital than ever today for creating resilience in a changing environment. Thanks to diversity, no matter what climate changes affect your environment, some organism or ensemble of organisms will know how to deal with it. When you have such a pluralistic system, adds Amory Lovins, "it automatically adapts to turn every form of adversity into a manageable problem, if not something advantageous." (He's paraphrasing his late mentor Edwin Land, who said, "A failure is a circumstance not yet fully turned to your advantage.")

Because "pluralism is not diversity alone, but the energetic engagement with diversity," explains the Pluralism Project at Harvard on its website, "mere diversity without real encounter and relationship will yield increasing tensions in our societies." A society being "pluralistic" is a reality (see Syria and Iraq). A society with pluralism "is an achievement" (see America). Pluralism, the Harvard Project also notes, "does not require us to leave our identities and our commitments behind . . . It means holding our deepest differences, even our religious differences, not in isolation, but in relationship to one another." And it posits that real pluralism is built on "dialogue" and "give and take, criticism and self-criticism"—and "dialogue means both speaking and listening."

Being able to embrace and nurture this kind of true pluralism is a huge asset for a society in the age of accelerations—and a huge liability if you cannot, for a host of reasons. Indeed, I would go a step further and say that the ROI—return on investment—on pluralism in the age of accelerations will soar and become maybe the single most important competitive advantage for a society—for both economic and political reasons.

Politically, pluralistic societies that also have pluralism enjoy much greater political stability. They have a much greater ability to forge social contracts between equal citizens to live together equally, rather than have to rely on an iron-fisted autocrat keeping everyone in line from the top down. In a world where all top-down command-and-control systems are weakening, the only way to maintain order is through social contracts forged by diverse constituencies from the bottom up. Syria, Libya, Iraq, Afghanistan, and Nigeria, for instance, are all stories of pluralistic societies today that lack pluralism and are paying a huge price for that—now that their diversity can no longer just be controlled from above. A working melting pot that melds diverse citizens who can then do big, hard things together will be a huge advantage in the twenty-first century, when so many more people will be on the move.

At the same time, in the age of accelerations, societies that nurture pluralism—gender pluralism, pluralism of ideas, racial and ethnic pluralism—tend to be more innovative, everything else being equal. A pluralistic country that embraces pluralism has the potential to be much more innovative, because it can draw the best talent from anywhere in the world and mix together many more diverse perspectives; oftentimes the best ideas emerge from that combustion. Even countries that are not ethnically or religiously diverse—think Korea, Taiwan, Japan, and China—can enjoy the fruits of pluralism if they have a pluralistic outlook; that is, if they develop the habits of reaching out to the best ideas anywhere in the world to adapt and adopt them.

As the social scientist Richard Florida observed in a December 12, 2011, essay on this subject on CityLab.com:

> Economic growth and development has long been seen to turn on natural resources, technological innovation and human capital. But a growing number of studies, including my own research, suggest that geographic proximity and cultural diversity—a place's openness to different cultures, religions, sexual orientations—also play key roles in economic growth.
>
> Skeptics counter that diversity is an artifact of economic development rather than a contributor. They argue that diverse populations flock to certain locations because they are either rich already or are fast becoming that way.
>
> An important new study by economists Quamrul Ashraf of Williams College and Oded Galor of Brown University should help put many of the skeptics' claims to rest. "Cultural Diversity, Geographical Isolation and the Origin of the Wealth of Nations," recently released as a working paper by the National Bureau of Economic Research, charts the role of geographic isolation, proximity and cultural diversity on economic development from pre-industrial times to the modern era.
>
> It finds that "the interplay between cultural assimilation and cultural diffusion have played a significant role in giving rise to differential patterns of economic development across the globe." To put it in plain English: diversity spurs economic development and homogeneity slows it down . . .

> The evidence is mounting that geographical openness and cultural diversity and tolerance are not by-products but key drivers of economic progress.

P. V. Kannan is the cofounder of 24/7 Customer, which began as a call center operation in India and, since 2007, has expanded into a customer service/analytics company with a thousand clients around the world. I have watched his company grow from a start-up in Bangalore, with a lot of people answering phones, into a global big data services firm where high-paid data engineers are working on screens. When I asked him what his clients look like today, Kannan responded: "I go into a client in Sydney and their data expert is sitting in California and they are talking about their call centers in the Philippines and India and their top management is spread around the world—and even those in Sydney [all] come from different countries. The whole stereotype of all white men in one place is gone. If you are running a smart company today, it is filled by people from everywhere . . . Pluralism allows you to be fast and smart."

This is only going to become more true as Moore's law and the Market move deeper into the second half of the chessboard. Lovins argues:

> Let's say you have two genomes. And genome A has one gene that is perfectly adapted for today's system of cold and genome B has twenty genes, only one of which is expressed as resistance for cold. Genome A has one option to mutate the gene until it randomly hits on the solution to the problem or it dies. Genome B might have twenty offspring. Genome B has twenty potential answers. It will express or modulate each of those twenty, and there is a very good chance that one of them will be the right solution to the problem it faces.

One of the most valuable tutorials on the virtues of diversity I received came in 2014, when I took part in Showtime's *Years of Living Dangerously* documentary series on the impacts of climate change and environmental degradation around the world. My contribution was to look at how climate change and environmental destruction had affected Syria, Yemen, and Egypt. The interview I learned the most from, though, happened in Salina, Kansas, and it underscored the close parallel between monocultures and polycultures in nature and politics. Our film crew came out to America's wheat-growing farmland to illustrate how the drought that hit wheat farms in central Kansas in 2010 ended up raising bread prices in Egypt and, as we've seen, helping to fuel its revolution in early 2011. Our visit was built around an interview with Wes Jackson, founder and president of the Land Institute, an experimental farm where his team of bioscientists is trying to develop a perennial variety of wheat, called Kernza, that would not require annual tilling and planting. Jackson, a bioscientist and MacArthur Foundation "Genius" awardee, started the interview by giving me a tutorial on the prairie, which I wrote up as a column along the following lines:

The prairie, Jackson explained, was a diverse wilderness, with a complex ecosystem that naturally supported all kinds of wildlife, as well as American Indians—until the Europeans arrived, plowed it up, and covered it with single-species crop farms: monocultures, mostly wheat, corn, or soybeans. Annual monocultures are much more susceptible to disease and pests and require much more fossil fuel energy—plows, fertilizer, pesticides—to maintain resilience, because in a monoculture, one pest or disease can wipe out the whole field. They also exhaust the topsoil, which is so vital for life. Polycultures, by contrast, noted Jackson, provide species diversity, which provides chemical diversity, which provides much more natural resistance to disease and pests and "can substitute for the fossil fuels and chemicals that we've not evolved with." They also naturally maintain the topsoil. That is why during the Dust Bowl years of the thirties, Jackson noted, all the monoculture crops died but the remaining parts of the polyculture prairie, with its diverse ecosystem, survived. The polyculture prairie stores water, cycles nutrients, controls pests, and becomes ever more diverse, productive, beautiful, and adaptive.

As I listened to Jackson explain all of this, I responded: Isn't it interesting that Al Qaeda often says that if the Muslim world wants to restore its strength, it needs to go back to the "pure" days of Islam, when it was a monoculture in the Arabian Peninsula, unsullied by foreign influences? But in fact the golden age of the Arab-Muslim world was between the eighth and thirteenth centuries, when it became arguably the world's greatest polyculture, centered in Spain and North Africa. That was a period of great intellectual ferment in the Arab-Muslim world, which became the place to study science, math, astronomy, philosophy, and medicine. And what drove this intellectual ferment was the way Islamic scholars of the day bridged and integrated the very best teachings from a wide variety of civilizations, from China and India to Persia and Greece. It defined polyculture and made the Arab world incredibly wealthy, healthy, and resilient.

Unfortunately, today in the Middle East Al Qaeda and the Islamic State, using funding from the sale of fossil fuels and cash donated by Sunni fundamentalists from the Persian Gulf, are trying to purge Iraq, Yemen, Libya, and Syria of any religious or ethnic diversity. They are trying to plow up all the polycultures of the region—think of Baghdad, Aleppo, Palmyra, Tripoli, and Alexandria, once great melting pots of Jews, Christians, and Muslims; Greeks, Italians, Kurds, Turkmen, and Arabs—and turn them into monocultures, making these societies much less able to spark new ideas. Al Qaeda and ISIS are effectively trying to opt out of evolution to become a specialized closed system. To put it another way, diversity and tolerance were once native plants in the Middle East—the way the perennial polyculture prairie was in the Middle West—and it gave that region enormous resilience and healthy interdependencies with so many other civilizations. Al Qaeda and ISIS, using high-density fossil fuels, are trying to wipe out all that diversity and create a monoculture that is enormously susceptible to conspiracy theories and diseased ideas. This has left that region barren, weak and unhealthy for all its inhabitants.

I would argue the same thing happened to the Republican Party in America. The G.O.P. used to be an incredibly rich polyculture. It gave us ideas as diverse as our national parks (under Theodore Roosevelt), the Environmental Protection Agency and Clean Air and Clean Water Acts (under Richard Nixon), radical nuclear arms control and the Montreal Protocol to close the ozone hole (under Ronald Reagan), cap-and-trade to curb acid rain (under George H. W. Bush), and market-based health care reform (under Mitt Romney when he was governor of Massachusetts). And for decades the party itself was a pluralistic amalgam of northern liberal Republicans and southern and western conservatives. But in recent years the Tea Party and other hyperconservative forces, also funded in large part by fossil fuel companies and oil billionaires, have tried to wipe out the Republican Party's once rich polyculture and turn it into a monoculture that's enormously susceptible to diseased ideas: climate change is a hoax; evolution never happened; we don't need immigration reform. All of this weakened the G.O.P.'s foundation and opened the way for an invasive species such as Donald Trump to make deep inroads into its garden.

A 2012 study by the Kauffman Foundation revealed that immigrants founded one-quarter of U.S. technology start-up companies. The study, entitled "America's New Immigrant Entrepreneurs: Then and Now," shows that "24.3 percent of engineering and technology start-up companies have at least one immigrant founder serving in a key role," Reuters reported on October 2, 2012. "The study paid particular attention to Silicon Valley, where it analyzed 335 engineering and technology start-ups. It found 43.9 percent were founded by at least one immigrant. 'High-skilled immigrants will remain a critical asset for maintaining U.S. competitiveness in the global economy,' wrote the authors of the study."

It is not true just for America. As George Yeo, the veteran Singaporean cabinet minister, told a conference of the Lee Kuan Yew School of Public Policy in October 2014, Singapore's "ability to work in dense networks and be able to connect to different cultural domains, and to turn it into our own economic advantage," is its secret sauce. "Ultimately, what drives Singapore, what gives Singapore our special advantage, is the ability to arbitrage across cultures."

## *Ownership Cultures*

There is no perfect human analog to the way nature unconsciously evolves a sense of belonging in ecosystems, but there is a rough parallel—and that is promoting a culture of ownership in human societies, which always creates more resilience.

"Ownership is the one thing that fixes more things so other things can be made easier to fix," argues the education expert Stefanie Sanford of the College Board. More often than not, she says, when citizens feel a sense of ownership over their country, when teachers feel a sense of ownership over their classrooms, when students feel a sense of ownership over their education, more good things tend to happen than bad. You get outcomes that are much more internally generated and therefore self-sustaining. And where ownership doesn't exist, where people feel like renters or transients, more bad things tend to happen.

When someone assumes ownership, it is difficult to ask more of them than they ask of themselves. In education, "there is nothing I can do for you if you don't own it yourself first," argues Sanford. Andreas Schleicher, who runs the Programme for International Student Assessment exams, a global evaluation of scholastic performance, observed that those scoring highest are Asian countries that have "ownership cultures—a high degree of professional autonomy for teachers . . . where teachers get to participate in shaping standards and curriculum and have ample time for continuous professional development." They are not disengaged from the tools of their own craft, like a chef whose only job is to reheat someone else's cooking.

When you are an owner, you care, you pay attention, you build stewardship, and you think about the future. If you build a house for a quick flip, how strong will you build its foundation? People always tend to cut corners in a place where they won't actually be living. And that is why I have so often over the years quoted the dictum "In the history of the world, no one has ever washed a rented car." Ownership focuses you on long-term thinking over short-term, and on strategy over tactics.

I have spent a lot of time, in America and abroad, covering the struggles of different groups to assert ownership over their societies and the consequences of their lack of it. And I am forever struck at how quickly ownership can change behaviors and enable adaptation, self-propulsion, resilience, and healthy interdependencies.

In February 2011 I was in Cairo's Tahrir Square for the climactic toppling of Egypt's president, Hosni Mubarak. The uprising in Tahrir Square was all about the self-empowerment of a long-repressed people no longer willing to be afraid, no longer willing to be deprived of their freedom, and no longer willing to be humiliated by their own leaders, who told them for thirty years that they were not ready for democracy. Indeed, the Egyptian democracy movement was everything that Hosni Mubarak said it was not: homegrown, indefatigable, and authentically Egyptian. On February 9, I spent part of the morning in the square watching and photographing a group of young Egyptian students wearing plastic gloves taking garbage in both hands and neatly scooping it into black plastic bags to keep the area clean. For centuries Arabs have just been renting their countries from kings, dictators, and colonial powers. So they had no desire to wash them. Now they did. Nearby hung a sign that said: "Tahrir—the only free place in Egypt." So I went up to one of these young kids on garbage duty—Karim Turki, twenty-three, who worked in a skin-care shop—and asked him for my column: "Why did you volunteer for this?" He couldn't get the words out in broken English fast enough: "This is my earth. This is my country. This is my home. I will clean all Egypt when Mubarak will go out."

Three years later, in April 2014, I found myself in Kiev's Independence Square, known in Ukrainian as the Maidan, shortly after the uprising there against the country's corrupt leadership. The barricades of piled cobblestones, tires, wood beams, and burned cars erected by Ukrainian revolutionaries were still there. The whole scene looked like a Broadway set of *Les Misérables*. People were still laying fresh flowers at the makeshift shrines for the more than one hundred people killed there. My local guide explained to me that in winter, when the revolution broke out, the square and its sidewalks were usually covered with a layer of ice that the city never managed to effectively clear. But after the protesters took the square, elderly women came with little picks and shovels, chopped the ice, and kept it clean. They did it themselves. They did it for free—exactly like those young students in Tahrir Square.

Ownership is also self-propulsive, which makes it an important ingredient for resilience. In February 2015, I was invited to speak at the U.S. Coast Guard Academy in New London, Connecticut. I stayed over on the campus and the morning after my talk I was given a tour by LCDR Brooke Millard, who taught writing there but had previously commanded a coast guard cutter. Since she was considerably shorter than me, I couldn't help but ask her how she managed to command all those male crewmen, which—rank or no rank, formal authority or no formal authority—could not have been easy out there on the blue ocean. She thought about it for a couple of days and then wrote me this e-mail, which is as good a description of leading by extending ownership as I could find:

> Thanks for asking. At the unit I worked at previous to my command tour, I was in charge of about 10 chiefs—guys with about

> 18+ years, all subject matter experts in their specialty field. I was a brand new LT with four years in [and] 26 years old. When I told them to jump, I was expecting a "how high?" response, but instead I got [a very hostile] attitude. The first 6 months there were tough. I had to come up with a different leadership technique. I knew that children are often given two choices for food: "Do you want carrots or apples for a snack?"—both options of which mom approves of, but giving a child the option gives him/her a choice and [makes] them learn to own that decision. I tried a similar technique with my chiefs. I introduced a problem/issue, solicited their advice/ideas, and ultimately came up with two options for action; one was usually a better option than the other, and they naturally chose the option I liked best. But it appeared—to them, at least—that they had a choice—and, therefore, buy-in. It worked at the training command, so I applied the same technique with many decisions as a captain of a ship. At 29 years old, I led a crew of 17 men—and at least five of them were older than me. I think getting my command cadre's buy-in for major decisions helped—it helped them feel empowered and listened to, and it also helped me to weigh options and get the support I needed to carry out the decisions.

Millard shared ownership of both her ship's problems and its solutions and thereby leveraged the full energies of her crew, making the whole ship more resilient. As the Mumbai-based McKinsey management consultant Alok Kshirsagar once remarked to me, if you want to solve a big problem, "you need to go from taking credit to sharing credit to multiplying credit. The systems that all work, multiply credit." Multiplying credit is just another way of making everyone in the system feel ownership, and the by-product is both resilience and propulsion.

## *Getting Federalism Right*

In both nature and politics it is very important to get the balance right between individual ecosystems and the larger whole, so each can nurture the other. There is no hard-and-fast rule for doing this, but resilience comes from having the right balance at the right time. In

American politics today, in the age of accelerations, the balance between the federal, state, and local levels needs rebalancing, noted Will Marshall, president of the Progressive Policy Institute.

For much of the twentieth century, he said, the "arrow of history pointed to the centralization of political power and the nationalization of policy solutions" for addressing the big problems of the day. The primary tools of politics at that time were seen as "the emerging national bureaucracy and the administrative state," Marshall explained. That was quite logical in early twentieth-century America, "since state and local governments needed the weight of the national government to deal with new, monopolistic economic actors who could buy legislatures and overwhelm the puny power of states," let alone localities. And then came the Great Depression and its aftermath.

The New Deal, added Marshall, "expanded the scope of federal power dramatically, by launching huge public works and relief programs; regulating prices and wages; nationalizing income support and labor protections; establishing Social Security; and multiplying federal agencies staffed by a new breed of college-educated technocrats. Washington also replaced laissez-faire with Keynesian spending designed to manage the business cycle." And that nationalizing impulse intensified after World War II, he noted, "reaching its peak in LBJ's Great Society. This period of expansive liberalism saw the federal government assume responsibility for problems that had previously been left mainly to states and local authorities: racial injustice, poverty, illness, gender inequality, urban decay, educational inequity and pollution." Geopolitics also pushed things up to Washington, D.C., which had to finance and sustain a global Cold War competition with the Soviet Union. Plus, you had the need for real expertise from the federal government for solving new, complex industrial-age problems.

This was the broad and defining trend of American politics in the twentieth century that shaped many of the key planks of the "left" and "right" political agendas we know today—with the conservative right tending to be more sympathetic to the interests of owners and capital, always looking for more market-based solutions and less federal government regulation, and the liberal left tending toward more government-led solutions that promoted not just equal opportunities but equal outcomes, particularly for minorities and the poor.

The fact is, however, that the age of accelerations poses a different set of challenges and opportunities than the industrial age, and it requires a different balance between the center and periphery, the federal and the local. Today we need to reverse the centralization of power that we've seen over the past century in favor of decentralization. The national government has grown so big bureaucratically that it is way too slow to keep up with the change in the pace of change. Meanwhile, states and many localities have grown more flexible and capable—living on the edge of the iceberg, they feel every change in temperature and wind first; they need to react quickly, and now they can.

Many businesses are now global and quite dynamic; many cities now sponsor their own international trade missions and create their own consortiums of local businesses, educators, and philanthropies to upgrade their workforces. And thanks to local think tanks and universities participating in public policy, there is plenty of localized expertise available. Very often I meet mayors who have a much better grasp of the world, and the requirements for competitiveness, than their congressmen. Meanwhile, the federal government is in no position to supplement the fiscal deficiencies of states and cities alike, and it won't be for at least a generation, until the baby boomers die off; so localities will have to figure out for themselves how to generate the growth and incomes to sustain their own pension obligations.

More and more in the age of accelerations it strikes me as wise to push government down to the platform where trust is the highest—at the state and local level—because where trust is high people can do big hard things fast, because big hard things can only be done together.

"Annual Gallup polls continue to show that a majority of Americans trust their state governments (62%) and their local governments (71%) to handle problems," note economist Laura Tyson and consultant Lenny Mendonca, in a Jan. 6, 2017 essay on Project Syndicate. "A 2014 Pew study found that while only 25% of respondents were satisfied with the direction of national policy, 60% were satisfied with governance in their own communities. And the U.S. Constitution allows individual states to function as what Judge Brandeis called laboratories of democracy by experimenting with innovative policies without putting the rest of the country at risk."

This doesn't mean we can get along without a federal government. Hardly. We still need it to manage the national economy, national security, national health care, taxation, and social safety nets. "But we live in a different world," said Marshall. "Power today flows out of Washington. Urban America—centers of economic and social dysfunction a generation ago—have now become the nation's laboratories for public innovation."

Therefore the real question, argues Marshall, is how can the states and cities and towns—and the federal government—become better partners? The short answer is: wherever possible, the thrust of federal government should shift from offering solutions driven by the national bureaucracy to incentivizing, enabling, and inspiring experimentation and innovation from the local and individual level upward.

We will look at this issue more closely in the next two chapters, but for now suffice it to say that national and state leadership should be about enabling the compounding acceleration of local start-ups in both the economic sector and the social sector to build resilient and prospering citizens who have the skills and institutional support to keep pace with the age of accelerations.

## *Mother Nature's Political Party*

And that leads to the last of Mother Nature's killer apps that we need to consciously translate into politics in the age of accelerations. We need an entrepreneurial mind-set, a willingness to approach politics and problem-solving with an utterly hybrid, heterodox, and nondogmatic mixing and matching of ideas, without regard to traditional left-right catechisms—letting all kinds of ideas coevolve, just as plants and animals coevolve in nature.

Unfortunately, as noted above, that is not the mind-set of our two parties in America today. For now their mind-set is to double down on their old ideas—tax cuts, deregulation, and opposition to immigration for Republicans; and more social welfare, more support for teachers' unions, more regulation, more identity politics, and more redistribution from a very slow-growing pie for Democrats. For reasons of identity and fund-raising these two parties cannot let ideas that are best paired together actually be paired together; and for legacy reasons they cannot take out that blank sheet of white paper and think wholly anew about innovating around the great accelerations. We can do better—and not by splitting the difference between the two parties we have, but by rising above both and going beyond both, until they crack up and reconstitute themselves entirely around the challenges of facing three climate changes at once, and using Mother Nature as a mentor.

Throughout American history, with every leap forward in our economic platform, we have understood the need to adapt our education, tax, regulatory, and investment policies to get the most out of the new technologies and to cushion the worst. When we all worked on farms, we mandated universal primary education. When we moved to factories, it was universal secondary education. As we've moved into services, it was the G.I. Bill and Pell Grants so as many people as possible had access to post-secondary education. As we move into the age of accelerations, we need to adjust again. This era requires its own unique mix of education, tax, regulatory, and investment policies—in this case, policies that would translate Mother Nature's strategies for building resilience and propulsion when the climate changes into policies.

If Mother Nature had a political party—let's call it the "Making the Future Work for Everybody" party—here are some of the policies I think would be part of her platform. Mother Nature has no problem being to the left of the left and to the right of the right *at the same time*. Whatever should coevolve must coevolve. Here's what I mean:

1. She would favor a single-payer universal health care system funded by a progressive value-added consumption tax (except on groceries and other necessities). The level of this tax would be annually adjusted to the cost of health care, so citizens could feel the connection between the cost of health care and the VAT they pay at the store. If a single-payer system can work for Canada, Australia, and Sweden and provide generally better health outcomes at lower prices, it can work for us, and it would get U.S. companies out of the health care business and Medicare out of payroll taxes.

2. She would extend and expand the Earned Income Tax Credit (EITC) and the Child Tax Credit, which are essential trampolines to bounce people out of poverty, by topping up wages of low-income workers and thereby providing an incentive to work. (Both are set to expire in 2017.) Explaining how the credits work, the group Network Lobby for Catholic Social Justice noted, "For a couple with two children, the [EITC] credit rate is 40% of the first $13,090 in earnings, with a maxi-

mum credit of $5,236 if earnings reach $22,300. Over that amount the credit rate drops substantially until it reaches zero for taxpayers over $47,162 . . . The Child Tax Credit allows a nonrefundable credit against income taxes of $1,000 per qualifying child under age 17." Trampolines that incentivize work—and the dignity, discipline, and learning that come from work—are the best mechanisms for sustainably lifting families out of poverty. Some recent research also suggests that topping up the wages of low-income working parents through the EITC results in more lasting benefits for their kids—in terms of school success and college enrollment—than family support programs like prekindergarten or Head Start.

3. She would pair support for free trade agreements—the Trans-Pacific Partnership (TPP) with the United States and eleven Pacific Rim countries and the Transatlantic Trade and Investment Partnership (TTIP) between the United States and the European Union—with wage insurance for workers impacted by free trade. Economic research has now proved that the surge of imports into America after China was invited to join the World Trade Organization in 2001 hammered a specific set of American workers, while benefiting a much wider public with cheaper imported goods. Rather than close off trade with China or any other country, we need to expand trade, which benefits the economy as a whole, while finally getting serious about protecting those among us who specifically have been harmed by trade. David Autor, an economist at MIT, co-wrote the widely discussed February 2016 study "The China Shock: Learning from Labor Market Adjustment to Large Changes in Trade," which details the very real job-destroying impact of Chinese imports on certain American communities. He told *The Washington Post* on May 12, 2016, that it is quite possible for the overall American pie "to grow by 3 percent, and some slices to contract by 40 percent, and we've seen that. We still have lots of displaced people, lots of angry people."

That is not fair or sustainable. "Many workers displaced by trade and offshoring haven't been able to find new jobs that pay as much as the previous ones," noted the Brookings Institution policy expert Bill Galston in *The Wall Street Journal* on May 10, 2016:

> These workers and their families are asked to make do with incomes that can be 40% lower than they once enjoyed. This is

why the mostly ineffective Trade Adjustment Assistance program should be bolstered by adopting a system of wage insurance. Under this system, displaced workers would receive a wage supplement amounting to half the gap between their current and previous earnings, up to an annual maximum of $10,000. The supplement wouldn't be permanent, but because it is tied to employment and is more generous than traditional unemployment insurance, it would give workers an incentive to find a job as quickly as possible. That would minimize the negative effects of extended unemployment while shoring up growth in the U.S. labor force.

4. She would make all postsecondary education at an accredited offline or online university or technical school fully tax deductible. If every person is going to have to be a lifelong learner, we need a tax environment that makes that economically as easy as possible for everyone. Moreover, it will create jobs. The more people who are lifelong learners, the more people will become lifelong teachers. Anyone with an expertise in any subject—baking, plumbing, column writing—will be able to create apps or podcasts to teach their specialty.

To this end she would also encourage states to follow the lead of Tennessee, which in 2014 decided to make tuition and fees free for high school graduates who wanted to enroll in any state community college or technical school—on the condition that they maintain at least a C-average, stay in school for consecutive semesters, contribute eight hours of community service each semester, and meet with a volunteer coach/mentor who would help them stay on track to get their degree. Starting in 2018, Tennessee adults who don't already have a two-year degree will be able to go to any state community college and earn one free as well.

At the same time, she would make Common Core education standards the law of the land, to raise education benchmarks across the country, so high school graduates meet the higher skill levels that good jobs will increasingly demand. But those higher standards should be phased in with sufficient funding, so that every teacher can have the professional development time to learn the new curriculum those standards require and be able to buy the materials needed to teach it.

Mother Nature would also use her bully pulpit to urge every university to move to three-year undergraduate degrees from four. If universities in Europe, such as Oxford, or in Israel, such as the Technion, can instill enough learning into young people in three years to merit a BA or BS degree, American ones can do so as well and save their families 25 percent of the cost of a college degree and all that student debt.

Additionally, she would create extremely generous tax incentives for companies to create in-house lifelong learning opportunities for their employees—the way AT&T has. These programs would encourage workers to constantly upgrade their skills in a particular industry, while also giving them an easily accessible and affordable platform to do so.

5. She would roll back the 2005 "reform" of the bankruptcy laws, which has hurt start-ups by making it much costlier for entrepreneurs to declare bankruptcy and start again, especially those who used their credit cards for seed capital. As *Business Insider* reported on March 8, 2011:

> Growing evidence exists that bankruptcy reform is promoting fear amongst entrepreneurs, retarding the growth of new ventures, delaying economic recovery, and preventing new and small businesses from doing what they've always done best: creating jobs . . .
>
> A 2010 USC study established a direct link between changes in U.S. bankruptcy law and reduced entrepreneurial activity. Its authors concluded, "Many entrepreneurs go through several business models before they are successful . . . The harsh provisions of the new law appear to discourage some potential entrepreneurs from starting new businesses, and to keep entrepreneurs who have a business failure from starting anew."

6. On immigration, she would be for a very high wall with a very big gate. That is, we need to tighten security on the 1,945-mile-long U.S.–Mexico border, with both more physical fencing and virtual fencing with sensors, drones, and televisions. Americans need to believe that they live in a country where the borders are controlled. But they also need to understand that to thrive as a country we need a steady flow of legal immigration. Our ability as a country to embrace diversity is one of our greatest competitive advantages. We need to control low-skilled immigration so our own low-skilled workers are not priced out of jobs,

while removing all limits on H-1B visas for foreign high-skilled knowledge workers. We should also double the research funding for all of our national labs and institutes of health to drive basic research. Nothing would spin off more new good jobs and industries than that combination of more basic research and more knowledge workers.

7. To ensure that next-generation Internet services are developed in America, she would put in place new accelerated tax incentives and eliminate regulatory barriers to rapidly scale up the deployment of superfast bandwidth—for both wire line and wireless networks. Numerous studies show a direct correlation between the speed and scope of Internet access in a country and economic growth.

8. She would also borrow $50 billion at today's almost-zero interest rates to upgrade our ports, airports, and grids and to create jobs.

9. She would ban the manufacture and sale of all semiautomatic and other military-style guns and have the government offer to buy back any rifle or pistol in circulation. It won't solve the problem, but Australia proved that such programs can help reduce gun deaths.

10. To provide sufficient government revenues to pay for these investments, she would support major tax reform. For starters, she would eliminate entirely America's corporate income tax, now 35 percent, the highest in the world. The global average is in the low 20s. John Steele Gordon, author of *An Empire of Wealth: The Epic History of American Economic Power,* pointed out the many benefits this would have in a December 29, 2014, essay in *The Wall Street Journal*: We would get rid of the legions of lobbyists and accountants wasting time trying to game the corporate tax system; with the increased profits this tax relief would bring to companies, many would "increase both dividends and investment in plant and equipment, with very positive effects for the economy as a whole and increased revenue to the government through the personal income tax." At the same time, "stock prices, which are a function of perceived future earnings, would rise substantially, inducing a wealth effect as people see their 401(k)s and mutual funds rising in value. That would lead to increased spending and thus increased tax revenues . . . The distinction between for-profit and nonprofit corporations would disappear. So nonprofit corporations would not have to jump through hoops to qualify for that status" and "much of the $2 trillion of foreign earnings, now kept abroad to avoid being taxed when repatriated, would

flow into this country." Finally, America would go from the highest corporate income tax rate to the lowest, which would attract many more foreign corporate investors to the United States.

At the same time, she would embrace an idea President Obama contemplated in his first term—changing the inflation formula used to determine cost-of-living increases in Social Security checks to slow the annual growth in Social Security benefits and thus ensure the solvency of the system for future generations. Otherwise, she wouldn't touch Social Security. In an age of zero interest rates, retirees will need it more than ever.

To generate the tax revenue sufficient to replace corporate taxes and other government income streams, she would use a carbon tax, a small tax on all financial trades (stocks, bonds, and currency) and a tax on bullets—with offsets for the lowest-income earners. She would also take away the preferential tax treatment for dividend income and for capital gains and tax them at the normal rate for income. We need a tax system that specifically incentivizes the things we want—investment, work, and hiring—and shrinks the things we don't want: carbon emissions, corporate tax avoidance, overregulation, climate change, and gun violence. We simply can't afford them any longer.

Think about this: on January 1, 2013, the U.S. Senate resolved its fiscal-cliff negotiations by agreeing on a $600 billion tax hike—$60 billion a year for ten years. Just a few days earlier, on December 28, 2012, the Senate approved a $60.4 billion aid package to help New York and New Jersey recover from the devastation caused by one storm—Superstorm Sandy—that raked across the eastern United States in October 2012. In other words, we spent on one storm all the new additional tax revenue for that year.

11. She would require labeling on all sugary drinks, candies, and high-sugar-content fast foods, warning that excess consumption can cause diabetes and obesity—just as labels on cigarette packages warn that they cause cancer. On April 6, 2016, a study published in the respected journal *The Lancet* found that the global cost of diabetes is now $825 billion per year. The press release noted that "diabetes results in a person being unable to regulate levels of sugar in their blood, and increases the risk of heart and kidney disease, vision loss, and amputations . . . Using age-adjusted figures, they found that in the last 35 years, global diabetes

among men has more than doubled—from 4.3% in 1980 to 9% in 2014—after adjusting for the effect of aging. Meanwhile diabetes among women has risen from 5% in 1980 to 7.9% in 2014." It added that "the largest cost to individual countries [was] in China ($170 billion), the US ($105 billion), and India ($73 billion)." Mother Nature would not be for telling anyone what to eat, but she would be for making sure they are fully aware of the consequences of excess.

12. She would appoint an independent commission to review the Dodd-Frank financial reforms and the Sarbanes-Oxley accounting regulations to determine which—if any—of their provisions are needlessly making it harder for entrepreneurs to raise capital or start businesses. We need to be sure we're preventing recklessness—not risk-taking.

13. She would also create a Regulatory Improvement Commission, as proposed by the Progressive Policy Institute in a May 2013 policy paper. The PPI argues that "the natural accumulation of federal regulations over time imposes an unintended but significant cost to businesses and to economic growth. However, no effective process currently exists for retrospectively improving or removing regulations." Often agencies are asked to review their own regulations, and that rarely results in meaningful change. The RIC proposed by the Progressive Policy Institute would "be modeled after the successful Defense Base Closure and Realignment Commission. The Commission would consist of eight members appointed by the President and Congress, who, after a formal regulatory review, would submit a list of 15–20 regulatory changes to Congress for an up or down vote. Congressional approval would be required for the changes to take effect, but Congress would only be able to vote on the package as a whole without making any adjustments."

14. She would copy Great Britain and limit national political campaign spending and the length of the national campaign to a period of a few months. In a world that is getting this fast, we in America cannot afford to govern the country for one hundred days every four years and spend the rest of the time preparing for the midterms and presidential campaigns. That is insane.

15. She would encourage every state to end gerrymandering, by following California's move to a nonpartisan commission of retired jurists who draw up congressional districts in the most balanced way possible. If you have nonpartisan boundaries, you are much less likely to have

safe Republican or Democratic seats, so elections will be more competitive around the center and candidates will have to appeal to independent voters. In safe districts a Republican, most of the time, can lose only to another, more conservative Republican and a Democrat can lose only to a more liberal Democrat. The result is a Congress made up of more people from the far right or far left than the true disposition in the country. With more center-left Democrats and more center-right Republicans, it should be possible to build more legislative coalitions from the center out rather than from the extremes in.

She would also introduce ranked-choice voting in all Senate and House elections. In this system, instead of voting for just one candidate, you rank each candidate in order of preference. If no one gets a majority, the candidate with the least number of first-preference votes is eliminated. Then his or her votes are redistributed to those voters' second preferences, and that process continues until somebody has a majority. This allows voters to embrace alternatives and take a flyer on someone outside the box, who might be from, say, a third or fourth party. You can take a chance on someone because if that person loses, your vote will be redistributed to your second preference. "These systems encourage innovation and the entry of new alternatives," explained the Stanford University political scientist Larry Diamond. We should also eliminate the sore loser ban. In forty-five states in America, if you lose your party's primary you are not allowed—by law—to run in the general election. This prevents a moderate who might lose his party's primary—to someone from the far right or far left—from running in a general election in which he would have a much better chance when all voters go to the polls.

16. On national security, she would ensure that our intelligence services have all the legally monitored latitude they need to confront today's cyber-enabled terrorists—because if there's one more 9/11, many voters will be ready to throw out all civil liberties. And with the world cleaving into zones of "order" and "disorder," we'll need to project more power to protect the former and stabilize the latter.

In regard to the latter, she would elevate and expand the Peace Corps to be an equal branch of service with the Army, Navy, Air Force, Coast Guard, and Marine Corps, including with its own service academy. If the Army, Navy, Air Force, Coast Guard, and Marine Corps constitute our "defense," the Peace Corps would be our "offense." Its primary task would be to work at the village and neighborhood levels to help create economic opportunity and governance in the World of Disorder, thus helping more people to live decently in their home countries and not feel forced to flock to the World of Order.

17. She would condition all U.S. foreign aid to developing countries on their making progress on gender equality, and on access for every woman who wants it to family planning technology. As a global community, and environment, we simply can't afford the population explosions that, in combination with climate change, desertification, and civil strife, are making more and more swaths of the world uninhabitable. The welfare burden on the World of Order and the stress on the planet generally will become increasingly disruptive and unmanageable. Family planning and poverty alleviation and climate mitigation are policies that have to coevolve and not be treated as separate.

18. She would initiate three "races to the top" from the federal level—with prizes of $100 million, $75 million, and $50 million—to vastly accelerate innovations in social technologies: Which state can come up with the best platform for retraining workers? Which state can design a pilot city or community of the future where everything from self-driving vehicles and ubiquitous Wi-Fi to education, clean energy, affordable housing, health care, and green spaces is all integrated into a gigabit-enabled platform? Which city can come up with the best program for turning its public schools into sixteen-hour-a-day community centers, adult learning centers, and public health centers? We need to take advantage of the fact that we have fifty states and hundreds of cities able to experiment and hasten social innovation.

For instance, as part of this initiative, Mother Nature would have the Federal Government install 3-D printing labs in every public high school in America and mandate that they be open until 10 p.m. every evening so adults can use them after school. Who knows what new jobs or industries people will come up with if we truly unleashed our society's full creative and experimental talents with 3-D printers. We might end up with a whole new generation of artisans and manufacturers of products and devices we cannot currently imagine.

19. Finally, Mother Nature would never be politically correct. So her party would unabashedly promote community, citizenship, and the core values of all thriving modern societies—tolerance, pluralism, eco-

nomic inclusion, respect for women, freedom of speech, etc. To grow and sustain a healthy democratic system, there has to be a rich topsoil of shared core values that everyone is expected to buy into. Political scientist Andrew A. Michta put it well in an April 12, 2017, essay in *The American Interest*, entitled "The Deconstruction of the West."

"Over the past two decades Western elites have advocated (or conceded) a so-called 'multicultural policy,' whereby immigrants would no longer be asked to become citizens in the true sense of the Western liberal tradition. People who do not speak the national language, do not know the nation's history, and do not identify with its culture and traditions cannot help but remain visitors. The failure to acculturate immigrants into the liberal Western democracies is arguably at the core of the growing balkanization, and attendant instability, of Western nation-states, in Europe as well as in the United States."

A democratic country will not thrive today unless people see themselves as citizens with rights but also with responsibilities to uphold, embrace and enrich "the commons"—that is, public spaces, public institutions, and core liberal values. These are the ingredients of "civic capital" and communities and countries rich in civic capital—where people don't just live behind their own walls, and mix with their own tribes—will always be more resilient and propulsive.

In sum, in an age of extreme weather, extreme globalization, extremely rapid change in the job market, extreme income gaps, extreme population explosions in Africa that are destabilizing Europe, extreme deficits, extremely low interest rates, and extremely unfunded pension liabilities, we need to get extremely innovative in our politics and extremely willing to give all our basic institutions a makeover. We need a dynamic, hybrid politics that is unafraid to combine ideas from across the traditional political spectrum and also to go above and beyond it. I am talking about a politics that can strengthen work-based safety nets, to catch those for whom this world is becoming too fast. I am also talking about a politics that can unleash accelerated entrepreneurship, innovation, and growth to sustain those needed safety nets. And I am talking about a politics able to stimulate more of the social technologies and institutional, political, and regulatory reforms we need to keep up with all the changes in our physical technologies spurred by the age of accelerations. Finally, I am talking about a politics that understands that in today's world the big political divide "is not left versus right but open versus closed," as the pollster Craig Charney puts it, and that therefore chooses open—openness to trade, immigration, and global flows, as opposed to closing them off.

And, finally, I am talking about a politics more focused on creating tools and trampolines, not walls; on fostering workers, not jobs; on creating taxpayers, not just tax cuts; and on forging citizens and neighbors, not political tribes that live isolated one to the next.

If the traditional left- and right-wing parties in America and across the globe can adapt themselves to this new agenda—which requires a much more heterodox approach to politics—well and good. But my guess is that many will implode as the pressure for adaptation, for mimicking Mother Nature, and for building resilience and propulsion in this age of accelerations becomes too great for their rigid orthodoxies.

There is no quick fix for adapting to the age of accelerations—no single tax cut or hike, no single wall or exit from a trade deal, will get it done. If you want your country or community to thrive in the twenty-first century—and sustain the dream of a decent middle-class life for all of its citizens—you need to rethink everything about how we tax, educate, govern, and incentivize, and more often than not stimulate from the bottom up, from the community and state level. My list of fixes above is my crack at that challenge. I welcome others. But if you don't have a list of mutually reinforcing adaptations for the age of accelerations, you don't have an answer and you don't have a meaningful political party.

Since we started with Mother Nature's wisdom, let's end with the same: biological systems that thrive all have one thing in common, notes Amory Lovins: "They are all highly adaptive—and all the rest is detail."

ELEVEN

# *Is God in Cyberspace?*

***There has never ever been a time when the human being was capable of doing something and yet, eventually, that something did not happen. That means one of three things: 1) the human psyche is going to change fundamentally (good luck with that!); 2) the worldwide social contract changes so that the "angry men" can no longer be "empowered" (good luck with that too!); or 3) boom!***

***—Garrett Andrews, online comment on my October 21, 2015, column on NYTimes.com***

***Love does not win unless we start loving each other enough to fix our [expletive] problems.***

***—Comedian Samantha Bee, commenting on the Orlando massacre on her TBS show,* Full Frontal, *June 13, 2016***

I have been on the road selling different books ever since I published *From Beirut to Jerusalem* in 1989. I've given several hundred book talks to different audiences. So what's the best question I ever got from someone in the audience on any book? That's easy to answer. It was at an event at the Portland Theater, in Portland, Oregon, in 1999, when I was promoting *The Lexus and the Olive Tree*. A young man stood up in the balcony and asked me this question: "Is God in cyberspace?"

I confess, I didn't know how to answer his question, which was asked with the utmost sincerity and demanded an answer. After all, mankind had created a vast new realm for human interaction—called cyberspace—somewhere out there between Heaven, Earth, and Hell. So who is in charge there? Amazon.com? God on high? The Devil? The question seized me. So I called one of my most cherished spiritual mentors, Rabbi Tzvi Marx, a great Talmudic scholar whom I had gotten to know at the Shalom Hartman Institute in Jerusalem and who now lives in Amsterdam. I hoped to enlist his advice on how I should respond.

I thought Rabbi Marx's answer was so good that I slipped it into the paperback edition of *The Lexus and the Olive Tree* and then more or less forgot about it. But the more I worked on the conclusion of this book, the more I found myself reflecting on that question, as well as Rabbi Marx's answer. Indeed, I occasionally took the opportunity to pose the same question to religious leaders and others. When I asked the archbishop of Canterbury, Justin Welby, "Is God in cyberspace?" he joked at first that God must be in cyberspace because every time he is in the London subway, "I hear people saying into their cell phones, 'Oh God, why doesn't this work!'"

Here is how Rabbi Marx originally answered: He began by suggesting that whenever I get the question "Is God in cyberspace?" I should start by responding: "That depends what your view of God is." If your view of God is that he is, literally, the Almighty, and makes his presence felt through divine intervention—by smiting evil and rewarding good—then He sure as hell isn't in cyberspace, which is full of pornography, gambling, blogs and tweets trashing different people from every direction, pop and rap music with suggestive lyrics and four-letter words, not to mention all manner of hate speech and now cybercrime and recruitment by hate-filled groups such as ISIS. Indeed, it used to be said that the most oft-used three-letter words on the World Wide Web were "sex" and "MP3"—the once essential protocol for the free downloading of music—not "God."

Rabbi Marx added, though, that there is a Jewish postbiblical view of God. In the biblical view of God, He is always intervening. He is responsible for our actions. He punishes the bad and rewards the good. The postbiblical view of God is that we make God present by our own choices and our own decisions. In the postbiblical view of God, in the Jewish tradition, God is always hidden, whether in cyberspace or in the neighborhood shopping mall, and to have God in the room with you, whether it's a real room or a chat room, you have to bring Him there yourself by how you behave there, by the moral choices and mouse clicks you make.

Rabbi Marx pointed out to me that there is a verse in Isaiah that says, "You are my witness. I am the Lord," adding that second-century rabbinic commentators interpreted that verse to be saying, "If you are my witness, I am the Lord. And if you are not my witness, I am not the Lord." In other words, he explained, unless we bear witness to God's presence by our own good deeds, He is not present. Unless we behave as though He were running things, He isn't running things. In the postbiblical world we understand that from the first day of the world, God trusted man to make choices, when He entrusted Adam to make the right decision about which fruit to eat in the Garden of Eden. We are responsible for making God's presence manifest by what we do, by the choices we make. And the reason this issue is most acute in cyberspace is that no one else is in charge there. There is no place in today's world where you encounter the freedom to choose that God gave man more than in cyberspace. Cyberspace is where we are all connected and no one is in charge.

So, as I wrote in the paperback edition of *The Lexus and the Olive Tree*, I started telling anyone who asked "Is God in cyberspace?" that the answer is "no"—but He wants to be there. But only we can bring Him there by how we act there. God celebrates a universe with such human freedom because He knows that the only way He is truly manifest in the world is not if He intervenes but if we all choose sanctity and morality in an environment where we are free to choose anything. As Rabbi Marx put it, "In the postbiblical Jewish view of the world, you cannot be moral unless you are totally free. If you are not free, you are really not empowered, and if you are not empowered the choices that you make are not entirely your own. What God says about cyberspace is that you are really free there, and I hope you make the right choices, because if you do I will be present."

The late Israeli religious philosopher David Hartman added an important point: In some ways cyberspace resembles the world that the prophets spoke about, "a place where all mankind can be unified and be totally free." But, he went on, "the danger is that we are unifying mankind in cyberspace but without God"—actually, without any value system, without any filters, without true governance.

Indeed, there are no stoplights in cyberspace, no police walking the beat, no courts, no judges, no God who smites evil and rewards good, and certainly no "1-800-Please-Stop-Putin-From-Hacking-My-Election." If someone slimes you on Twitter or Facebook, well, unless it is a death threat, good luck getting it removed, especially if it is done anonymously, which in cyberspace is quite common.

And yet this realm is where we now spend increasing hours of our day. Cyberspace is where we do more of our shopping, more of our dating, more of our friendship-making and sustaining, more of our learning, more of our commerce, more of our teaching, more of our communicating, more of our news broadcasting and news seeking and news making, more of our selling of goods, services, and ideas. It's where America's president can communicate through Twitter directly with tens of millions of followers—without an editor, fact checker, libel lawyer, or any other filter. But what is even more frightening is that the leader of ISIS can do the same thing from his bunker in Syria or Iraq.

At the same time, as we have discussed earlier, the age of accelerations is also making men, women, and machines vastly more powerful and capable, in an interdependent world, of impacting large swaths of the globe all at once. And this, too, has moral implications.

Because when we weaken all top-down authority structures and strengthen bottom-up ones; when we create a world with not only superpowers but also super-empowered individuals; when we put so many distant strangers into proximity; when we accelerate the flow of ideas and innovation energy; when we give machines the power to think, alter DNA to remove diseases, and design plants and new materials; when Greeks not paying taxes can undermine bond markets and banks in both Bonn, Germany, and Germantown, Maryland; when a Kosovar hacker in Malaysia can break into the files of an American retailer and sell them to an Al Qaeda operative who can go on Twitter and threaten the U.S. servicemen whose identities were hacked; when all of this is happening at once, we've collectively created a world in which what every single person imagines, believes, and aspires to matters more than ever, because they can now act on their imaginations, beliefs, and aspirations so much faster, deeper, cheaper, and wider than ever before.

*To put it bluntly, we have created a world in which human beings have become more godlike than ever before. And we have created a world with vast new territories—called cyberspace—that are law-free, values-free, and, seemingly, God-free unlike ever before.*

Put those two trends together and you understand why I found more people in recent years asking me not only if God was in cyberspace but about values in general—how we can anchor more people in communities and contexts governed by values of decency, honesty, and mutual respect. In their own ways they were asking that we rethink ethics just as we have to rethink the workplace, politics, and geopolitics. In short, people are looking for moral innovation. And they are right to do so. If there was ever a time to pause for moral reflection, it is now. "Every technology is used before it is completely understood," Leon Wieseltier wrote in *The New York Times Book Review* on January 11, 2015. "There is always a lag between an innovation and the apprehension of its consequences. We are living in that lag, and it is the right time to keep our heads and reflect. We have much to gain and much to lose."

While it is obvious how much we are now living our lives in cyberspace, a realm that is God-free, how exactly are we becoming Godlike? Think about it: At 8:15 a.m. on August 6, 1945, when an American B-29 bomber dropped an atomic weapon on the Japanese city of Hiroshima, triggering the nuclear arms race that followed, we entered a world in which one country could kill all of us—could conceivably destroy the whole planet. If it had to be one country, I am glad that it was America.

But now we are entering a world where one person can kill all of us. It used to take a person to kill a person. Then one person could kill ten people. Then one person could kill thousands. Now we are approaching a world where it is possible to imagine a single person, or small group, being able to kill everyone. That used to take a country or an organization. Not anymore. How long before you read that ISIS has acquired 3-D printer technology and designs to put together a suitcase bomb with a little fissile material? How long before some terrorist or disturbed loner tries to get hold of a virus, like Ebola, and tries to turn it into a bioweapon? In March 2016, it was reported that ISIS militants were plotting to take a Belgian nuclear scientist hostage in order to get access to Belgium's nuclear research facility.

Today, "if you can imagine it, it will happen," argues Eric Leuthardt, the neuroscientist. "It is just a matter of how much it will cost. If you can imagine mass chaos or a mass solution to poverty or malaria, you can make it happen more [easily] than ever before." The scalability of individual behavior is both a problem and a solution today. "Individual behavior can now have global consequences. My behavior scales to the world now—and the world scales to me."

This applies to biology as well. "In the past only Mother Nature controlled the evolution of the species, and now man is inheriting that capability at scale," notes Craig Mundie. "We are beginning to manipulate the biology on which all life is based." For instance, today people are asking: Should we wipe out that species of mosquito that carries the Zika virus, because the technology exists to do that through computing and data collection? It's called a "gene drive." *MIT Technology Review* reported on February 8, 2016:

> A controversial genetic technology able to wipe out the mosquito carrying the Zika virus will be available within months, scientists say.
>
> The technology, called a "gene drive," was demonstrated only last year in yeast cells, fruit flies, and a species of mosquito that transmits malaria. It uses the gene-snipping technology CRISPR to force a genetic change to spread through a population as it reproduces.
>
> Three U.S. labs that handle mosquitoes, two in California and one in Virginia, say they are already working toward a gene drive for *Aedes aegypti,* the type of mosquito blamed for spreading Zika. If deployed, the technology could theoretically drive the species to extinction.

The supernova facilitates the use of synthetic biology to create organisms that did not exist before, it's imbuing existing ones with attributes they did not have before, and it's eliminating organisms that were problematic or nonproductive that Mother Nature herself evolved. All of that used to be Mother's Nature's work through natural selection. Soon, though, you'll be able to play this game at home.

On the most positive side of the ledger, we are approaching a world where, acting together, we could sustainably feed, clothe, and shelter every person, as well as cure virtually every disease, increase the free time of virtually every person, educate virtually every child, and enable virtually everyone to realize their full potential. The supernova is enabling so many more minds to work on solving all the world's great problems. "We are the first generation to have the people, ideas, and resources to solve all of our greatest challenges," argued Frank Fredericks, founder of World Faith, a global interfaith movement.

That is why I insist that as a species, we have never before stood at this moral fork in the road—*where one of us could kill all of us and all of us could fix everything if we really decided to do so.*

And that is why, properly exercising the powers that have been uniquely placed in the hands of our generation will require a degree of moral innovation that we have barely begun to explore, in America or globally, and a degree of grounding in ethics that most leaders lack.

"Maybe this is overly romantic, but I think leadership is going to require the ability to come to grips with values and ethics," remarked Jeffrey Garten, the former dean of the Yale School of Management:

> Education will need a strong dose of liberal arts. How will we think about privacy or genetic experimentation? These are areas where there's no international framework at all. In fact, there's barely a national framework. China has embarked on large-scale genetic engineering in certain animals. Where is that going? What should be the legal and ethical principles on which such activity should be based? And who has the wherewithal to even establish the right principles? How do you balance technological progress with this sense of humanity? You're not going to get that if you went to MIT and all you did was study nuclear physics. This is the supreme irony. The more technological we get, the more we need people who have a much broader framework. You'll be able to hire the technologist to make the systems work, but in terms of the goals, that takes a different kind of leader.

Amen.

## *I'll Have Some Beer with That ISIS Video*

That we are creating vast new ungoverned spaces—free from rules, laws, and the FBI, let alone God—is indisputable. Consider a couple of unusual news stories that broke in the last two years. The first concerned the exposure of the fact that YouTube was running commercial advertisements before videos posted by ISIS and other terrorist groups.

On March 3, 2015, CNNMoney.com reported: "Jennifer Aniston lauds the benefits of Aveeno, Bud Light shows off beer at a concert, and Secret sells its freshly scented deodorant. Pretty standard commercials, but what's different is the content that comes after. In this case, they're all followed by ISIS and jihadi videos."

When YouTube sells advertising slots to companies, the ads are automatically inserted by algorithms before a video plays. As CNNMoney noted, "Advertisers don't directly control where their ads are placed although they can specify the demographics they'd like to target." The story quoted the legal analyst Danny Cevallos as saying, "From a contract perspective, these corporations that are paying lots of money to get YouTube clicks may not be that pleased when they find out that their video is placed right before an ISIS recruitment video."

There probably aren't a lot of beer drinkers among ISIS followers. Maybe the algorithm detected that a lot of young men were coming to these sites and assumed there would be a lot of beer drinkers among them! However it happened, the advertisers were not aware or amused.

After reviewing one of the videos, a vice president of consumer connections at Anheuser-Busch told CNNMoney, "We were unaware that one of our ads ran in conjunction with this video." YouTube removed the ISIS-related video in the wake of the CNNMoney report.

The website Bustle.com picked up the story from there:

> The way that advertising works on YouTube is this: After the brand pays for a slot, the video site's algorithm will randomly place the ad before a video, but neither YouTube nor the company will know which video exactly unless they watch it. Even though companies can't request specific videos for their ads,

> they can request certain demographics to target. It's certainly a mystery, then, how ads for Bud Light, Toyota, and Swiffer ended up airing before videos produced by ISIS, because it's safe to assume none of these companies chose to target extremist militants between the ages of 18 and 55 who want to incite terror on the world.

Or consider this story from Sydney, Australia. On December 24, 2014, the mobile taxi-booking app Uber had to apologize for instituting surge pricing during a terrorist incident at a café, in which three people plus the gunmen were killed during a sixteen-hour siege. BBCNews.com reported that after a gunman took over the café and people started fleeing from the area by foot and by car, Uber's "surge pricing" algorithm "raised fares by as much as four times its normal rate."

> On the day of the Martin Place siege in Sydney, Uber came under heavy criticism on social media for raising its fares, so it started offering free rides out of the city.
>
> It also said it would refund the cost of the rides that had been affected by the higher fares . . .
>
> "We didn't stop surge pricing immediately. This was the wrong decision" [Uber said in a blog post] . . .
>
> The company said that its priority was to help as many people get out of the central business area safely, but that was "poorly" communicated, and led to a lot of misunderstanding about its motives.
>
> Uber has defended its surge pricing strategy in other cities, but reached an agreement with regulators in the U.S. to restrict the policy during national emergencies.

What all these stories have in common is that the algorithms were in charge—not people, not ethics, and certainly not God. What all of these stories also have in common is the fact that a number of technological forces came together to create an exponential step change in the power of men and machines—much faster than we have reshaped ourselves as human beings, much faster than we have been able to reshape our institutions, our laws, and our modes of leadership.

"We are letting technology do the work that human beings should never abdicate," argued Seidman. "Someone made the decision to let the YouTube algorithm put these commercials on these videos. But that was never a job of technology before." That was always the job of people. "Technology creates possibilities for new behaviors and experiences and connection," he added, "but it takes human beings to make the behaviors principled, the experiences meaningful and connections deeper and rooted in shared values and aspirations. Unfortunately, there is no Moore's law for human progress and moral development. That work is messy and there is no linear program for it. It goes up and down and zigs and zags. It is hard—but there is no other way."

This is especially challenging as cyberspace enters the home. Recall the November 2015 story from Cañon City, Colorado, where more than one hundred students in the local high school were caught trading nude photos and hiding them in secret photo-vault smartphone applications. After taking nude pictures of themselves and sharing them, the students used "ghost apps" on their cell phones to store and hide them. Ghost apps look like any normal application—one of the most popular is a calculator—so if your parents or a teacher get ahold of your phone, that is all they would see. But if you type in a secret code on the keypad, you are transported to a hidden page where you can store pornography, videos, and sexting messages. It sounds like something that Q might have installed on James Bond's cell phone ten years ago. Now every high school kid has it. Private Photo Vault is among the most downloaded photo and video apps on the Apple App Store. This is a technology designed to keep out parents, police, and anyone peddling sustainable values.

"In the old days a parent would catch their kid doing something bad, and what would they do? They would say, 'Go to your room,'" remarked Seidman. "As long as they knew where their kids were physically in the house, they could control them—so they sent them up to their room, where there was no TV." Now you send your kid to their room, he added, and they are still connected to the whole world with secret apps that Mom and Dad cannot penetrate—where it looks like they are calculating but are actually sexting.

You gave your kids a cell phone so you could better track them down after midnight or have them brought home from a party by Uber. But that Apple iPhone, rather than being just an extended leash, turns out to also be the key to a world of forbidden apples. So, "Go to your room" now has to be "Hand over your smartphone, your tablet, your iPod, your Apple Watch, your wireless card, and the code to your vault apps—and then go to your room."

## *All Flows Are Not Created Equal*

What is the point here? John Hagel, who explained earlier why flows have become so much more important than stocks, answers that question best: in this age of increasingly fast and voluminous digital flows, we must never forget that "all flows are not created equal."

As Hagel explains on his blog Edge Perspectives: right now, we're beginning to realize that all this connectivity enabling richer flows on a global scale also has its downside. Bad stuff, fake news, slander, computer viruses, hacking scandals, flash crashes in markets, rumors—all circulate farther, faster, deeper, and cheaper than ever before.

"It turns out that too much flow can make our systems more fragile," explains Hagel.

So, how do we take advantage of a world of flows without weakening our societies? Hagel answers with one word: "Friction—it requires friction—institutional arrangements and personal practices that tend to slow down flows and reduce the likelihood that these flows will cascade into the breakdown of systems . . . For example, in the case of trading markets, it might take the form of buffers that require some reflection or more analytics before acting upon information. In the case of individuals interacting with each other, productive friction might come from welcoming others with diverse perspectives and experiences as an opportunity to challenge our own beliefs and evolve to much more creative approaches than would be likely if we just interact with others who have similar perspectives. Sure, it might take us longer to come to some agreement or resolution, but the outcome would be much more innovative and help us to learn faster. The key to keeping friction productive is to foster mutual respect for all the participants even though we might disagree on the topics or challenges under discussion."

This book is not called *Thank You for Being Late* for nothing.

Hagel's lesson is particularly apposite for parents and educators. It is vital that we teach digital civics to young people, starting from kindergarten onward. The lesson can be very simple, but should be pounded into every young person: the Internet is an open sewer of untreated, unfiltered information, where they need to bring skepticism and critical thinking to everything they read and basic civic decency to everything they write. A November 22, 2016 study published by the Stanford Graduate School of Education found "a dismaying inability by students to reason about information they see on the Internet . . . Students, for example, had a hard time distinguishing advertisements from news articles or identifying where information came from . . . One assessment required middle schoolers to explain why they might not trust an article on financial planning that was written by a bank executive and sponsored by a bank. The researchers found that many students did not cite authorship or article sponsorship as key reasons for not believing the article."

Professor Sam Wineburg, the lead author of the report, said: "Many people assume that because young people are fluent in social media they are equally perceptive about what they find there. Our work shows the opposite to be true."

Therefore, teaching every child how to read the Internet, how to check with other sources whether what they read is true and anchoring them in a value system that promotes mutual-respect and tolerance matters more than ever. It also means—if we are now living so much of our lives in cyberspace, in this God-free realm—that our privacy laws have to be updated. We cannot give in to privacy absolutists who would argue that the police have a right to search your terrestrial bedroom with a proper warrant, but not your virtual one in cyberspace. Terrorists will have a field day with that way of thinking.

CNN.com reported on December 17, 2015, in the wake of the Paris jihadist suicide attacks that "investigators of the Paris attacks have found evidence they believe shows some of the terrorists used encrypted apps to hide plotting for the attacks . . . Among the apps officials found used by the terrorists were WhatsApp and Telegram, both of which boast of end-to-end encryption that protects the privacy of their users and are difficult to decrypt."

Consider, too, the April 2016 case, when the FBI demanded that Apple give it the keys to a cyberlocker in the iPhone used by Syed Rizwan

Farook, the gunman in the December 2, 2015, shooting in San Bernardino, California, that killed fourteen people. Apple refused to help the FBI, citing privacy concerns for the iPhone's users all over the world. The FBI eventually managed to break into the phone and extract the data by purchasing "a tool" from a third-party cybersecurity outfit that the then-FBI director, James Comey, would not identify. This arms race between the principles of privacy and the necessities of security has only just begun. It demands a serious rethinking by the U.S. Congress of how privacy in cyberspace should be governed and balanced against the rising impact of super-empowered angry men and women.

## *Time for Everyone to Go Back to Sunday School*

To be sure, there will always be evil in the world, there will always be criminality, there will always be swindlers who use the fruits of technological progress or the freedom of cyberspace to cheat the community or their neighbor or a stranger. To talk about how to better govern such realms is always, at best, to talk about increasing the odds of restraining more bad behaviors than not—because they will never be eliminated.

The first line of defense for any society is always going to be its guardrails—laws, stoplights, police, courts, surveillance, the FBI, and basic rules of decency for communities like Facebook, Twitter, and YouTube. All of those are necessary, but they are not sufficient for the age of accelerations. Clearly, what is also needed—and is in the power of every parent, school principal, college president, and spiritual leader—is to think more seriously and urgently about how we can inspire more of what Dov Seidman calls "sustainable values": honesty, humility, integrity, and mutual respect. These values generate trust, social bonds, and, above all, hope. This is opposed to what Seidman calls "situational values"—"just doing whatever the situation allows"—whether in the terrestrial realm or cyberspace. Sustainable values do "double duty," adds Seidman, whose company, LRN, advises global companies on how to improve their ethical performance. They animate behaviors that produce trust and healthy interdependencies and "they inspire hope and resilience—they keep us leaning in, in the face of people behaving badly."

When I think of this challenge on a global scale, my own short prescription is that we need to find a way to get more people to practice the Golden Rule. And it doesn't matter which version you were taught. It can be "Do unto others as you would wish they would do unto you," or its variant from the Babylonian Talmud, where the great Jewish teacher Rabbi Hillel famously said, "That which is despicable to you, do not do to your fellow. This is the whole Torah. The rest is commentary. Go and learn it." Or any other variant enshrined by your faith.

When one of us can kill all of us, when all of us can fix everything, and when more others can do unto you from farther away and you can do unto more others from farther away, the Golden Rule has never been more important and more in need of scaling.

What is so special about the Golden Rule is that while it is the simplest of all moral guides, "it produces the most complex of all behaviors—it's ever adaptive, it applies to every imaginable situation in a way that no rulebook ever could," argues Gautam Mukunda, a professor of organizational behavior at Harvard Business School. When the world is already complex, you don't want to make it more complicated. Make it simple. And no moral edict packs more punch simply than the Golden Rule—everything else really is commentary.

I know—to even talk about scaling the Golden Rule to more people in more situations sounds utterly unrealistic. But the simple truth is: If we can't get more people doing unto others as they would want others to do unto them, if we can't inspire more sustainable values, we will be "the first self-endangered species," argues Amory Lovins.

Is that realistic enough for you?

Changing what people believe is hard. Universal acceptance is not in the cards. Even broaching the notion sounds naïve today. But I will tell you what is really naïve: ignoring this challenge—this need for moral innovation—in this age of super-empowered angry men and women. Thinking that's automatically going to end well is the essence of naïveté, not to mention recklessness. For my money, naïveté is the new realism.

Nearing the end of his second term in office, President Obama gave voice to exactly this sentiment in the speech he delivered as the first American president to visit Hiroshima, on May 27, 2016: "Science allows us to communicate across the seas and fly above the clouds, to cure disease and understand the cosmos, but those same discoveries can be turned into ever more efficient killing machines," said Obama. "The wars of the modern age teach us this truth. Hiroshima teaches this truth. Technological progress without an equivalent progress in human institutions can doom us. The scientific revolution that led to the splitting of an atom requires a moral revolution as well."

Our calling today, Obama added, is "to see our growing interdependence as a cause for peaceful cooperation and not violent competition. To define our nations not by our capacity to destroy but by what we build. And perhaps, above all, we must reimagine our connection to one another as members of one human race."

I don't have the words to say it better—and it is not naïve. It's the essence of cold, hard realism today. I repeat: *naïveté is the new realism*—naïve is thinking that we are going to survive as a species in the age of accelerations without learning to govern our new realms in new ways and our old realms in new ways. And, yes, that is going to require some very rapid moral and social evolution.

Where to even begin?

## *The Martian*

One practical way to begin is to anchor as many people as possible in healthy communities. Beyond laws and guardrails, police and courts, there is no better source of restraint than a strong community. Africans didn't coin the phrase "It takes a village to raise a child" for nothing. Communities also do double duty. They create a sense of belonging that generates the trust that has to underlie the Golden Rule, and also the invisible restraints on those who would still think of crossing redlines.

I was in Israel on September 11, 2001, and interviewed Israeli intelligence experts the next morning about what they had learned about suicide bombers, having confronted so many in their fight with the Palestinians. I never forgot what they said. They said that while Israel, with its deeply embedded intelligence networks, could stop some of these bombers before they headed out of their home villages in the West Bank or Gaza and blew themselves up in a bus or restaurant, a few would always get through—unless the Palestinian village said "no," unless the village said that this is not martyrdom that we approve of but murder that we don't approve of.

In a healthy community people are not only looking out for each other; they are getting out of Facebook and into each other's faces. Healthy communities shame and mobilize against destructive and abusive behaviors. When family, community, and cultural and religious restraints are removed, or never present, suicide bombers can much more easily flourish.

Here is another story about the truck-driver terrorist in Nice who killed eighty-five people. It's from AFP:

> Neighbors of the man suspected to have killed scores of people in a truck attack on the Nice seafront described him Friday as a loner with no visible religious affiliation, as forensic experts searched his flat. AFP reporters interviewed about a dozen neighbours of the man, named by police as 31-year-old Franco-Tunisian Mohamed Lahouaiej-Bouhlel, whose identity papers were found in the truck. They portrayed him as a solitary figure who rarely spoke and did not even return greetings when their paths crossed in the four-storey block, located in a working-class neighborhood of Nice.

Hal Harvey, the environmental strategist, once remarked that "what keeps me up at night is the thought of some guy in a dark room eating pizza out of a delivery box, staring into a computer figuring out how to open the gates of the Hoover Dam—stuff you would only think of if you are morally and socially disconnected. It is much easier to break a dam than build one." In a world of super-empowered individuals we need to redouble our efforts to ensure that in as many ways as possible we are creating moral contexts and weaving healthy interdependencies that embrace the immigrant, the stranger, and the loner, and inspire more people in more places to want to make things rather than break things.

There is no restraint stronger than thinking your friends and family will hate or disrespect you for what you do—and that can be generated only by a community. "All over the country there are schools and organizations trying to come up with new ways to cultivate character," my colleague David Brooks noted in his November 27, 2015, column in *The New York Times*. "The ones I've seen that do it best, so far, are those that cultivate intense, thick community. Most of the time character is not an individual accomplishment. It emerges through joined hearts and souls, and in groups."

One way to reinforce and scale the character-building norms of healthy communities is by showing people the joys and the fruits that can come from joining hearts, souls, and hands—what happens when we don't just not do unto others but actually do with others in ways that are big and hard and make a difference.

For instance, I really loved the movie *The Martian*—but not only for the wonderful acting and the plot about a U.S. astronaut, played by Matt Damon, who gets marooned on Mars. My favorite scene was when NASA has to quickly assemble a rocket to ferry critical supplies to its stranded astronaut, but the rocket explodes shortly after takeoff because the time frame did not allow for proper inspections and preflight testing. As NASA scrambles for another solution (it takes a long time to build a rocket), the movie suddenly cuts—as *China Daily* noted in its September 12, 2015, review—to "the inner sanctum of the China National Space Agency. Two ranking officials . . . discuss what China could possibly do to help the hopeless situation, and how it might play out for China politically, diplomatically and financially if they did. They just so happen to have a rocket all ready to go, but because China's program is so secretive, no one else in the world knows about it, so if they don't offer to help, no one would be the wiser."

But the Chinese, in an unprompted act of international collaboration, decide to help save the U.S. astronaut on Mars from starving to death and offer their delivery rocket to get "the Martian" his desperately needed care package. We see the Chinese and American space experts collaborating to solve the problem, and at the end of the movie we see CNSA's leadership side by side with NASA's, rooting together—along with people all over the globe—for what becomes a successful rescue mission.

Alas, it could only happen in Hollywood. It was political science fiction because, as *China Daily* also noted, "since 2011, NASA has been banned by Congress from collaborating with China, because of human rights issues and national security concerns. The ban was slipped into the 2011 budget by then congressman Frank Wolf, a long-serving Republican from Virginia, who chaired the subcommittee overseeing NASA. 'We don't want to give [China] the opportunity to take advantage of our technology and we have nothing to gain from dealing with them,' he told *Science Insider* then."

But the author of the book and the producers of the film *The Martian* were on to something. That make-believe scene of international cooperation touched me, and I was not alone in feeling that. It was reported that in many theaters, audiences applauded at the ending, with its Hollywood depiction of international cooperation. The beauty of the film, though, was in how the director made it all look so normal, so logical, so right—leaving you to think: "Why don't we always behave that way? How much better off would we all be?"

The fact is, for our survival as a species, our very notion of "community" has to expand to the boundaries of the planet. That is a big statement, but it is true: if Mother Nature is treating us all as one, and if the power of one, the power of machines, and the power of flows can touch all of us at once, then we are a community whether we like it or not, whether we admit it or not. And if we are a global community, we have to start to act like one.

"Interdependency is a moral reality," explains Seidman. "It is a reality in which we rise and fall together; we affect each other profoundly from great distances in ways we never could before. In such a world, there is only one strategy to survive and thrive: it is to forge healthy, deep, and enduring interdependencies—in our relationships, in our communities, between businesses, between countries—so that we rise, and not fall, together. It's not complicated, but it's hard." Our motto in today's world, adds Tom Burke, the British environmentalist, should be: "It takes a planet to raise a child."

Why is that so hard? Because "the one big bug we have as humans is that we are tribal," answers Marina Gorbis, executive director at the Institute for the Future. "We always need the group to give us identity. We are wired that way. From the first campfire, human beings evolved as tribal beings."

And therein lies the challenge, and the need for moral innovation: in a much more interdependent world we have to redefine the tribe we are in—we have to enlarge the notion of community—precisely as

President Obama advocated in his Hiroshima speech: "What makes our species unique [is that] we're not bound by genetic code to repeat the mistakes of the past. We can learn. We can choose. We can tell our children a different story, one that describes a common humanity, one that makes war less likely and cruelty less easily accepted. The world was forever changed here, but today the children of this city will go through their day in peace."

Gorbis is right that we are wired to be tribal, but we are not hard-wired to view our tribe in the narrowest way possible. Unlike animals, we can adapt, and we can learn that in order to survive we have to widen the circle of the campfire. The opera star Carla Dirlikov Canales, thirty-six, is the product of a Mexican mother and a Bulgarian father and an upbringing in the state of Michigan. She has sung Carmen more than eighty times all around the world. We first met at an arts festival at the Kennedy Center, and she articulated this challenge in a compelling way that I had never heard before. She said that growing up in America as a non-WASP, she spent her life "checking the box labeled 'other.' It made me feel like I didn't belong in any box. It made me feel like an alien. And I didn't like that feeling. Because I think, as humans, we yearn to belong, and as I started to think about it on a broader level, I began to think that I am human, so I do belong. I belong to the 'all' box. We all belong to the 'all' box . . . and we all need to go from 'the other' to 'the all.'" At a time when America is becoming a "minority-majority" country, Canales has started her own little organization to widen the campfire—"to help others make the journey from the other to the all."

On May 3, 2016, National Public Radio's *Morning Edition* carried a story by its social science reporter Shankar Vedantam, who specializes in the unseen patterns in human behavior, about new research on the health benefits of dancing with others. "Psychology researchers at the University of Oxford," Vedantam explained, had "recently published a study in the journal *Evolution and Human Behavior*. They brought volunteers into a lab and taught them different dance moves. They placed the volunteers in groups of four on the dance floor and put headphones on them so they could hear music. Some of them were taught the same dance moves, and others were taught different dance moves. Before and after the volunteers danced to music, the researchers measured their pain threshold by squeezing their arms . . . with a blood pressure cuff."

What did they find? Said Vedantam:

> There were huge differences in pain perception before and after the volunteers danced together . . .
>
> When the volunteers were taught the same dance moves and heard the same songs as the others, their movements synchronized on the dance floor . . . Afterwards, these volunteers were able to withstand significantly more pain—their threshold for pain increased.
>
> By contrast, the volunteers who heard different songs, or were taught different dance moves to the same music, and didn't synchronize their movements, these volunteers experienced either no change in their pain perception or an increase in their pain perception, they actually felt more pain than they did before.

What's the explanation? What researchers think is going on, Vedantam said, is this:

> When experiences feel good that's usually a signal that they have served some kind of evolutionary purpose—so the brain evolved to find certain kinds of food tasty because eating those foods had survival value for our ancestors.
>
> As a social species, being part of a group has survival value. Evolution also may have adapted the brain to experience a sense of reward when we did things with and for other people—dancing together especially in synchrony can signal that you're actually simpatico with lots of other people. The researchers think this is why so many cultures have synchronized dancing and why it might have health benefits.

In an interview I did with the U.S. surgeon general, Vivek Murthy, he instinctively echoed that finding: "We have such a fascination with new medicines and new cures, but if you think about it, compassion and love are our oldest medicines and they have been around for millennia. When you practice medicine you learn very quickly how much they are a part of the healing process."

I have no illusions about how difficult scaling that kind of medicine is—or how many people will still be inclined to run away from the all and seek shelter as the other. The European Union was born out of a realization that after rivalries and tribal hatreds ignited two world wars, Europeans would be better off acting as a "common market." But that insight seems to be wearing off lately—see the British vote to exit the EU. And it's not only happening there. In the Middle East, a place that I have covered as a reporter/columnist my whole adult life, Israelis and Palestinians, Shiites and Sunnis, Iraqis and Iraqis, Syrians and Syrians are running, not walking, in the wrong direction. And, most sadly, many of them know it.

As I was completing this book, on May 2, 2016, *The New York Times* carried a story from Syria about the horrors of life there after five years of civil war. At the end, it quoted a mosque caretaker in Damascus, Salim al-Rifai, eighty-five, as saying that even the worst calamities did not last forever and "this, too, will pass." But before it could pass, Mr. Rifai added, his countrymen needed to change: "We need to believe in God and do what he asks of us," he said. "And we need to help each other to be human again."

When 250,000 people are killed in a civil war, roughly one-tenth of a country's population, it is safe to say that Syrians forgot how to be human in Syria. That is true of a lot of people in Iraq, Libya, Somalia, Yemen, the Congo, Rwanda, Ukraine, and Bosnia as well—way too many of them reached a point where they hated each other more than they loved their own children. That is what forgetting how to be human actually looks like. It means killing another person on the basis of their sect, their religion, or the hometown listed on their ID card or revealed by their accent, even knowing that it means sowing seeds of hatred that will burn the very ground under all their children's feet and their future. It is the opposite of building community.

There are countertrends worth noting. For instance, on April 22, 2016, Earth Day, world leaders from 175 countries signed the Paris climate agreement. While that agreement achieved the lowest common denominator of self-imposed emission constraints, it was impossible to ignore how high that lowest common denominator had become. Nothing as global as this long-sought accord to slow the dangerous rise of greenhouse gases had ever been reached before. Indeed, it may be this challenge from Mother Nature's acceleration that finally gets humanity to shift its thinking from the other to the all. There is no better example of the choice we now have to either destroy everyone or fix everything than the choice of whether we rise to the challenge of climate change or not, noted Hal Harvey. With the steady drop in the price of renewable energy and efficiency, "it now costs the same to destroy the climate or save it," said Harvey. "The price is basically the same, but at the micro scale there will be different winners and losers." Coal and oil companies and traditional utilities will lose out. Wind, solar, hydro, nuclear, and efficient and distributed energy purveyors will win. "At the macro scale, though, the whole world will win or the whole planet will lose. The impact will hit every generation going forward and will not respect national boundaries in the least."

It is our call. To repeat what President Obama said at Hiroshima: "We can tell our children a different story." And we must. And it is not naïve. It is strategic. I am not yet ready to go as far as Israeli historian Yuval Noah Harari, the author of *Homo Deus*, who predicted in a March 25, 2017, interview with *WorldPost*: "Human history began when men created gods. It will end when men become gods." I still believe we can make choices. And it is a job for everyone—parents and politicians, teachers and spiritual leaders, neighbors and friends. If you're looking for a story to start with, I can recommend the one told by the rabbi at my home synagogue, Kol Shalom, in Maryland, Jonathan Maltzman, as his opening sermon on the Jewish New Year in 2015. It went like this:

> A rabbi once asked his students: "How do we know when the night has ended and the day has begun?" The students thought they grasped the importance of this question. There are, after all, prayers and rites and rituals that can only be done at nighttime. And there are prayers and rites and rituals that belong only to the day. So, it is important to know how we can tell when night has ended and day has begun.
>
> So the first and brightest of the students offered an answer: "Rabbi, when I look out at the fields and I can distinguish between my field and the field of my neighbor, that's when the night has ended and the day has begun." A second student offered his answer: "Rabbi, when I look from the fields and I see a house, and I can tell that it's my house and not the house of

my neighbor, that's when the night has ended and the day has begun." A third student offered another answer: "Rabbi, when I see an animal in the distance, and I can tell what kind of animal it is, whether a cow or a horse or a sheep, that's when the night has ended and the day has begun." Then a fourth student offered yet another answer: "Rabbi, when I see a flower and I can make out the colors of the flower, whether they are red or yellow or blue, that's when night has ended and day has begun.

Each answer brought a sadder, more severe frown to the rabbi's face. Until finally he shouted, "No! None of you understands! You only divide! You divide your house from the house of your neighbor, your field from your neighbor's field, you distinguish one kind of animal from another, you separate one color from all the others. Is that all we can do—dividing, separating, splitting the world into pieces? Isn't the world broken enough? Isn't the world split into enough fragments? Is that what Torah is for? No, my dear students, it's not that way, not that way at all!"

The shocked students looked into the sad face of their rabbi. "Then, Rabbi, tell us: How do we know that night has ended and day has begun?"

The rabbi stared back into the faces of his students, and with a voice suddenly gentle and imploring, he responded: "When you look into the face of the person who is beside you, and you can see that person is your brother or your sister, then finally the night has ended and the day has begun."

Hastening that heavenly day is the moral work of our generation. I don't know where it ends, but I know where it has to start—by anchoring people in strong families and healthy communities. It is impossible to expect people to extend the Golden Rule very far if they are unmoored, unanchored, and insecure themselves. How to build strong families is beyond my skill set, but I know something about strong communities, because I grew up in one. And so I hope you'll indulge me if I end this journey by taking you back home with me to discuss the final kind of innovation we need to promote resilience and propulsion in this age of accelerations—innovation in the building of healthy communities.

TWELVE

# *Always Looking for Minnesota*

***Anyone who has grown up in the hills or used to sit by the spring to drink, or played outdoors in the neighborhood square; going back to these places is a chance to recover something of their true selves.***

***—Pope Francis's encyclical on climate change,***
***"Laudato Si'," May 24, 2015***

One afternoon in the fall of 2015, while I was writing this book, I was driving in my car and listening to SiriusXM Radio. On the folk music station the Coffee House, a song came on with a verse that directly spoke to me—so much so that I pulled off the road as soon as I could and wrote down the lyrics and the singer's name. The song was called "The Eye," and it's written by the country-folk singer Brandi Carlile and her bandmate Tim Hanseroth and sung by Carlile. I wish it could play every time you open these pages, like a Hallmark birthday card, because it's become the theme song of this book.

The main refrain is:

I wrapped your love around me like a chain
But I never was afraid that it would die
You can dance in a hurricane
But only if you're standing in the eye.

I hope that it is clear by now that every day going forward we're going to be asked to dance in a hurricane, set off by the accelerations in the Market, Mother Nature, and Moore's law. Some politicians propose to build a wall against this hurricane. That is a fool's errand. There is only one way to thrive now, and it's by finding and creating your own eye. The eye of a hurricane moves, along with the storm. It draws energy from it, while creating a sanctuary of stability inside it. It is both dynamic and stable—and so must we be. We can't escape these accelerations. We have to dive into them, take advantage of their energy and flows where possible, move with them, use them to learn faster, design smarter, and collaborate deeper—all so we can build our own eyes to anchor and propel ourselves and our families confidently forward.

The closest political analogue for the eye of a hurricane that I can think of is a healthy community. When people feel embedded in a community, they feel "protected, respected, and connected," as my friend Andy Karsner, whose father grew up in Duluth and mother in Casablanca, likes to say. And that feeling is more important than ever, because when people feel protected, respected, and connected in a healthy community, it generates enormous trust. And when there is more trust in the room, citizens are much more likely to mirror Mother Nature's killer apps. When people trust each other, they can be much more adaptable and open to all forms of pluralism. When people trust each other, they can think long-term. When there is trust in the room, people are more inclined to collaborate and experiment—to open themselves up to others, to new ideas, and to novel approaches—and to extending the Golden Rule. They also don't waste energy investigating every mistake; they feel free to fail and try again and fail again and try again.

"Collaboration moves at the speed of trust," argued Chris Thompson, who works with cities for the Fund for Our Economic Future, in an essay on its website. When people trust each other, they take ownership of problems and practice stewardship. The political scientist Francis Fukuyama, who wrote a classic book in 1996 on why the most successful states and societies exhibit high levels of trust—*Trust: The Social Virtues and the Creation of Prosperity*—noted that "social capital is a capability that arises from the prevalence of trust in a society or in certain parts of it. It can be embodied in the smallest and most basic social group, the family, as well as the largest of all groups, the nation, and in all the other groups in between." Where trust is prevalent, he explained, groups and societies can move and adapt quickly through many informal contracts. "By contrast, people who do not trust one another will end up cooperating only under a system of formal rules and regulations, which have to be negotiated, agreed to, litigated, and enforced, sometimes by coercive means," wrote Fukuyama.

It's for all these reasons that Dov Seidman argues that trust "is the only legal performance-enhancing drug." But trust cannot be commanded. It can only be nurtured and inspired by a healthy community—between people who feel bound by a social contract. "Trust is something that emerges from how people interact politically, for mutual benefit, through institutions," adds the Harvard University political philosopher Michael Sandel. "Healthy communities build civic muscles that lead to greater trust."

Indeed, the best explanation I ever heard for the emotional effect that trust has on a person or community came from U.S. surgeon general Murthy, who offered a beautiful analogy between the way trust breathes life into a community and the way our bodies pump oxygen into the heart:

> The heart pumps in two cycles—systole, when it contracts, and diastole, when it relaxes. And one of the things we often think is that contraction is the most important phase, because that is what gets the blood pushed out everywhere around your body. But you realize when you study medicine that it's in diastole—when the heart relaxes—that the coronary blood vessels fill and supply the heart muscle with the lifesaving, sustaining oxygen that it needs. So without diastole there can be no systole—without relaxation there can be no contraction.

In human relations, trust creates diastole. It is only when people relax their hearts and their minds that they are open to hear and engage with others, and healthy communities create the context for that.

Fortunately, America today is blessed with many healthy communities. It is why I often tell foreign visitors that if you want to be an optimist about America, "stand on your head," because our country looks so much better from the bottom up than from the top down. What has been saving us at a time when our national politics has become increasingly toxic and unable to produce the social technologies we need to keep pace with the accelerations in the Market and in Moore's law is the dynamism coming from our cities, towns, and communities, from the bottom up. They have given up waiting for Washington, D.C., to get its act together. Many of them are forging local public-private partnerships—involving businesses, educators, philanthropists, and governments—to put in place the tools they know their citizens and kids will need to dance in the hurricane.

And thank goodness—because the healthy city, town, or community is going to be the most important governing building block in the twenty-first century.

One of my teachers on this subject is Israel's Gidi Grinstein, president of the Reut research and strategy group, which has been focused on reimagining communities in Israel. In this age of accelerations, he argues, we need to "reinvent the basic organizing unit of society." Of course, as we discussed earlier, we still need federal and state governments to maintain the foundations of the national economy, and welfare, security, and health care. But more and more it is becoming clear, says Grinstein, that "the basic architecture of a resilient and prosperous twenty-first-century society must be a network of healthy communities."

At the other end, though, the single-family unit is too weak to stand alone in the face of the hurricane-force winds of change, especially since many families, particularly single-parent ones, are living so close to the edge—without savings, pensions, or homeownership. Just one health, auto, or employment crisis can derail them. At the same time, such families lack the time and financial resources essential for ensuring their own employability and productivity in an era that demands lifelong learning to sustain lifelong employment and income.

The ideal adaptive political unit for producing resilience and propulsion in the age of accelerations is the healthy community. That's because it is much easier to generate trust at the community level, and trust enhances flexibility and experimentation. Also, mayors and city council members have to be more accountable and responsive to voters—and less ideological than the typical U.S. congressman—because they see their voters every day, their voters see them, and both have to live with the immediate effects of what their city does right or wrong. It makes for much more pragmatic and responsive decision-making.

But just because you have a community doesn't mean it will be healthy. We have plenty of failing communities in America and around the world. What do all healthy communities have in common today? Broadly speaking, they have all created what I call *complex adaptive coalitions*. These adaptive coalitions are able to take advantage of trust to move fast and, as Grinstein puts it, "reinvent" all the core institutions of the community—from schools to community centers to city hall to public spaces—to enhance "the employability, productivity, inclusion, and quality of life of its members." These adaptive coalitions help to foster resilience and propulsion for their citizens at a time when more and more families need a local hand up to keep pace with the accelerating pace of change.

A community cannot be healthy, Grinstein explains, unless its citizens have the education and skills to be employable and productive their whole lives—and with people now living longer that means creating lifelong learning opportunities for everyone from ages 18 to 70. Today, a community also cannot be healthy unless it is inclusive—and that means inclusive socially of different races and religions, and the disabled and the elderly, but also inclusive economically, with some minimum sense of the shared prosperity that characterized the American middle class at its height. Therefore the public schools in these healthy communities, adds Grinstein, have to remain open after normal school hours to act as adult lifelong learning and day-care centers, serving children, parents, and the elderly, and hosting social service groups that can truly make sure no family or child is left behind.

And, finally, argues Grinstein, the healthy community creates and expands "public spaces"—parks, trails, transportation networks, swimming pools, baseball diamonds, hockey rinks, community gardens, basketball courts, and mixed-use gathering places for living and shopping and dining, as well as art and music opportunities. Not only do rich public spaces build inclusion and a sense of place so vital to a healthy community, they also enable families who cannot afford to join a private club or pay for private cultural opportunities for their kids to enjoy these things. In doing this, they enhance inclusion.

And so the healthiest communities today are the ones that are building complex adaptive coalitions to leverage their local assets precisely to create this kind of inclusive growth. What do the best complex adaptive coalitions have in common? What you always see is that the business community has become deeply involved in the public school system and local community college, injecting into them in real time the constantly evolving skill demands of the global economy. It doesn't wait for the public schools to try to guess what business needs by way of skills but partners with them directly. Local business also draws on the local community colleges and universities for talent and innovations. What you always see are local philanthropies and civic groups working alongside business and government to create supplemental learning opportunities and internships for both young people and adults to enhance their lifelong learning opportunities and chances to build new skills.

What you also always see in these coalitions is that they ask their unsung heroes to sing. That is, some of the most important actors in a healthy community, explains Grinstein, are "the leaders of key local institutions—the school principals, counselors and superintendents, the heads of libraries and community centers—they are not that sexy or hip, but when you involve them as change agents you get tremendous bang for your buck. They want to play a role and are usually highly ethical and can really pull a community forward. I call it 'extending the yoke.' The longer the yoke the more horses you can harness to pull your wagon"—and the faster and farther you can go.

What you also always see are the local mayors and city councils working to bring all these complex actors together into a coalition—always seeking to lengthen their yoke and making sure the horses pull in the same direction to create the total reinvention the community needs.

Finally, what you always see with the best adaptive coalitions is that they are constantly looking to leverage their local assets. That is, rather than hoping and praying that Ford Motor comes to their town with a 25,000-person factory, they start by cataloging their local assets—which may be climate, geography, a university, a coastline, ports, diversity, artisans, an arts culture—anything that cannot be easily outsourced or moved and are unique to them. And then they try to use technology and globalization to build on these assets and nurture local industries, tourism, manufacturing and artisanship. The best adaptive coalitions also hire development recruiters to seek out investors nationally and internationally to invest in expanding those local assets.

In short, the best adaptive coalitions start every day by asking, "What world am I living in? How do I enhance the employability and productivity of my citizens and the inclusive growth of my community in this world?" They are not waiting for the 1950s to come back.

When you make a community work, Grinstein says, "you can really affect the quality of life for the vast majority." And here is the good news: traveling around America today, you can find so many communities leveraging these kinds of assets in innovative ways. Indeed, this is true to such an extent that nothing new needs to be invented today—whatever you can think of, someone is already doing it. It is the opposite of what is happening in Washington, D.C. "The innovation needed to address the challenges facing our society is already sprouting among us bottom-up," concludes Grinstein. "It just needs to be highlighted, modeled, and scaled."

And the country that builds the broadest and richest network of healthy communities—each sharing ideas with the other—will be the country with the most resilience and propulsion to thrive amidst all the climate changes we're now going through.

Eric Beinhocker, executive director of the Institute for New Economic Thinking at Oxford, calls this focus on building resilience and propulsion from the bottom up through healthy communities the "new progressive localism." For too long, he argues, "progressives have been so focused on Washington they've missed the fact that most of the progress on the issues they care about—environment, education, economic opportunity, and workforce skills—has happened at the local level. Because that is where trust lives."

## *The St. Louis Park Story*

I know a lot about this subject because I saw a healthy community get built, brick by brick, block by block, neighbor by neighbor, close up. It was the one that I grew up in: St. Louis Park, Minnesota, a suburb of Minneapolis. And that's why I am going to conclude this book with two chapters about where I began—literally: in the midwestern community I called home from the mid-1950s to the early 1970s.

This is not an exercise in nostalgia. Returing to St. Louis Park is the appropriate way to close this book for two simple reasons. First, as I explained at the outset, a column has to combine three things: your own value set, how you think the Machine works, and what you have learned about how the Machine affects people and culture and vice versa. Well, my value set and my affinity for a politics that embraces inclusion, pluralism, and always trying to govern with Mother Nature's best ideas—a mix of center-left and center-right—was instilled in me by the community where I grew up. And second, because those values seem more relevant today than ever in America as a whole, and in the world at large. At a time of rising racial tensions and political debates tearing at the fabric of our country, I grew hungry to understand what made that little suburb where I came of age politically such a vibrant community, anchoring and propelling me and many others. I found myself hungry to reexamine whether the inclusive tapestry I saw woven while growing up there a half century ago was just something I dreamt, or was real. And I wanted to assess how well those civic engines were still working today—with a much more diverse community—and whether those lessons could be shared and scaled.

Hint: Yes, it was real. Yes, those engines are still working. Yes, the challenges are a lot harder now. And yes, the end of the story has yet to be written, but it matters now more than ever. Let me explain . . .

Long before I met Ayele Bojia in the Bethesda parking garage, I was aware that I brought a heterodox value set to column writing. I could put my core values on a bumper sticker, but I would need your whole bumper: I am a socially liberal, deeply patriotic, pluralism-loving, community-oriented, fiscally moderate, free-trade-inclined, innovation-obsessed environmentalist-capitalist. I believe that America at its best—and we're not always at our best—can deliver a life of decency, security, opportunity, and freedom for its own people, and can also be a bulwark of stability and a beacon of liberty and justice for people the world over. How did I come to this worldview? As I said, not by reading any particular philosophers. Rather, it emerged bit by bit from the neighborhood, the public schools, and the very soil of the community where I spent my first nineteen years.

I grew up in a time and place where being in the middle class was a "destination," somewhere you could actually arrive and stay. In the 1950s my mom and dad got on an elevator and pressed the button labeled "MC," got off on the middle-class floor, and stayed there their whole lives. I also grew up in a time and place where politics, though still partisan, worked, where, at the end of the day, the two major parties and community leaders collaborated and forged compromises to do big, hard things together. I grew up in a time and place where big businesses helped to pioneer corporate social responsibility by donating 5 percent of their gross revenues to the arts and education.

I grew up in a time and place where my parents bought their first house on the GI Bill, thanks to my mother's service in the U.S. Navy during World War II, where my dad never made more than twenty thousand dollars a year before he died in 1973, but where we still could afford to belong to a local golf club and just about all my friends lived in the same size rambler house as we did, went with me all the way through the same public school system, and drove the same kinds of cars—and if anyone was richer than anyone else, it didn't seem to make that much difference. That is, it was not yet true what Dorothy Boyd, the secretary played by Renée Zellweger in the movie *Jerry Maguire*, tells her son about flying first-class: "It used to be a better meal, now it's a better life."

I grew up in a time and place where the word "public" had deep resonance and engendered the highest respect as a source of innovation—as in public schools, public parks, public deliberations, and public-private partnerships. I grew up at a time and place when I was anchored in concentric communities and where the American Dream—"my parents did better than their parents and I will do better than mine"—seemed to be as certain as spring following winter, and summer following spring.

And I grew up in a time and place where Jews were the biggest "minority" but gradually integrated themselves and were integrated by the dominant white, non-Jewish society and culture, and while it wasn't always easy or pretty, somehow it happened.

So where was this place over the rainbow and when was this time?

The Land of Oz that I speak of was the state of Minnesota, and, for me, its Emerald City, where I grew up, was, as I said, a small suburb/town just outside of Minneapolis called St. Louis Park. The time (I was born July 20, 1953) was the 1950s, 1960s, and early 1970s. Growing up in that community at that time was a gift—a gift of enduring values and optimism—that has kept on giving my whole life. Three decades of reporting from the Middle East tried to leach that out of me. So, today, mine is not a naïve optimism that everything will turn out well; I've learned better. But it is an enduring confidence that things can turn out well, if people are ready to practice a politics of compromise and pursue an ethic of pluralism.

I know it sounds corny, but there really is something called "Minnesota nice." In August 2014, I was back in St. Louis Park for a wedding and sitting with my childhood friend Jay Goldberg. Jay told me that his wife, Ilene, had come home that day very flustered and angry. She had been driving on one of the main highways around Minneapolis and another driver cut her off, nearly forcing her off the road.

Ilene told Jay when she got home: "Jay, I was so mad, I almost honked."

When Jay told me that story, I said to him: "Is there a better definition of 'Minnesota nice' . . . 'I was so mad at another driver who nearly ran me off the highway, I almost honked!'" That is Minnesotan for road rage. Ilene's reaction was that of a fundamentally decent person shaped by a fundamentally decent place.

This story of St. Louis Park is the story of how an ethic of pluralism and a healthy community got built one relationship, one breakup, one makeup, one insult, one welcoming neighbor, one classroom at a time—from bricks and logs that were not automatically destined to fit together easily. And I tell it here because St. Louis Park is a microcosm of the "ordinary miracles" that make America what it is when it is at its best. I tell this story because we are going to need these ordinary miracles more than ever—communities whose inhabitants feel connected, respected, and protected and that can both anchor and propel their citizens in the age of accelerations.

And that's why all these years later, as a reporter-columnist, I am still always looking for Minnesota, always looking for ways to re-create that spirit of inclusion and civic idealism that was imbued in me in the time and by the place where I grew up. In short, ever since I left in 1973 for college and then a career in journalism, I've just been trying to get back home.

## *Something in the Water*

Whenever I look back on the impact growing up in St. Louis Park, Minnesota, had on me, I can't help but recall that opening scene in the musical *Jersey Boys*. The group's founder, Tommy DeVito, locates their beginning. DeVito comes onstage, following a French rendition of the Four Seasons' classic "Oh What a Night," and declares: "That's our song. 'Oh What a Night.' 'Ces soirées-là.' French. Number one in Paris, 2000. How'd that happen? You ask four guys, you get four different versions. But this is where all of them start—Belleville, New Jersey. A thousand years ago. Eisenhower, Rocky Marciano, and a few guys under a streetlamp singing somebody else's latest hit."

That riff always transports me back to my roots. It was a long journey from that little town to the op-ed page of *The New York Times*. How'd that happen?—Minnesota. Sixty years ago. Hubert Humphrey. Walter Mondale. The Minnesota Vikings. Target. The State Fair. And a few guys and girls growing up in a one-high-school suburb called St. Louis Park.

St. Louis Park was incorporated as a village in 1886 and assumed the status of a city in 1955. By the late 1950s and 1960s, something was in the water—both figuratively and literally. The literal part is explained on the Minnesota Department of Health website: From 1917 to 1972, the Reilly Tar & Chemical Corporation, known as Republic Creosoting Company in St. Louis Park, "distilled coal tar and made a variety of products, including creosote that was used to treat rail ties and other lumber at the site. The area was sparsely populated initially, but as the community grew following World War II, the appearance and odors of the site became a cause of increasing concern to residents as well as city and state officials."

They sure did. According to the U.S. Environmental Protection Agency, "Reilly disposed of waste on-site in several ditches that flowed to an adjacent wetland. In 1972, the facility was dismantled and sold to the City of St. Louis Park . . . The main contaminant was polycyclic aromatic hydrocarbons, or PAHs, which contaminated soil at the site, a nearby wetland and groundwater beneath the site." In September 1986, St. Louis Park became one of the first enforcements of the 1980 federal Superfund law, after a settlement was reached requiring Reilly to clean up contaminated groundwater and pay $3.72 million to the city, state, and federal governments. During the 1980s, the peat bog was replaced with clean soil and the site was redeveloped into a city park and multifamily housing. As the EPA put it: "It is estimated that approximately 47,000 people use the groundwater from aquifers near the site, which are now treated to meet all required health standards."

I, my parents, my two sisters, and all of our neighbors grew up drinking that water.

But there seemed to be something else in that water besides PAHs.

During the late 1950s, 1960s, and early 1970s, the 10.86-square-mile township of St. Louis Park, with its roughly forty-five thousand residents, was the childhood home of the movie directors the Coen brothers—Joel and Ethan Coen; the political scientist Norm Ornstein; the senator and former comedian Al Franken; the two-time Grammy Award–winning classical guitarist Sharon Isbin; Bobby Z (aka Bobby Rivkin), the drummer for the late R&B megastar Prince; and the former Chicago Bears head football coach Marc Trestman (our high school quarterback, who formed a neighborhood band with Bobby Z when they were in junior high). It was also the home of the feminist historian Margaret Strobel and the Grammy Award–winning songwriter Dan Wilson, who cowrote, with the British singer Adele, her hit song "Someone Like You." The *New York Times* bestselling authors Peggy Orenstein, the author of *Girls & Sex* and *Cinderella Ate My Daughter*, and the environmental journalist Alan Weisman, the author of *The World Without Us* (named the best nonfiction book of 2007 by *Time* magazine) both went to St. Louis Park High. So, too, did the Hautman family. Pete Hautman's book *Godless* won the 2004 National Book Award for Young People's Literature, and Joe, James, and Robert are nationally renowned wildlife artists who won ten Federal Duck Stamp Contests and inspired the duck stamp subplot in the Coen brothers' movie *Fargo*. The Coens and Hautmans were childhood friends. One of Harvard's most popular professors, the philosopher Michael Sandel, was raised just across the St. Louis Park boundary in Hopkins but attended the St. Louis Park Talmud Torah Hebrew school (in my class), and the same was true for Oprah's favorite interior designer, Nate Berkus, also a graduate of the St. Louis Park Hebrew school.

All of us, and there were many others propelled by this small town, either grew up in St. Louis Park or went through its public schools or Hebrew school in roughly the same fifteen-year span. The Coen brothers based their 2009 movie *A Serious Man* on St. Louis Park, circa 1967, and our Hebrew school. When they were young, the Coen brothers often hung out at Mike Zoss Drugs on Minnetonka Boulevard, a few miles from my house. If you look closely at their classic film *No Country for Old Men*, you'll see that the pharmacy just across the Mexican border that the lead character, Chigurh, played by Javier Bardem, enters to steal medicine after he blows up a parked car is called "Mike Zoss Pharmacy"—one of many homages in Coen brothers movies to our hometown and its unlikely Jewish community who settled in these wintry midwestern plains and called themselves "the Frozen Chosen."

To this day I am not sure any of us knows what was the dynamic that unlocked all this human energy, but in my mind it had something to do with pluralism—with the combustion that happened when a new generation of American Jews got unlocked from their Minneapolis ghetto in the mid-1950s and were thrown together with a bunch of progressive Scandinavians in one little suburb. If Israel and Finland had had a baby, it would have been St. Louis Park.

After he left government, Vice President Walter Mondale once invited the Coen brothers, Franken, and Ornstein to submit letters for a dinner event Mondale and I were doing together back in Minneapolis—letters that would try to explain what they thought was going on in St. Louis Park back in the 1950s and 1960s. He later published them in the Minnesota *Star Tribune* on December 5, 1999. Here are some excerpts of what they wrote:

*Dear Mr. Vice President,*

*It is my honor to write a letter that you can read aloud in your introduction of my friend Tom Friedman. I understand that it will save you from having to write something yourself and give you more time to take on your enormous workload at [your law firm] Dorsey . . . When people hear that the five of us all grew up in the same suburb, they are astonished. "What's in the water?" they sometimes joke. But it's not a joke. During our childhood, St. Louis Park was home to a large creosote plant, which leached tons of the toxic chemical into our groundwater. Studies have shown that ingesting large quantities of creosote can lead to two things: increased intellectual creativity and/or prostate problems. This is why Tom insists that we all get regular prostate exams, and why neither Norm nor Tom nor I drink a large Diet Coke before watching one of the Coen brothers' movies.*

*Have a great lunch (dinner?),*
*Al Franken*

*To: Walter F. Mondale*
*From: Norm Ornstein*
*Re: St. Louis Park*

*I didn't know Al Franken, Tom Friedman or the Coen brothers growing up (although my sister did go out on a date with Tom). They were a couple of years younger than me . . . We are tied not just by our similar backgrounds and experiences, but by our love for politics and government. We all feel the link that St. Louis Park, and more generally Minnesota, creates for us. And frankly, I attribute a major share of it to you and your contemporaries. We are all children of the Humphrey/Mondale/Fraser/Freeman era—an era when Minnesota politicians were all substantially above average, when they aspired to do something for the disadvantaged and for world stability . . . Not that you aren't good-looking, but you and your contemporaries were not chosen because you were blow-dried TV anchormen, but because of your ideas and your passion. Because of the Humphrey/Mondale/Fraser/Freeman connection, we felt that Minnesota was special, and so we must be special too. That, and the creosote.*

*Norm Ornstein*

*Dear Tom,*

*It is an oddity often remarked upon that at the turn of the century, one small, obscure provincial area of Hungary, then under the benign tutelage of Emperor Franz Josef, spawned several towering figures in the fields of physics and mathematics—among them Edward Teller, George de Hevesy, Eugene Wigner, Leo Szilard and John von Neumann. This group, many of them Nobel Prize winners, all of them products of the Jewish middle class, were referred to in their diaspora as the "Men from Mars" because of their obscure provenance and their thick Finno-Ugaric accents. What explosive tinder in this remote corner of the Carpathians had nourished such a forest fire of genius? No one knows. Many years later, the Jewish middle class of remote and obscure St. Louis Park, Minn., produced a group of people who also emigrated and overcame funny accents to achieve their own measure of success. Their goyishe tutelary spirits were not Emperor Franz Josef but Don Fraser, Hubert Humphrey and, yes, Walter Mondale. What created this oddly local flowering of intellectual activity? Why, indeed, is St. Louis Park commonly called the City of Flowers? Because there's a "Rosenbloom" on every corner? Coincidence? We doubt it . . . Maybe St. Louis Park, like the cosmos itself, defies easy explanation—even though, unlike the cosmos, it is next door to Hopkins. Maybe [the local commentators] George Rice or Al Austin could have explained it—or, if not them, Roundhouse Rodney. But they are gone. Maybe you, Tom, who have explained so much, could turn your attention to it.*

*We wish you all the best.*

*Joel Coen, Ethan Coen*

In and between the lines of those letters, there is not only an affection for this place we all called home but also an appreciation for the fact that the community that emerged from this mix of cultures didn't happen by accident—we were blessed with extraordinary local and state leaders, school principals, and parents, who time and again made decisions about the kind of inclusive place they wanted to build and who fought for those values against, at times, entrenched opposition. Pluralism doesn't just happen because diverse people are thrown together. Like other communities in America in this period, these local leaders had their blind spots—Jews could be welcomed or at least tolerated, but African Americans were a bridge too far for many back then; and some came along slower than others, but in time they built a community that became unusually welcoming for its era to misfits, different ideas, and different people with funny accents.

## *The Frozen Chosen*

Let's start at what for me is the beginning: How did all these Jews get out to the Minnesota prairie and then gather in this unlikely town called St. Louis Park, where the biggest industry was a creosote plant? Minnesota was not the most natural or obvious place for Jews to settle. Indeed, in the press kit for the Coen brothers' movie *A Serious Man*, Ethan Coen observed to the website MinnPost.com on September 25, 2009, that "to us the [flat midwestern] landscape with Jews on it is funny, you know? Maybe this is part of why we put in that little story [set in a shtetl] at the beginning of the movie, to kind of frame it. You look at a shtetl, and you go, 'Right—Jews in a shtetl.' And then you look at the prairie in Minnesota and you kind of think—or we kind of think, with some perspective on it, having moved out—'What are we doing there?' It just seems odd." Joel Coen added: "Mel Brooks once had a song called 'Jews in Space.' I guess that's sort of the idea."

That space where they settled was not originally St. Louis Park, but the inner city of North Minneapolis, where many Jewish immigrants—our grandparents—took root between 1880 and the early 1900s. That, in fact, is where I was born, as were my parents, Margaret and Harold Friedman. Minneapolis North High School, which both my parents attended, was a heavy mix of blacks and Jews. One of my predecessors at both UPI and *The New York Times*, Harrison Salisbury, was also part of that Jewish community and graduated from North High a few years before my parents, in 1925.

My grandfather on my mother's side was a junk dealer and on my father's side a photographer, although both saw their businesses crushed during the Great Depression. My dad was vice president of a ball bearing distribution company, United Bearing, started by a friend, and my mom was a housewife and part-time bookkeeper. When I was born we lived in a duplex on James Avenue North in Minneapolis with my mom's sister's family, who owned a cigar store with a three-stool dinette—Burt's Smoke Shop—where my uncle and his partner made breakfast and lunch out front and augmented their income by running a little betting operation out back.

There was quite an active Jewish Mafia in Minneapolis, which reached its zenith during Prohibition, led by the notorious Isadore Blumenfeld, better known as "Kid Cann." My dad, who was not in this Mafia but grew up with many of these characters, would occasionally tell me stories about them. Indeed, one of my earliest childhood memories was my dad telling me about a friend of his who had been sentenced to a jail term. This was quite a shock to a young boy. I couldn't imagine my father knowing anyone who actually went to jail. So I asked my dad: Why? And in one of the greatest euphemisms I ever heard—so good it stuck with me all these years—my dad told me that his friend was sentenced to jail because "he was shopping in a store before it was open."

Breaking and entering has never been described so benignly.

The largest minority groups, Jews and blacks, settled in North Minneapolis "because it was one of the few areas that would rent to them when more blatant discrimination abounded in housing practices," noted the June 2, 2013, essay "A Brief History of Jews and African Americans in North Minneapolis," by Rachel Quednau on The-City-Space.com:

> Jews from Russia and Eastern Europe moved into the Northside during a general period of immigration—the early 1900s. They built a Hebrew school in 1910—Talmud Torah—which is still well known today. At that time, it offered social services in addition to education at a nearby community center. Jewish businesses also cropped up in the area. Meanwhile, African Americans made their homes throughout North Minneapolis before this time, but they appeared in larger numbers after World War II . . .
>
> Both Jews and African Americans provided a significant portion of residents in a North Minneapolis public housing project, Sumner Field Homes, which was built during the New Deal. These projects were racially segregated, but interviews with past residents [and this was very much the story my parents told me] . . . suggest that children from different backgrounds played together and other mingling occurred.

The biggest social problem for my grandparents' and parents' generations in Minneapolis was not relations with blacks but with anti-Semitic whites. An essay on the St. Louis Park Historical Society website, written by Jeanne Andersen, quoted an article entitled "Minneapolis: The Curious Twin," by Carey McWilliams, that was published in *Common Ground* magazine in September 1946. McWilliams proclaimed: "Minneapolis is the capital of anti-Semitism in the United States. In almost every walk of life, 'an iron curtain' separates Jews from non-Jews in Minneapolis." The article went on to say that "although only 4 percent of the population, Jews were publicly and unapologetically excluded from membership in private country clubs and also Rotary, Lions, and Kiwanis Clubs and groups like the Toastmasters. Jews were even barred from the Minneapolis chapter of the American Automobile Club." I remember growing up being told by my parents that there was a time they could not join Triple-A. "In 1948," McWilliams wrote, "frustrated Jewish doctors started their own hospital, Mt. Sinai, after being denied access to Minneapolis medical facilities." I was born there. The article also noted that Jews were blocked from joining local chapters of labor unions that had been started in New York by Jewish organizers and that "summer resorts on Lake Minnetonka advertised that they catered to 'Gentiles only.' Department stores such as Montgomery Ward refused to interview Jewish job applicants. Many neighborhoods were 'restricted,' barring Jews, Blacks, and even Catholics and Italians. Jewish teachers were few and far between." The discrimination was "much more pronounced in Minneapolis than in St. Paul," according to McWilliams.

So, with the first chance to get out, after World War II, the Jews left the Northside urban core of Minneapolis en masse for St. Louis Park. As Quednau put it, "Many of them had ascended in class and wealth since they or their parents immigrated to America, and this allowed them more control over their housing along with more opportunity to be treated fairly in the housing market."

It wasn't quite that easy, though, to just move to any suburb. Many of the western suburbs adjacent to Minneapolis were not platted for many starter homes, or had large agricultural tracts or had a legacy of refusing to sell homes to blacks or "Hebrews." St. Louis Park, though, had been platted for small forty-foot lots since the early part of the twentieth century, explained Jeanne Andersen of the Historical Society. "For some reason, the community's early developers and factory owners always had a growth mind-set," she told me. So there was a lot of available housing stock and the real estate developers "were perfectly happy to sell it to Jews," in contrast to other suburbs at the time, such as Golden Valley or Edina. Speaking of the suburbs that grew up around Minneapolis after World War II, Andersen remarked, "Besides St. Louis Park, I didn't find anyone else holding out the welcome mat for Minneapolis's unwanted Jews."

My parents and virtually all my Jewish friends' parents were part of that great 1950s exodus. When I was three, in 1956, my parents packed up our Buick and joined the migration of Jews westward—from North Minneapolis to St. Louis Park, seven miles away. We lived in an aluminum-sided, three-bedroom rambler—I had two older sisters, Shelley and Jane—and like everyone else on West Twenty-Third Street, we all went to the local public elementary school, junior high, and high school with nearly the same classmates for all twelve years. Our house cost my parents the grand sum of $14,500.

It's hard to believe that this little town that looked just like all the towns around it and was not separated from them by a wall or moat could develop such a unique liberal culture of its own, but it did. "From the very beginning, Park has had a welcoming attitude" toward strangers, oddballs, bars and bartenders, observed Andersen. Where other suburbs resisted bars, and gas stations, said Andersen, "Park just couldn't say 'no.' An unfortunate proclivity for dirty industry was part of this 'progressive' attitude, bringing us plants that processed lead, lithium, concrete, and, of course, creosote, but those industries also provided jobs."

My family and faith community formed the first ring of the many concentric, reinforcing communities in which I grew up. My dad's brother lived in an apartment house 250 yards away from our home; my mom's sister and brother-in-law three doors away. We celebrated every Jewish and non-Jewish holiday together with our extended family, with our moms taking turns as to who made the matzoh balls and popovers on Passover and the turkey on Thanksgiving.

The Minneapolis Jewish community was very tightly knit. About 25,000 strong, it was also extremely philanthropic. In the 1960s, you could not join the old-money Jewish country club, Oak Ridge, unless you also gave a minimum amount to the Jewish Federation to support the Jewish old age home, community center, and Israel. At the annual Federation dinner for the biggest givers, you had to stand up in front of the whole community and announce publicly your family's gift for that year. My parents were not wealthy enough to make that dinner, but they always gave within their means; they considered it as much an obligation as federal income taxes.

My parents' generation of Jews built their wealth mostly as entrepreneurs, since so many professions in the 1940s and 1950s were closed to them. So the term "never pay retail" was a staple in our household. When the school year began, my sisters would go with my dad to Ed Neff's showroom—he and his brother sold women's clothing—and they would stock up on samples for the year. (My mother, who loved to sew, also made a lot of their clothes, including both my sisters' wedding dresses and some of my suits.) When I needed a new winter coat, we went to Irv Baumel's house—he was a jacket designer—and I would choose from samples off the clothing racks in his basement. Needed steaks? Call Marvin Greenstein, who owned a meat packing business. Needed a new washer or dryer? We only went to Benis Koval, my dad's childhood pal, who had the General Electric franchise, Koval's Appliances. And when we needed anything else wholesale, there was Marshall Lifson. Marshall was a liquor salesman and had connections everywhere—baseball tickets, football tickets, Las Vegas shows, the right doctor—you name it—Marshall always knew the guy who knew the guy who could get you in. "Call Marshall" was an all-purpose mantra in our house.

I am keenly aware that my generation was a transition generation between the era of my parents—who always felt that life was a suitcase with a false bottom, so you should never get too comfortable—and my daughters' generation, for whom anti-Semitism is something they largely learned about through history books. Most of our grandparents were immigrants from various European pogroms, and our parents were born into the Depression and then World War II. So even though we had found our own little *goldene medina* in St. Louis Park, they were always wary. Our parents and grandparents were a generation of Jews who were at home in America and Minnesota—but tensely at home. They were always concerned that things were too good to be true. They had seen the Holocaust; they had touched the bottom of the Depression. They knew demons always lurked below. The acceptance of Jews and the existence of Israel seemed a startling departure to them—not a feature of nature.

It came out in little ways and phrases that often stuck in the mind of my generation. My childhood friend Howard Carp used to say of his Minnesota Jewish grandmother: "A chicken to my grandmother was like a Buffalo to the Sioux—no part was left unused: the neck, the butt. We used to say: 'Grandma, what are we eatin' here?'" Howard's grandmother never knew when she would be able to afford the next one, so she knew she'd better use all of this one.

For most of my years growing up, the established golf clubs in the Minneapolis area did not accept Jews. We had our own Jewish golf club, Brookview, and it was another community within the Jewish community—the members put on theatrical plays that they wrote each summer; they had regular summer Sunday dinners and bingo games, a swim team, family talent contests, and a poker club where all the winnings each week went into a pot, and when it got big enough the husbands all took their wives on a trip to Acapulco. Brookview was a real anchor in our lives. During the winter, the golf club organized a bowling league every Sunday morning, with each player betting with the other using a bowling handicap system. As a young boy I always accompanied my father to the bowling alley on Sundays to watch and root for my dad. To borrow the Harvard political scientist Robert Putnam's image, in those days there was no one in my St. Louis Park community "bowling alone."

I grew up caddying at Brookview for my dad and his friends and learning to play golf from the age of five. Some of my best friends today are still the guys I played with and caddied with back then. And because most of these men I caddied for owned small businesses, I was exposed, through their golf-course patter, to the world of business, and from that developed a respect for entrepreneurs and risk takers. I would overhear them talking about their deals, their successes, hot stocks, and, yes, losses. The first time I was exposed to the concept of bankruptcy was on the golf course. There was a guy at our club who my dad informed me one day had to drop out because he went "bankrupt." I did not know exactly what that meant, but I could see that he was out of money and out of golf balls and out of the club and it was not something I wanted to ever happen to my dad. Caddying teaches you many things, but most of all it offers an insight into character. We caddies all knew who cheated. We all knew who had integrity. We all knew who blamed their caddy for a bad shot. And most of all, we knew, as the great amateur player Jimmy Dunne observed in the September 8, 2011, *Golf Digest*, that at the food shack between nines "there were some guys who would let you get a soda. There were some guys who let you get a soda and a hot dog. And there was the rare guy who would let you get a soda and a hamburger. And you knew who those guys were. You knew."

My dad and I used to play golf in the summer after he got home from work—six or seven holes after dinner and before the sun went down. To get to the club we had to drive through the intersection of Louisiana Avenue and Highway 12. Every so often we would go past it and my dad would remind me that during the Great Depression he worked at a CCC—Civilian Conservation Corps—camp near there when he was a teenager. The CCC was the public works relief program established by the Roosevelt administration from 1933 to 1942 to provide employment for young, unmarried men, who built public buildings and parks. More than once my dad told me he made one dollar a day working there, most of which he saved for his family, and that he could only afford to buy a loaf of bread to eat—"and I can still feel it stuck in my throat," he would say. Every so often, when we would pass that intersection, I would say, as only a smart-ass teenager could, "I know, I know, you can still feel that loaf of bread stuck in your throat." He never forgot it, but neither did I. Fortunately my daughters will never know such a feeling.

Brookview eventually relocated and built a new course in Hamel, a more westerly suburb, and my dad died there from a heart attack on the par-four fifteenth hole, when I was nineteen. He lied three. After he passed away in 1973, I was walking down the fairway at Oakridge Country Club, where the older Jewish money belonged, playing with a friend of my father's. It was a beautiful summer day and the course was in magnificent condition—bright green grass and flowers everywhere—when out of the blue this family friend put his arm around my shoulder and whispered, "Tommy, if the goyim [Gentiles] knew we had something this nice, they would take it away from us."

Since I experienced childish versions of anti-Semitism in my high school—kids throwing pennies at the Jews because they were supposedly so cheap they would pick them up—I was not innocent about such matters, but his remark jarred me. That was the abiding ethic of my parents' generation of Jews—things were always too good to be true.

If the Jewish community of St. Louis Park had a beating heart, a holy of holies, it was not a synagogue or the Jewish Community Center. It was the Lincoln Delicatessen, mostly just known as "the Del," and the competitor to my aunt and uncle's Boulevard Del. My mom worked at the Lincoln Del as a bookkeeper to put my sister Jane through Bryn Mawr College, and as a little boy I used to play on the baker's wooden tables, sometimes braiding challahs. The Lincoln Del was owned by my parents' dear friends Morrie and Tess Berenberg. Morrie held court there at a table in the dining room every afternoon and evening, entertaining customers while keeping an eye on the deli counter.

Berenberg's granddaughter Wendi Zelkin Rosenstein and Kit

Naylor, who have been drafting a history entitled *Memories and Recipes from the Lincoln Del*, observed in their book proposal that "for Jewish and non-Jewish customers alike, it was the Minneapolis version of 'Cheers,' except that at the Del everybody really did know your name." What made the Del so beloved, added Rosenstein, "was that it was the true center of life for the Minneapolis Jewish community—it was the place for everyone to meet after school or before a movie, to tailgate before a bus ride to a sports event, to get engaged, or to celebrate life after a funeral. From graduations to business meetings, the Lincoln Del remains a touchstone for people who grew up in Minneapolis and St. Louis Park." The Del was also a vitally important mixing spot for the whole St. Louis Park community, a place where non-Jews got comfortable eating Jewish food and experiencing Jewish culture. People drove from all over the region to buy the Del's bagels.

What was The Del's secret recipe? Why is it so fondly and indelibly etched in the memories of so many of us who grew up in Minneapolis in the 1950s, 1960s, and 1970s? Was it just the food? Of course not. The Del "sold" something so much more compelling that kept its customers constantly coming back, something that is increasingly rare these days, to which Rosenstein alluded. It wasn't knishes—it was community; and it wasn't the bagels, it was the bonding. Anybody could learn how to make knishes—a doughy Eastern European delight—and sell them. But what I, and so many others, actually remember is the times we squeezed into a booth with family, friends, or a first date, and forged friendships that endure to this day—or that we miss to this day. The Del was the place where we did things together. And if you hopped from booth to booth you could be together with so many others doing things together, and together it all wove the warm tapestry of community. The knishes just made it extra special.

I confess that living in Washington, D.C. for over a quarter century, I still have no such place to go where you know lots of people and people call you by your first name, and where you can go from table to table and shake hands, gossip, and have a good laugh with other customers. The delicatessens in D.C. do sell matzo balls, but without memories. That is why whenever I eat at one of them, no matter how much I eat, I always leave hungry. My stomach is full but my soul is empty. Nobody ever left The Del hungry. It filled body and soul.

I also confess here—for the first time—that when I was working for the competing Boulevard Del, I would drive down to the Lincoln Del bakery in our truck every day to pick up the bagels, which the Boulevard bought wholesale from the Lincoln. Occasionally, temptation got the better of me; the smell of those warm bagels wafting forward from the back of that truck became too great to resist. So more than once I could not help stealing a plain bagel from the back of the truck and wolfing it down while it was still warm. I can still taste it to this day.

When my dad died suddenly, my widowed mom couldn't afford my college tuition, so Morrie and his friend Jake Garber, my dad's boss, and my aunt and uncle, all pitched in. Morrie was the driving force behind it all, though. I did not come to him for help. He just came to me one day and said, "You can't afford this," and that he would make it happen. It was a powerful lesson in community for me: When you are in a real one, never, ever say to someone in need: "Call me if you need help." If you want to help someone, just do it.

A few years before he died, Morrie had a stroke and lost his speech. So he would use a little white notepad instead. I would bring my girls with me to see him when we visited Minneapolis, and without fail he would write on that pad that my two lovely daughters "take after your wife." He had a way of keeping you off-balance. Anyone who put on airs around him got cut down to size. It was from him I first learned the lesson: "Remember the people you meet going up, you may meet the same people coming down"—and always, always leave a big tip for the waitress. If you leave big tips your whole life it may add up to $5,000; you'll be helping someone who needs it—and your service in Heaven will be amazing.

Our Hebrew school was serious, even if we weren't. From third through seventh grade, Monday through Thursday, we walked out the door of our public elementary schools at around 3:00 p.m. and got directly onto the Hebrew school bus, which drove us to the St. Louis Park Talmud Torah. There, we got chocolate chip cookies and chocolate milk and had ninety minutes of Hebrew classes on four weekdays, plus Sunday mornings. That was the after-school program for my generation until we were thirteen and had our bar and bat mitzvahs. Virtually every Jewish kid I knew growing up went through the local Hebrew school. Its vitality ended up attracting even more Jews from North Minneapolis.

As a result, by the 1960s, roughly 20 percent of St. Louis Park's residents, and public school population, was Jewish. As Al Franken told *The New Yorker* on July 20, 2009: "Not exactly a shtetl, but by Minnesota standards, a lot of Jews."

## *St. Jewish Park*

And thus began a great accidental mini-experiment in American pluralism.

It was as if America's Founding Fathers reconvened and said, "Let's have some fun. Let's test just how well we can make 'out of many—one': Let's mix dark-haired third-generation Jews, newly liberated from the inner city and energized by the postwar era—named Goldberg, Coen, and Friedman—with blond Protestant and Catholic Swedish, Norwegian, Finnish, and German-descended Americans, named Swenson, Anderson, and Bjornson, and do it almost overnight in one very small town in Minnesota, and see what happens!" It was no wonder people started calling it "St. Jewish Park." The Coen brothers captured this cultural clash and synthesis well in *A Serious Man*, when in the synagogue bar mitzvah scene an elderly man is asked to lift the Torah scrolls, a tradition in every Jewish service, but the load is too heavy for him. As he starts losing control of the Torah scrolls, he exclaims, "Jesus Christ!"

Building pluralism, making one out of many, a great American tradition, does not happen automatically or easily. Real pluralism never comes easy, because it has to be built not just on tolerance of the other but also on respect of the other, trust of the other. Like all such cultural encounters that have happened around America over the centuries, there was a fascination with and rejection of the "other," attraction and repulsion, beautiful moments of understanding and painful moments of misunderstanding; there were crushes and breakups, intermarriages, divorces and remarriages. In any given week, I saw biases melt away and biases displayed. We dated each other, slighted each other, tolerated each other, quietly mocked each other and embraced each other—all at the same time. We worked on yearbooks and newspapers and sports teams and student councils together, and although we prayed at different buildings to different gods on different days in different ways, somehow through trial and error, we built a community—but not without some broken emotional bones along the way.

In other words, real pluralism doesn't get built through Facebook posts, instant messages, or Twitter encounters. You can't download the values that undergird real pluralism. They have to be uploaded, the old-fashioned way—by real encounters over the fence in the backyard, at a first Bar Mitzvah party, through integrated public schools, through group car washes for the yearbook or midnight broomball games in winter or when Jewish kids in the chorus had to sing "Silent Night," while the gentiles had to learn the words to "Hanukkah, Oh Hanukkah." Real pluralism is built one matzo ball, one soy sauce bottle, and one slice of venison shot by your neighbor's dad during hunting season at a time—and by learning not to recoil, but rather to taste—and maybe even to like—the unknown, the exotic.

As in all such cultural encounters, there was fascination with and rejection of the "other," attraction and repulsion, beautiful moments of understanding and painful moments of misunderstanding, there were crushes and breakups and intermarriages and divorces and re-marriages. In any given week, I saw biases melt away and biases displayed. We dated each other, slighted each other, tolerated each other, quietly mocked each other, and embraced each other—all at the same time. We worked on yearbooks and newspapers and sports teams and student councils together, and while we prayed at different buildings to different Gods on different days in different ways, somehow through trial and error, design and accident, we built a remarkable community—a community that changed us and mostly for the better. We became more tolerant of each other and others in general—but not without some broken emotional bones along the way, because it was not always love at first sight.

According to the St. Louis Park Historical Society, "As usual, the High School Prom of the St. Louis Park Class of 1949 was to be held at the Automobile Club in Bloomington. A manager there found out that a Jewish student planned to attend, and banned him. St. Louis Park School Superintendent Harold Enestvedt personally told the Club that if all of his students were not welcome, the Prom would be held somewhere else. The Club reversed itself and everyone went to Prom as scheduled." That was the kind of leadership that made St. Louis Park special: ordinary men and women in positions of authority acting decently—not heroically, just decently.

One of my earliest memories was playing basketball on the asphalt court behind Eliot School, my elementary school. I am guessing I was seven or eight years old. And there was a boy, not Jewish, getting beaten up for some playground violation by one of my neighbors, Keith Roberts, who also was not Jewish. The boy, who was getting the bejesus beaten out of him, started shouting at Keith, "You dirty dhew!" He had a lisp. He wanted to deliver the worst insult he knew to Keith—"dirty Jew." Keith just laughed at him and said, "I'm not Jewish."

I'm sure they both quickly forgot about the incident. I never did. Even then I could figure out that this kid didn't know a Jew from a Gentile, but he clearly had learned this at home and dragged it out as a kind of all-purpose insult on the playground.

I quietly applauded Keith's beating the crap out of him.

Paul Linnee, who graduated from St. Louis Park High in 1964 and later became a local policeman, reminded me that Toledo Avenue, which ran right alongside Highway 100, was known when he was growing up as the "Gaza Strip," with a large percentage of the population east of there being Jewish and a large percentage west of there being Gentile.

It would take much longer for African Americans to have even a block of their own. Paul Linnee's sister Susan, who would later become the Associated Press bureau chief for East Africa, West Africa, and Spain, reminded me of that by retelling a remarkable incident from the summer of 1962, at their home at 2716 Toledo. She recalled:

> Toledo Avenue was unusual in that the houses did not all meet the sidewalk from the same distance. Some were far back, some were flush, some had been built long before anyone thought of sidewalks. But it was—as all St Louis Park neighborhoods were at the time—very white. The Sperlings—our neighbors to the south—were the first Jewish family in the neighborhood, and there had been a campaign to muster all the homeowners to block the sale of the house—owned by super devout Christian Scientists who could not have cared less to whom it was sold—to Jews. When a resident from across the street called on

my mom, Jane, and asked her to sign a petition, she was shown the door, with my mom telling her: "Finally the block is going to be interesting."

Her dad, said Susan,

was raised a Lutheran; my mother was a non-Scandinavian Episcopalian; they switched to the Congregational Church because it was more "liberal." Both were lifelong Democrats and my dad really agonized over voting for a Catholic for president, fearing, as many Protestants did, that "the Pope would run the White House" . . . We had a set of *World Book* encyclopedias next to the dinner table to settle any disputes that might arise—basically between my dad and me. Paul and I have talked about this several times. Neither of us can remember, other than my father's fear of the Pope, which soon dissipated, a single racist or intolerant word, other than against Republicans, at home—or even at school.

Susan graduated from St. Louis Park High in 1960 and attended the University of Minnesota. Through her boyfriend at the time, she met an African student from Macalester College in St. Paul at a party. She remembered that "he was wearing a trench coat and a hat and looked like someone in one of the French crime movies." So one day in the summer of 1962 she invited this exotic African and some other friends to their home in St. Louis Park—when her parents weren't home.

One of the neighbors, seeing black men enter the house, called the police. A few days later her dad sat her down to find out what happened. Susan sent me her recollection of the dialogue:

My dad came to my room in the early evening, quite uncomfortable and hesitant. He was an old-school Swede.

Dad: "Ummmm . . . have you had any . . . err . . . black . . . visitors recently?"

Susan: "What's this all about? Why are you asking me this?"

Dad: "Ummm . . . ummmm . . . errrr . . . someone called the

police to say black men were visiting the house while we were away, and the police called me."

Susan: "What? Who was it? Who called?"

Dad: "Errr . . . ummm . . . they wouldn't say . . ."

Susan: "Okay, here are all the 'men' who came while you were gone: Fred, Kofi, David, etc., etc. . . ."

Dad: "Wait, who's Kofi?"

Susan: "He's Ghanaian, from Africa . . ."

Dad: "Africa, so is he black?"

Susan: "I guess so . . ."

Later. Mom: "Let's invite them all for dinner!"

And so on a summer evening in 1962, the Macalester College student Kofi Annan and several of his friends drove back to St. Louis Park in a tomato soup–colored Studebaker. *Yes, that Kofi Annan,* who would later become a Ghanaian diplomat and then the seventh secretary general of the United Nations—but at the time was finishing his economics degree at Macalester on a Ford Foundation scholarship.

"Many people were out mowing their front lawns," Susan recalled. "Kofi leads the group. Mom and Dad walk out to meet him. Hands are shaken. All enter the house and eat corn on the cob. My mother died at one hundred, on April 7, 2013. I asked Kofi if he would prepare something for her memorial, which he did with great aplomb. We have kept in touch over the years and would meet when he was in Nairobi after the post-election violence in 2007 and 2008. He would always ask after my mother, although they never met again."

Fifty-four years later I asked Annan if he recalled the incident, which he did in vivid detail.

"I was quite a young student—there was a group of us, an Indonesian, an Indian, and others, and we all hung out together" at Macalester, said Annan. "On the whole, people in Minnesota were very nice and hospitable. My wife is from Sweden and she likes to say that Minnesota's Swedish immigrants prepared me for her!" As for Susan's mother, Annan added, "She had a lot of spirit—she had this attitude of 'I'll be damned if anyone is going to tell me who can come to my house or whom I can receive.'" For foreign students from Africa, India, or Indonesia, newly independent countries, this kind of racism, that a neighbor would call the cops because they saw a black man enter a house, was a bit of a shock, said Annan. "For a young Ghanaian, whose country was just a few years out of independence and [was] so proud of his country, it takes you a while to register and understand. We all came from cultures where we were the majority, where we never had this experience. When I sometimes hear societies say, 'We have no discrimination'—I am sure they don't, until they have someone to discriminate against." Because of that, said Annan, "You have to respect the courage of the individual who stands up against it, and they go up in your esteem and the bond and friendship strengthens, and that is how I felt about Susan and her family. It is incredible because somebody else would have reacted differently"—and some certainly did. All in all, though, looking back on his adventures in St. Louis Park and Minnesota, Annan concluded, "It was a community, and those of us who came in from outside could feel it."

Kofi was the second black man Susan's brother Paul Linnee ever met. In 1962, he recalled:

> I was pumping gas at Norm's Texaco at 5125 Minnetonka Blvd., when a dingy '62 Chevy Bel Air with Kansas plates pulled in for a tank of Fire Chief gasoline. While I was using my whisk broom to sweep out the passenger compartment, the rather large black man—who was, literally, the first black person I had ever spoken to in my life—who was driving, asked me if there was a pet hospital in town. I pointed him towards Fitch's Pet Hospital up behind the Pastime Arena. Not too long after that I heard that he had bought the pet hospital from Dr. Fitch, opened up his practice there, and became quite successful.

That man was Dr. B. Robert Lewis, a veterinarian, who went on to run for and serve on the St. Louis Park School Board and then became the first African American elected to the Minnesota State Senate. He was also the first African American to serve on a Twin Cities school board, and one of the founders of the St. Louis Park Human Relations Council. "He also went on to become a regular customer at Norm's Texaco. I like to think that I was the first St. Louis Park person he ever met, and I know he was the first black person I ever met," said Linnee.

Norm's Texaco was a most unusual hotbed of ecumenism, recalled Linnee:

> [The owner,] Norm Walensky, was an oddity. Most Jews I was aware of in the Park were professional persons or persons involved in trades, where they did not "get their hands dirty." Norm was different . . . Norm hired me and about a half dozen other non-Jewish teenagers and slightly older "gearheads" to be his service station attendants–cum–mechanics. His was probably the only Jewish gas station in the Park, and, as such, seemed to be the preferred place for most Jews to have their cars fixed, buy gas, and call for a jump start on cold winter mornings. A few days before every Christmas, Norm and "his boys" would spiff up the two-stall garage to be spic-and-span, and then lay out tablecloths on the drive-on hoist's ramp, go out and buy a real holiday spread and a few bottles of the good stuff and throw a holiday bash for all the customers and employees. I will always remember sharing holiday cheer with all of the successful Jewish doctors, dentists, and lawyers each year.

It was intriguing to me, forty years later, to hear what St. Louis Park, and this influx of Jews, looked like to non-Jews. Jane Pratt Hagstrom was in the St. Louis Park class of 1978, and grew up in the Westwood Hills neighborhood, one of the newer developments in St. Louis Park, where homes were a little bigger. "My family moved there in 1960," she remembered. "And I still recall the Realtor telling my parents, 'There won't be any Jews in the neighborhood.' My parents, who were from South Dakota and Iowa, thought she said 'trees.' Anyway, within a few years it was a predominantly Jewish neighborhood and my parents used to joke that I was turning into a Jew because I said 'Oy' all the time . . . But I remember going to college seeing bigotry. People said, 'You are from St. Jewish Park?'"

The discrimination wasn't all one-way, though. I had non-Jews recall for me hearing a Jewish friend's grandmother warning that he must never marry a *shiksa*, Yiddish for a Gentile girl. It is always striking to me how these little tribal asides, which you hear as a child and cannot fully understand, are remembered decades later by the person at whom they were directed.

Yes, we Jews could also be, well, annoying at times. Back in the 1960s and 1970s, a local television station sponsored something called *Quiz Bowl*, where brainy local high school teams would compete against each other, answering math, science, literature, and history questions. It was a local version of *GE College Bowl* and a big deal around the state. My AP history teacher, Marjorie Bingham, was the longtime coach of the St. Louis Park team, and she told me this story:

> We were doing well, but had to play St. Thomas Military Academy, which had been local champs. All the other teams would socialize before each contest, but before we went on the air, the adviser to the St. Thomas team, who was a priest, got his uniformed team in a circle and led them in prayer. Our team was predominately Jewish, and as St. Thomas finished their prayer session, the Park team spontaneously gathered in their circle and chanted—I can't remember what—probably something from Monty Python. I've repressed that part! The priest glared at me, like "Can't you keep your students under control?" But honestly, I couldn't help feeling St. Thomas deserved it and I didn't apologize. We won. You never knew what Park students would come up with.

Margaret Strobel was born in North Dakota, but her family moved to St. Louis Park in the 1950s, where she went through junior high and graduated from Park in the class of 1964, just before my sisters. She went on to become director of the women's studies program at the University of Illinois, and the author or editor of six books on feminism, race, and African history. She was also the director of the Jane Addams Hull-House Museum. She recalled:

> I know that being in school with so many Jewish kids also influenced me. It was so—relatively—soon after the Holocaust. My Presbyterian youth group did a weekend overnight with a Conservative Jewish youth group that I still remember—learning the hora—and it was the same youth group that got me to go around St. Louis Park knocking on doors to raise money for some civil rights cause—what I think now must have been the Freedom Summer of 1964, though I can't nail it down. I remember

> knocking on a door and the woman said something to the effect, "We should let them solve their own problems down there." It was one of those moments in my life around race that still sticks out.

Strobel reminded me that we also had a few Japanese Americans in our school, whose parents had been sent to internment camps during World War II. "My first experience of being the only white person in a room came when my friend Diana Shimizu invited me to the Japanese American youth group she was a member of," Strobel reminisced. "I also remember being outraged when she told me that her parents, when they moved to [St. Louis Park] from some internment camp, went around to homes in the block where they were considering buying a house and asked if neighbors would mind if they moved into the neighborhood." There is no better teacher of pluralism than a visit to the "other's" dinner table. "Experiential learning is so important," said Strobel:

> When I would eat over at Diana's house they had soy sauce on the table and I thought, "Soy sauce on the table, what is that about?" My family never went out to eat . . . I also remember going to Judy Light's on Friday nights and they would have a Sabbath dinner. And I remember it as sumptuous. And I remember her mom putting a napkin over her head and lighting candles and saying a prayer. And I am sure these things [all contributed to my] learning to live with the other. It was not a context of hostility. You were an invited guest into these homes.

But not everybody was.

Debra Stone, an African American woman who was a year ahead of me—her brother Melvin was in my class—gave an interview on June 22, 2012, to Jeff Norman, as part of the Jewish Historical Society of the Upper Midwest's Oral History Project. Stone's family moved, like the Jews, from North Minneapolis to St. Louis Park. She spoke eloquently of her own adventures in pluralism there.

In 1963, said Stone, her family moved out of North Minneapolis to 1637 Idaho Avenue South, not far from where I grew up, and she and her brother attended the same elementary school as me—Eliot School. Norman asked her how they ended up in St. Louis Park and not somewhere else in Minneapolis.

"They looked in Northeast Minneapolis," said Stone of her parents, "and because of housing discrimination and racism, they said it was easier to move to St. Louis Park." In Northeast Minneapolis, she said, "the Realtor wouldn't even show [my parents] a house . . . So they dropped that Realtor, and they found another Realtor who showed them the houses in St. Louis Park. [That second Realtor] was Jewish . . . As I understand it, we were the first family in that community . . . There were no other African American people living in St. Louis Park. We could walk down Cedar Lake Road; we could travel all over St. Louis Park; we could go to Knollwood [a shopping mall]; we wouldn't see another dark face, other maybe than some Sephardic-looking Jews who were dark. Other than that, it was us . . . It wasn't until I was in eighth grade that another African American family moved into that neighborhood."

What was the response of their neighbors? Norman asked.

"My parents were very protective," said Stone. "I heard that someone came and knocked on the door and said, 'Are you the African American family moving in?' My father said, 'Yes.' He said, 'Would you consider moving?' And my father said, 'No.' Then there was a little conversation, and my mother said, 'Your father said, "If you don't get out of here, I'm going to shoot you!"' [*Laughs*] That was it. We never had anyone else come. Everybody else was pretty nice to us after that . . . I played with the children in the neighborhood, played in their yards, played dolls in their houses—that was it. Jewish and non-Jewish families."

Did she ever attend a bar mitzvah or a bat mitzvah?

"Yes, I have," said Stone. "My good Jewish friend, Pam Russ—I went to her son's and her daughter's bat mitzvah. She grew up in Robbinsdale. I didn't get to be friends with her until a little bit later. We were at Temple Israel, and I attended bar mitzvahs and the parties for the kids. Pam is much more down-to-earth, so it was really kind of a nice affair—grandparents, and relatives, and friends, both Gentile and Jewish. It was a really moving ceremony, I felt—could understand why the kids would go through this process."

So all in all, what was it like to be the only African American kid in your classes back then? Norman asked Stone.

"I didn't have a problem with it," said Stone. She added: "Whenever something did come up, though, we could always rely on my mom to have our backs. There was one incident in which [someone] had done this poster of this "Mammy" figure. It played a very prominent feature in the school display. I don't remember what it was for, but I remember some of the kids snickering and stuff. I went home and I told my mom. I said, "There's this poster," and she said, well, that's totally inappropriate. So she went up to the principal, and it was gone the next day . . ."

Stone recalled that in high school she was voted to be a cheerleader and was part of the student council for a couple of years in a row. "So, yes, it was okay," she said. "There weren't any hot racial incidents or anything like that. We went to school. A couple of kids might have called us 'niggers' and we beat them up, and that was it."

Looking back on it all from the perspective of today, she concluded, "My experiences of living in St. Louis Park were really—as far a child growing up in an all-white and Jewish community—I would say because of the strength of my family, it was a good experience for me. I benefited from it in many ways . . . I was able to go to college; travel the world; do many, many things that a lot of my African American women peers have not been able to."

Precisely because there were so few blacks, "we Jews thought we were the minority," recalled one of my closest childhood friends, Fred Astren, now head of the Department of Jewish Studies at San Francisco State University. "There were three Chinese kids, three Japanese kids, two blacks, and everyone else was either Scandinavian or Jewish. We could afford to be liberals [on civil rights] because we never really met 'the other.'"

For me at least, that wasn't the only reason. My parents taught me not to be racist from a young age, though it was not by anything they ever said. As I noted, I played golf with my dad at our little club, Brookview, from about the moment I could walk. The first black people I ever met, I met there. One was—as you might predict back in the 1960s—the locker room attendant, Jimmy. The other, though, was the bartender: A slender handsome man named Victor, who went on to be president of the PTA at my elementary school. Victor liked to play golf, and several times each summer, when he had playing privileges on his Mondays off, my dad would invite Victor to play with the two of us in the evening. This happened over several summers. I was probably nine and ten at the time.

My dad was modeling something for me—insisting that Victor not play alone. So there were the three of us, pulling our carts together, playing golf until the sun went down. End of lesson. Not a word was ever spoken about it between my dad and me—he never said "this is how you should treat black people"—but I never forgot it. I am still hoping to show up at a book signing somewhere some day and have a handsome gray-haired black man come through the line and just say, "Sign it to Victor!"

## *Public Spaces*

The quality of the public schools in St. Louis Park and the pride we took in them were part of a larger respect for and celebration of public spaces and institutions. These public spaces were both a product of and an engine of trust, pluralism, and social capital generally. Each was a Mixmaster that brought people from diverse economic, religious, and racial backgrounds together. Virtually every person I knew went to public school. Growing up, in fact, I thought private school was only for kids with some kind of social or emotional problem—a place your parents sent you as a kind of punishment. The mere idea that someone would pay extra money, over and above their tax bills, to send their kids to private school because it was superior was simply not in our consciousness.

Even if we did not realize it at the time, a lot of us have since realized how good this public school was.

When I arrived at St. Louis Park High in September 1968, I took journalism as a sophomore from our then legendary high school journalism teacher, Hattie M. Steinberg. People often speak about the teachers who changed their lives. Hattie changed mine. I took her introductory journalism course in tenth grade, in room 313, and have never needed, or taken, another course in journalism since. It was not that I was that good. It was that she was that good. As I wrote in a column about her after she died, Hattie was a woman who believed that the secret of success in life was getting the fundamentals right. And she pounded the fundamentals of journalism into her students—not simply how to write a lede or accurately transcribe a quote, but, more important, how to comport yourself in a professional way and to always do quality work. I once interviewed an advertising executive for our high school paper who used a four-letter word. We debated whether to run it. Hattie ruled yes. That ad man almost lost his job when it appeared. She wanted to teach us about consequences.

Hattie was the toughest teacher I ever had. After you took her journalism course in tenth grade, you tried out for *The Echo*, which she supervised. Competition was fierce. In eleventh grade, I didn't quite come up to her writing standards, so she made me business manager, selling ads to the local pizza parlors. That year, though, she let me write one story. It was about an Israeli general who had been a hero in the Six-Day War, who was giving a lecture at the University of Minnesota. I covered his lecture and interviewed him briefly. His name was Ariel Sharon. First story I ever got published while on the staff. Little did I know how much our lives would intersect fifteen years later in Beirut.

Those of us on the paper, and the yearbook that she also supervised, lived in Hattie's classroom. We hung out there before and after school. Now, you have to understand, Hattie was a single woman, nearing sixty at the time, and this was the 1960s. She was the polar opposite of "cool," but we hung around her classroom like it was a malt shop and she was Wolfman Jack. None of us could have articulated it then, but we enjoyed being harangued by her, disciplined by her, and taught by her. She was a woman of clarity in an age of uncertainty. Her high school newspapers and yearbooks won top national honors every year. Among the fundamentals Hattie introduced me to was *The New York Times*. Every morning it was delivered to room 313 (a day late). I had never seen it before then.

Besides Hattie, I had other remarkable teachers who remain cherished friends to this day—particularly Miriam Kagol, my English teacher, and Marjorie Bingham, who taught me AP American history and indulged my fascination with Kennedy assassination conspiracy theories and my early obsessions with Israel, the Six-Day War, and the Middle East. I was intrigued to hear them both reminisce about the high quality of the public schools. "I had money to buy whatever books I wanted," Bingham remarked. "There were NSF [National Science Foundation] grants, there were national conferences you could go to. You never felt isolated in your classroom. [You felt as though] you were on a bigger stage and could talk to people from Illinois or California. [Today] a teacher will spend four or five hundred dollars of their own money on supplies. That just didn't occur" back then. St. Louis Park "teachers were encouraged by the administration to be creative."

This was an era in which Title IV-C Education Department grants enabled public school teachers to apply through their districts to create new curricula that other districts could purchase for low or no fees. It made high school teaching, for those who aspired to it, a very creative job—it wasn't just a matter of rinsing and repeating what was handed down from the central school's office. For instance, my World Studies teacher, Lee Smith, and his colleague Wes Bodin created a World Religions curriculum, stimulated by the multireligious nature of the student body in St. Louis Park and the desire by the St. Louis Park School Board to set some guidelines in 1971–1972 on what could and what could not be done religiously in the local schools. Their curriculum was adopted by schools all over the country. Bingham recalled that in 1977 she and Susan Gross, a teacher from Robbinsdale, one suburb over, won a Title IV-C grant to create an area studies program called Women in the World, to introduce high schoolers to women's history. They ended up distributing the curriculum they wrote nationally—more than a hundred thousand books.

There was a special kind of confluence taking place. An immigrant Jewish community that suddenly found the world open to it combined with a progressive Scandinavian civic ethic and the 1960s ferment all across America to produce an explosion of creativity out of this obscure place. Bingham's husband, Tom Egan, who also taught at St. Louis Park High and coached the cross-country team, put it this way: "There was a combustion that happened—and it had some rough edges. You cannot cook anything unless the water boils." Ah, yes, there were some rough edges. "In a classroom, you could not expect everybody to be quiet," said Bingham. "In most cases your Jewish students would challenge you and ask 'Why?' And they were always talking—and you thought they weren't listening, but they were. I remember a teacher who came in from out of state saying to me: 'You go to a synagogue and everyone is conversing. That is not a Lutheran thing!'"

I first met Miriam Kagol when I was a senior in high school. She taught my survey course on British literature and was the adviser to our literary arts journal, which she also founded. She taught me to enjoy Byron and Shelley and Keats and Yeats and great fiction in general—no easy task. She remembers that I often inquired of these great romantic poets, "Why don't they just say what they mean?" Kagol came from southern Minnesota, where her family had a hobby farm. She was twenty-two at the time and it was her first teaching job. "When I was hired at Park in 1967," Kagol recalled, "I signed a contract for fifty-six hundred dollars, and I told my dad and he looked at me in shock and said, 'You better earn every penny of it.'"

Kagol and I have been friends ever since. Reflecting on the role of the community as a value setter, Kagol mused:

> I remember I had one kid who plagiarized a poem for *The Mandala* [the high school literary arts magazine] and we published it, not knowing it was plagiarized, and found out after, and when we took him to the principal he told that student: "You have disrespected what this teacher stood for." I knew the principal would back me up no matter what the situation, and I had no worry that a parent would call up and ask for my head if I chastised their child. There was a community respect for the system and for the teacher—even if the parent thought the teacher was wrong.

There were no gated neighborhoods in St. Louis Park back then, or today. Before my friends and I could drive, we all used the same public bus system. When we were as young as ten or twelve years old—well before anyone had their driver's license—our great weekend treat was "going downtown." We'd catch the bus from St. Louis Park to Hennepin Avenue in the heart of Minneapolis, which cost ten or fifteen cents at the time. We would go shopping at Dayton's, which never involved purchasing anything. We'd just window shop, buy some caramel corn, and eat lunch at the Nankin, the most famous Chinese restaurant in Minneapolis. Then we'd take in a movie and catch the bus back home to St. Louis Park. I can't imagine what the waiters thought of these four twelve-year-old boys ordering off the menu at the Nankin. We must have looked so small in the big stuffed chairs. We were just kids, but our parents never seemed to worry about our freely roaming the city. Apart from the schools, the great public Mixmaster in Minneapolis was its string of lakes. The lakes were surrounded by some of the wealthiest homes in the city, but each lake was ringed by walking trails, bike paths, and public beaches open to all. I grew up walking with my mother around those lakes, and you inevitably saw everyone you knew there.

Sharon Isbin, born in 1956, was three years younger than me. I played clarinet—badly—in the St. Louis Park school band. Sharon became one of the world's greatest classical guitarists out of the same music room. Her father taught chemical engineering and nuclear physics at the University of Minnesota and served on the board of the Atomic Energy Commission. Her mother, a lawyer, had a lot of musicians and actors on her side of the family and insisted that Sharon and her two brothers and sister all take music lessons growing up.

"When I was nine, we all went to Varese, Italy, near Milan, for my dad's sabbatical year," Isbin recalled for me. "My brother wanted guitar lessons and my parents learned that there was a great classical guitar teacher in the area. But my brother had been influenced by the Beatles and had no interest in classic guitar, so I took his place."

When she returned home to St. Louis Park, Isbin found that her elementary school, Fern Hill, had a teacher who'd started a model rocket club. She was the only girl who joined, and on the side practiced guitar twenty minutes a day. "My father bribed me by saying 'If you practice guitar an hour a day you can launch your rockets.' Sharon eventually won a competition in 1970 to perform with the Minnesota Symphony Orchestra, "and walking onto the stage, I decided this was definitely more exciting than launching my worms and grasshoppers into space." Isbin also attended the St. Louis Park Hebrew School but eventually gave it up for French, which she felt would be more useful. She persuaded her parents by saying that for every Hebrew word she learned she forgot one in French.

"St. Louis Park High prided itself on being adventurous, innovative, and open-minded," Isbin recalled. "At one point they initiated modular scheduling, so I rigged up a system where in my junior and senior years I only went to school three hours a day and the rest was independent study at home. It meant I could practice guitar five hours a day. Without those hours, I would not have been able to win my first international competition." Isbin eventually earned degrees at Yale, became the founding director of Juilliard's guitar department, won two Grammys, and was described in a *New York Times* profile as "a trailblazer both for female musicians and for the guitar's place in the classical music world."

Looking back, she recalled, "this was a public school but the quality of the teachers was so high you could compare them to a college level. It was serious about learning and it nurtured a love of learning. My mentors Marj Bingham, who taught history, and Barb Smigala, who taught philosophy of literature, stimulated my thinking, questioning, and curiosity. It gave me the courage to believe there were no boundaries, and the same [came] from my parents: you can be anything you want to be. I did not have to battle the prejudices my father did for being Jewish and so I could focus my battles on being a woman in the guitar world, where it was assumed you would be a man, and on being a guitarist in the classical world, where it was assumed the instrument was inferior."

She also had to fight for her identity. Isbin came out as a gay woman when she was eighteen, in 1974. "My dad was okay," she recalled. "My mother took time to adjust that I would not marry a doctor and they eventually both became supporters." Her brother, a gay activist, died of AIDS.

During a chat one afternoon, as we both hit our sixties, Isbin said something to me that I think applied to so many of our generation that grew up then in St. Louis Park: "The freedom to ride your bike around all these lakes in the middle of this community or to walk around or cross-country ski—there was a sense of beauty and freedom combined. I felt no fear there and I felt no boundaries."

Of course that powerful spirit of community that generally characterized Minnesota was also a by-product of the harsh climate—subzero temperatures with frozen roads and slippery sidewalks, broken water mains, and snow that had to be removed—that characterized every winter. They all "made cooperation a necessity—not just a nicety," remarked Fred Astren, who now lives in the San Francisco Bay Area. "You may not like your neighbor, but you will always help him get his car going in the morning. He will help you up after you have fallen on the ice. And your boss will let you go home to deal with a winter emergency. As a result, people always know their neighbors, unlike the Bay Area, where there are few circumstances to bring neighbors together. The uncooperative neighbor in Minnesota is rare, because everyone knows that there will come a cold day when help is needed: Winter is always coming . . ."

Our next-door neighbor, Bob Bonde, moved to suburbia from his family's farm, and every winter he banked up the snow on the periphery of his backyard to form an ice rink that he filled with his garden hose. That is where I learned how to skate and play hockey. And then in the summer he plowed the same space into a garden plot, with perfect rows of corn, carrots, lettuce, and tomatoes. My sister Jane would lie on the ground and literally watch the carrots grow and constantly pester Bob about when it would be time for her to pull one out to eat. Our house was also at the top of the street, and our backyards and those on the parallel street all faced each other, without any fences between them. This arrangement created a long, natural football field of green space. As soon as most of the snow melted in April, I would take out my golf clubs and stand at the top of our backyard, which was slightly elevated, and hit five-irons 175 yards all the way down the block through the six backyards of our neighbors, sometimes reaching the open field at the end. I never once knocked out anyone's windows, and none of the neighbors ever complained. I have been very straight with a five-iron ever since!

That undeveloped open field at the end of our block was our wild frontier, where we played hide-and-seek among the tall shrubs, weeds, and trees. We did not know it, but it separated us from a giant factory owned by the Lithium Corporation of America. In researching this book, I learned from a story in *Twin Cities Business* on November 1, 2006, that "from 1942 to 1960, two firms, Metalloy and Lithium Corporation of America, produced lithium carbonate for the U.S. military, primarily for use in batteries and lifesaving equipment, from a property located at the end of Edgewood Avenue just off Cedar Lake Road in St. Louis Park. [That was only a few blocks from our house.] But while the company was carrying out its patriotic duty, it was also leaking lithium, fuel oil, and miscellaneous metals into the soil and groundwater beneath its plant." John C. Meyer III, in his memoir *Don't Tell Douglas,* claims that Metalloy used that factory in World War II to produce "an ingredient necessary for the Atomic bomb that was used on Hiroshima." My God! Between that plant and the creosote factory a few miles away, it is amazing that my sisters and I don't glow in the dark today. Two blocks in the other direction was a big public park filled with baseball diamonds in summer and hockey rinks in winter. There was a warming hut with wooden floors where we changed from boots to skates and came in for shelter from the cold when temperatures plunged far below zero. I can still smell that gas heater. You could pick up a hockey game any afternoon or evening, and the city provided lights to play by night. Marc Trestman was the quarterback of the St. Louis Park High football team, and three years younger than me. He went on to play for the University of Minnesota and then had a distinguished career as an offensive coordinator or quarterback guru for two college teams and ten NFL teams, capping his career as head coach of the Chicago Bears from 2013 to 2014. Being middle-class in Minnesota back then meant that almost everything you could imagine was accessible. After midnight on weekends, Trestman and his fellow jocks would occasionaly rent out the rink at the Metropolitan Sports Center, where the local NHL professional hockey team, the Minnesota North Stars, played. "There were no social messaging services at the time, no cell phones, nobody had credit cards," he recalled. "There were no ATMs. It was a hundred fifty dollars an hour after midnight. So, I think back now, 'How did we get twenty guys out to the [professional] ice arena at four in the morning on a Saturday, scrape together the cash, with no coaches, and just walk in and there was the Zamboni [ice-smoothing machine] and we just split up and played hockey for an hour?' "

A middle-class life then included a lot more spontaneity; money had not completely taken over public spaces, as it has today. "I remember one day my mom just said out of the blue, 'Let's go to the baseball game,'" Trestman told me. "The Twins were playing the Red Sox. So we just showed up and got tickets in the first row of the second deck. And Reggie Smith hit a foul ball and my mom reached up and caught it with one hand."

In those days, the Vikings played in the open-air Metropolitan Stadium and there were no skyboxes where the elites could find relief from the December chill with heaters and catered meals. We all froze together. I have a picture in my scrapbook that the then–*Minneapolis Tribune* took of my dad and me watching a Vikings game outdoors in winter, totally bundled and wearing knitted facemasks; it was truly an exercise in Arctic endurance. In high school, my friends and I had season tickets to the NHL North Stars. They cost $2.75 a game and put us up in the nosebleed section, where we sat next to the same four Polish guys from St. Paul every year. We cheered together, high-fived together, suffered losses together. We were "the skybox"—sky high near the rafters. I once came to a hockey game early and was interviewed by the North Star's announcer, Al Shaver, and won a transistor radio when the player whose name I picked from a hat actually scored the first goal of the game. Shaver was one of the great announcers of his day and imparted to me my first political science lesson that I never forgot: He ended every broadcast by saying, "Remember, when you win say little. When you lose say less—goodnight and good sports." It is still a rule I live by.

In 1970, the U.S. Open golf tournament was played at Hazeltine National Golf Club, in Chaska, Minnesota, forty minutes from downtown Minneapolis. I was in eleventh grade in high school and caddied regularly during the summer at our club, Brookview. Most of the clubs around Minneapolis were invited to nominate four caddies to caddy in the U.S. Open, and I was one of those selected by my club. In those days—and this is the real point—professional golfers were not allowed by the U.S. Golf Association (USGA) to bring a professional caddy to an "open," because amateurs were also invited and it was seen as giving the pros an advantage. A couple of weeks before the tournament, all of us local caddies gathered at Hazeltine and walked all eighteen holes of the course with its then head pro, Don Waryan, filling in a yardage notebook they gave each of us, detailing the distances to the greens from different trees and traps. Then we retreated to the clubhouse dining room. There, in the middle of the room, was a big silver bowl that had the names of every player in the tournament folded up on little pieces of paper inside. They called your name, you walked up to the bowl, stuck your hand in, and pulled out the pro for whom you would caddy. Talk about egalitarian! Some kid picked Jack Nicklaus, someone else picked Arnold Palmer, someone else picked the eventual winner, Tony Jacklin, and I picked . . . Chi Chi Rodríguez, the great Puerto Rican golfer and showman. He was tied for second place after the first day, made the cut, came in twenty-sixth, paid me $175, and gave me all the balls and gloves in his bag. I had the time of my life—and it could never happen to a seventeen-year-old kid today.

Because a few years later the USGA rescinded the ban on professional caddies at the Open, no high school junior would ever again get the chance to pull Jack Nicklaus's or Arnold Palmer's name out of a silver bowl and walk inside the ropes with him. Outside the men's locker room at Hazeltine today there is a picture of us crew-cut high school boys plucking our professional's name from that bowl—a lovely but distant memory of an age when, to paraphrase my childhood friend Michael Sandel, the Harvard philosopher, there were still things that "money couldn't buy."

Chi Chi, incidentally, gave me my first class in sex education. One day before our round, he was on the practice tee hitting balls, with a crowd watching from behind the ropes. He said to me loudly, "Tommy, I gave up giving golf lessons." I said, "Why, Chi Chi?"—not realizing I was his straight man.

"Well," he said, "there was this woman who came for a lesson and she was a slicer and I turned her into a hooker . . ." The crowd all started laughing. I didn't get it. I didn't know what a hooker was. My dad explained it to me later.

Some twenty years later, family friends of ours were visiting Chi Chi's home course in Puerto Rico and ran into him in the pro shop. As family friends will do, I am embarrassed to say, they asked him: "Do you remember who caddied for you at the U.S. Open at Hazeltine?"

And without missing a beat, they reported to me that Chi Chi answered: "Tommy." They then said to him, as family friends will do, "Do you know that he's more famous than you are today?" And without missing a beat, Chi Chi answered: "Not in Puerto Rico!"

## *Middle-Class Minnesota*

But what made those public spaces possible was the confluence of two things: a generally rising American and Minnesotan economy that elevated a rising middle class, and a unique generation of progressive politicians. And the two reinforced each other. There is nothing like a growing pie to sustain both public works and a politics that works, a politics of inclusion. I now fully appreciate that those of us who grew up in the middle class from the end of World War II to the early 1970s grew up during an extraordinary moment of American history. Or, as the Stanford University historian of America David Kennedy put it to me: "It was the greatest moment of collective inebriation in American history—the country was giddy with pride and opportunity." It was a time of "great compression of incomes and great shared prosperity—high growth and high equality."

The February 2015 annual report of the White House Council of Economic Advisers (CEA) looked at productivity growth in America post–World War II and labeled the years from 1948 to 1973 "The Age of Shared Growth," because of the way that

> all three factors—productivity growth, distribution, and participation—aligned to benefit the middle class from 1948 to 1973 . . . Income inequality fell, with the share of income going to the top 1 percent falling by nearly one-third, while the share of income going to the bottom 90 percent rose slightly. Household income growth was also fueled by the increased participation of women in the workforce . . . The combination of these three factors increased the average income for the bottom 90 percent of households by 2.8 percent a year over this period . . .
>
> This period illustrates the combined power of productivity, income equality, and participation to benefit the middle class.

I came of age exactly during that era. No wonder growing up then left me and so many others with an optimism bias and with an expectation that this kind of broadly shared prosperity should and would continue. It was a virtuous cycle of ascension. You felt the wind at your back—not in your face. Indeed, Representative Rick Nolan, a congressman from Minneapolis, likes to say that for my middle-class generation growing up then in Minnesota, "you had to have a plan to fail."

Senator Al Franken went to both public elementary school and junior high in St. Louis Park, but his parents shifted him to Blake, the main private school in Minneapolis, for high school. Franken was a rare exception in going to a private high school. In a February 28, 2015, speech to the Colorado Democratic Party, Franken elaborated on this moment:

> I remember back in 1957, when the Soviets launched Sputnik. They had nuclear weapons, and suddenly, they were ahead of us in space. Americans were terrified. I was six. My brother,

Owen, was eleven. And my parents sat us down in our living room in St. Louis Park, Minnesota, and said, "You boys are going to study math and science so we can beat the Soviets." Now, I thought that was a lot of pressure to put on a six-year-old. But we were obedient sons. And so Owen and I studied math and science. And we liked it! And we were good at it. My brother became the first in our family to go to college. He graduated with a degree in physics from MIT. And then he became a photographer. I also got into a really good college. And graduated. And then I became a comedian. My poor parents! But—we beat the Soviets. You're welcome!

As I mentioned, I grew up in St. Louis Park, a middle-class suburb of the Twin Cities. My dad was a printing salesman. We had a two-bedroom, one-bath house. And I felt like the luckiest kid in the world. And I was. I was growing up middle-class in America at the height of the middle class, back when being in the middle class meant real security. It meant you could put a roof over your family's head and food on the table. It meant you could send your kids to a good public school and take them to a doctor if they got sick. It meant you could take a vacation once in a while—although our vacation was always driving to New York to visit my uncle Irwin, my aunt Hinda, and my cousin Chuck. It meant you could count on your pension, and your Social Security, being there for you when you got older so you could have a comfortable retirement. And it meant you could take a chance on yourself. It meant that, if I worked hard and played by the rules, a kid like me could have a chance to do anything I wanted. Including being a comedy writer and a senator. In that order.

What resonated with me most about Franken's observation was his point that we had the economic security and psychological sense of being anchored in a community. As he put it, "you could bet on yourself—and you didn't worry that if I became a comedian, it was not as safe as the careers of other Harvard grads. I felt like it was ridiculous to believe that you could not make a living somehow."

I didn't become a comedian like Al, but I did start taking Arabic as a freshman at the University of Minnesota, and that got a lot of laughs from friends and family. There weren't a lot of Jewish kids studying Arabic at the university back then. My parents' friends used to ask them, "How in the world is Tommy going to get a job studying Arabic?" It beat the heck out of me, but it also never occurred to me that something wouldn't come out of it, so not to worry. No one was warning me that if I didn't get a STEM—science, technology, engineering, and math—education I would never feed myself again.

The leadership of the Minnesota business community was a big driver of the Minnesota way: they understood that government is there to compromise, make decisions, and support the private sector, and the private sector is there to create jobs and contribute to the public good, noted Lawrence Jacobs, director of the Center for the Study of Politics and Governance at the University of Minnesota's Humphrey School of Public Affairs. "In Minnesota the business community historically has been a real partner in building the state and keeping the two parties close to the center."

The roots of this run deep, as explained by a December 22, 2007, article in *The New York Times* entitled "Emerald City of Giving Does Exist." In the mid-1970s, it noted, Minnesota's leading businesses formed

> the Five Percent Club—in which Minneapolis–St. Paul corporations agreed to set aside 5 percent of their pretax income for philanthropy. Believe it or not—and it is a little hard to believe, given the modern emphasis on maximizing profits and pleasing Wall Street—the club still exists. Now known as the Keystone Club, it has 214 members, and 134 of them donate at the 5 percent level . . .
>
> The Guthrie Theater, the city's fine regional theater, recently moved into a sparkling new building by the river—one of five major arts organizations that have recently built new buildings or major additions. All of them were built, in no small part, with corporate contributions.

So it was no wonder that two years after I graduated from high school, the August 13, 1973, cover of *Time* magazine was devoted to a picture of a smiling Minnesota governor Wendell Anderson holding up a northern pike. The headline read, "The Good Life in Minnesota." At a time when the rest of the country was going through the agonies of Watergate, high inflation, and the Vietnam War, Minnesota was singled out as a "state that works." I remember that cover well. My dad had just died, and I had been accepted as a transfer student to Brandeis, where I would relocate a few weeks later, never again living permanently in Minnesota. But the state would also never leave me. Whether I lived in Boston, London, Oxford, Beirut, Jerusalem, or Washington, when people asked me, "Where do you live?" I always answered, "I live here, but I am from Minnesota."

## *Our Political Forefathers*

As we explained earlier, Minnesota wasn't always so nice, and so politically and economically inclusive—especially to blacks and Jews and other minorities. It is important to understand that it became more inclusive not simply because the economy improved after World War II, but because of some courageous political choices, by a unique generation of moderate Republican and Democratic-Farmer-Labor Minnesota politicians, namely Hubert H. Humphrey (a mayor of Minneapolis, senator, and vice president), Walter Mondale (a senator and vice president), Don Fraser (a congressman and mayor of Minneapolis), Eugene McCarthy (a senator), Arne Carlson (a Republican speaker of the state legislature and a governor) and Bill Frenzel (my congressman from St. Louis Park when I grew up and also a Republican), among others.

According to the St. Louis Park Historical Society website, in March 1936, "a lunatic fringe group called the Silver Shirts descended on Minneapolis, preaching anti-Semitism and paranoia to what they claimed were 6,000 followers in the state." The legendary CBS News editorialist Eric Sevareid, then a young reporter from Minneapolis, using his real name, Arnold, "was a journalist for the *Minneapolis Journal* and published a six-part expose of this group starting on September 11, 1936. The organization was led by William Dudley Pelley of Asheville, North Carolina, who chose to blame all of his problems on communists and Jews . . . Some of their most ridiculous ideas include: President

Roosevelt's real name was Rosenvelt, a Jew." Blacks suffered a similar and often worse fate, the Historical Society website noted. "In July 1947, the Governor's Interracial Commission of Minnesota issued 'The Negro and His Home in Minnesota.' Polling revealed that 63 percent would not sell their property to a black person, even if offered a higher price."

In the late 1940s and early 1950s, things started to change. Hubert Humphrey was a hero in our house in large part for the way he took on anti-Semitism when he became mayor and appointed a task force to eradicate it in city government. "The task force confirmed the allegations, and also shone light on discrimination against Blacks and American Indians. Humphrey turned the task force into a permanent Mayor's Council on Human Relations," the Historical Society reported. "Ordinances were passed in the next two years that outlawed anti-Semitic and racist practices in housing and employment."

We think of Hubert Humphrey today as a great civil rights crusader in the realm of black-white relations, but he got his start combating anti-Semitism among whites, explained Lawrence Jacobs: "One of the things that defined Minnesota was that the civil rights movement started here—but it was not about blacks. It was about Jews. Before Humphrey gave his famous speech calling for equality for blacks at the 1948 Democratic convention, he fought anti-Semitism in Minneapolis. The St. Louis Park you grew up in would never have been possible in the Minnesota of the 1930s and 1940s . . . You grew up in a period when it was possible to grow up and live life based on merit—and be Jewish," but that was not true in the 1930s and 1940s, when there were barriers all over America and in Minnesota to Jews. "Before Humphrey declared war on racism," Jacobs added, "he declared war on anti-Semitism, and that allowed this group of people in St. Louis Park to be unleashed on merit so their creativity and inspiration had space to grow."

Humphrey's transition from fighting anti-Semitism to fighting racism generally was defined by his speech to the 1948 Democratic Convention in the Philadelphia Convention Hall on July 14, 1948. The writer Thomas J. Collins described the scene in a retrospective fifty years later for the Hubert-Humphrey.com historical website: "Sweating through his plain black suit, his thin black hair matted to his head, Humphrey looked over the crowd that included national party leaders who advised him not to speak but desperately wanted him to—and those who threatened to walk out of the hall if he did. During the next eight minutes, Minnesota's happy warrior would for the first time engage a national political party in the civil rights battle that continues in fits and starts today." Making his case in that speech, Humphrey famously declared: "My friends, to those who say that we are rushing this issue of civil rights, I say to them we are one hundred seventy-two years late. To those who say that this civil rights program is an infringement on states' rights, I say this: the time has arrived in America for the Democratic Party to get out of the shadow of states' rights and to walk forthrightly into the bright sunshine of human rights."

It is hard to recall now what radical fightin' words those were. Several dozen southern delegates stormed out of the convention, led by South Carolina's governor, Strom Thurmond. The southerners would eventually back Senator Richard B. Russell of Georgia as a protest candidate against Harry Truman, and Thurmond himself would run for president on the Dixiecrat ticket. These events marked the beginning of the end of the Democratic Party as a coalition of southern conservatives and northern liberals—ultimately setting the stage for the Civil Rights Act of 1964.

Humphrey was a dyed-in-the-wool progressive, and he helped to infect a generation of Democratic politicians, and even many Republicans, in Minnesota. Growing up, my two congressmen in St. Louis Park, which was part of the Third Congressional District—the most Jewish, Democratic, and liberal district in the state—were both liberal Republicans: Clark MacGregor, who served from 1961 to 1971, and Bill Frenzel, who served from 1971 to 1991.

I interviewed Frenzel in 2014, shortly before he died at age eighty-six, about the evolution of Minnesota politics as I was growing up there in the 1950s and 1960s. He epitomized a now-extinct species—the liberal Republican. Frenzel was first elected during my junior year of high school, in 1970. Wherever I traveled in the world, I always referred to him as "my congressman." Sitting in the cafeteria at the Brookings Institution in Washington, where he was scholar in residence, Frenzel mused about those early days:

> We call them the kinder and gentler days. I was born in St. Paul and came back from the Korean War and worked for a family

company in Minneapolis. I didn't know whether I was a Republican or a Democrat. My family was upper middle class and doing just fine, but politics was not a big deal in the family. A lot of my father's friends were un-Rooseveltian, but my father never let me pick that up. FDR came to Minnesota once and my father took me down there, and my father was cheering and I was cheering. I was on my father's shoulders, and I said: "Why is everyone cheering? I thought we didn't like this guy," and he said, "No, son, you only get one of these at a time."

Frenzel recalled that when he went to the legislature,

there was a fair amount of camaraderie. It was the farms versus the central cities, and we suburbanites had a lot in common in supporting the central cities, and we worked a lot together. Hubert [Humphrey] was doing his thing from the Senate. He was also nice. If you met him on the street, he was nice to you. Everything about it was different from today. You tried to cooperate, and if you couldn't, you voted against. In the Minnesota legislature we probably had half a dozen party line votes out of five hundred or six hundred in the first term I was there. This was 1963 to 1969. It was just the way you expected to do business. You were expected to be able to make a deal. Our family business was transportation and distribution. If you needed a contract, you should not do business with that guy—you just needed a handshake. We were personally conservative, but in general [we were] liberal. Minnesotans paid their bills, saved their money, taught their children to save, but they also wanted to take care of their neighbors and build a good community. Today [Minnesota] is not like it was—but it is still better than anywhere else. The politics has gone sour, but the people have not.

When did politics start to change? I asked Frenzel:

It started to change when Reagan came in and challenged Democrats who controlled the House. Then the name-calling started. Then the House Bank scandal emerged. That got into campaigns that were really personal. Over the years, campaigns

> got progressively nastier. When I started in politics my mentors said, "For Heaven's sake, never mention your opponent's name," and now you begin by telling everyone what a skunk your opponent is . . . My campaign was run by guys in my kitchen. Now you hired a guy from Baltimore or L.A., and he didn't have to live in your district and didn't care about the wreckage he created. When the Republicans took the House of Representatives in 1994, they did not know how to be a majority, and Democrats did not know how to be a minority. When I was in the House, Republicans knew "their place." We had never been a majority, and it looked like we never would be, so we had no choice but to try to work hard and compromise, and we had to decide whether it was half a loaf or a third of a loaf—and so we made those deals.

It helped, though, Frenzel added, that "I had a tolerant constituency. It was also a consistent constituency. Things did not change." Most of the households in his district were consistent two-earner families, fairly well-to-do, he added, "who had moved to the suburbs and put their kids through good schools and knew why they were there and wanted to stay there. They wanted a congressman to be handsome, brave, and true and they wanted me to pay attention to them. I never felt that other than a lunatic fringe of the left and right that people were really pushing me. The principal thing for people was: were you paying attention to them. I don't think people thought that I was a Republican or Democrat or cared much."

Indeed, my mom, who was a dyed-in-the-wool liberal, always voted for Frenzel. When he would run for reelection, Frenzel said he used to buy "a big highway sign that just said: 'Frenzel for Congress.' It didn't say Republican."

After retiring, Frenzel, not surprisingly, served as a special adviser to a Democratic president, Bill Clinton, to help win passage of the North American Free Trade Agreement.

Walter Mondale, who was a Minnesota senator when I was growing up—1964 to 1976—told me a similar tale from the other side of the aisle:

> I grew up in a small town on the Iowa border. My dad was a minister in several different churches. Every five years we would

move. My mom was a musician and taught almost every young person to play piano and ran the choir in Elmore. My family and parents always expected us to be involved in the community and to be for things. Dad was an old Farmer-Labor guy. And Hubert's dad was a big social activist and his mom was, too. Don Fraser—the same thing. We took Minnesota, which had been an isolationist state—and Minneapolis was one time called "the capital of anti-Semitism"—and we changed all that. We changed the political culture.

In terms of state politics, added Mondale,

It was an optimistic time. We were all going to make something out of our lives. Education would get us there and it was available to everyone. The GI Bill gave everybody a chance to go to the university or beyond to get professional training. All over the state people became professionals. There was an equality of income and opportunity. Things just kept getting better as we went along, and as we went along you could see it was working. People were getting ahead and the economy was getting better. You could get an education, and you could see it and feel it. And Minnesota had this bipartisanship. We had progressive Republicans. We used to fight over which party did the most for the University of Minnesota, and [the Republican leader] Arne Carlson and the Republicans would never cede that to us. And the university loved it, of course, and would encourage the competition. It was not to slash and burn. And if someone showed up with that kind of politics, they would get turned down right away.

Speaking of those golden years after World War II and up to the mid-1970s, Mondale concluded:

We all just expected things to just keep getting better . . . The GI Bill babies were making a new life for themselves and then a lot of us spread our wings and went to Washington and we took a lot of Minnesota with us . . . I often think about the fact that racial tensions, so central to many states, were not as strong in

> Minnesota. Oregon and Washington were like us. It allowed us to be very progressive on civil rights early. And they used to criticize Hubert and say that he did not know what he was talking about because ours was an all-white state.

But Minnesota was not all white, Mondale added, and

> we always worked on making the community work, and getting the minimum wages up and spending on early childhood education. Nationally, [we need] to reclaim this momentum . . . I am depressed about how this paralysis has changed all that. Today, instead of community, you have this great sorting, and people are divided out . . . That is hurting us. I have seen big money in these campaigns and nobody knows where it is coming from. Does the [Supreme] Court have any idea what that Citizens United case has done to the public life of this nation? Money—you needed a little bit, but it wasn't important. Now it is everything.

What a blessing it was to come of age politically under the tutelage of such politicians. The experience shaped the political outlook of many of my friends, including Michael Sandel. Sandel is now a renowned political philosopher at Harvard, where his courses attract as many as a thousand students a semester. The titles of his books—including *Democracy's Discontent*, *Public Philosophy*, *Justice: What's the Right Thing to Do?*, and *What Money Can't Buy*—reflect a persisting concern with the fate of democracy, community, and civic virtue in our time. I asked him to reflect on how Minnesota helped to shape the civic sensibilities that inform his writing and teaching. He explained:

> Although we barely perceived it at the time, the civic idealism of our Minnesota upbringing shaped our view of what it means to be a citizen. The Minnesota we knew as young boys was a place that cultivated a democratic sensibility, though not in an explicit, heavy-handed way. Civic sensibilities were imparted through well-supported local and municipal institutions—strong public schools, public libraries, public parks, and recreational facilities. We imbibed our civic education from the

> landscape of everyday life. Without quite realizing it, we imbibed the conviction that politics and civic activism can make the world a better place . . . These were stable middle-class communities that nurtured the belief that politics can be about the common good. The Democratic Party in Minnesota was called the DFL—the Democratic-Farmer-Labor Party. It grew out of a progressive-era alliance between farmers and workers that pushed for agrarian reform, strong unions, social security, and public ownership of railroads and utilities. This progressive tradition still infused Minnesota politics when we were growing up. It encouraged us to care about the wider world. Its representative figures—Hubert Humphrey, Orville Freeman, Walter Mondale—were remarkable politicians, full of optimism and idealism. Today, we think of Humphrey, tragically, as the establishment politician he became as Lyndon Johnson's vice president during the Vietnam War. But he began his career as a bold proponent of civil rights.

During our childhood, Sandel continued, "this Midwest populist tradition left its mark—on national politics, and on us. We were eleven years old when [President Lyndon] Johnson signed the Civil Rights Act, and fourteen years old when Eugene McCarthy, another Minnesota senator, challenged LBJ in the New Hampshire primary to protest the war in Vietnam."

Sandel also pointed to another, subtler source of our civic education: "The Minnesota of our day offered a rich array of class-mixing public spaces and experiences. In the suburbs at least, the public schools were strong. Public parks and recreation facilities were plentiful and widely used by people from all social backgrounds. The Minnesota State Fair attracted people from all walks of life. So did Metropolitan Stadium, where baseball fans gathered to root for the Minnesota Twins."

Going to a baseball game "was a more democratic experience in those days," Sandel observed:

> Of course, seats behind home plate were always more expensive than the cheap seats in the bleachers. But the difference was not as vast as it is today. A bleacher seat cost about a dollar, and

> a box seat about three dollars and fifty cents. As a result, going to a baseball game was a class-mixing experience. Business executives sat side by side with teachers and mail carriers. Everyone drank the same stale beer, ate the same soggy hot dogs, and waited in the same long lines for the restrooms. And when it rained, everyone got wet. Of course, we didn't go to Metropolitan Stadium for a civic experience; we went to root for the Twins, and to see Harmon Killebrew hit home runs. But the class-mixing conditions at the ballpark made for a shared democratic experience. These conditions also obtained—not perfectly, but for the most part—in the neighborhoods, in the public schools, and in most of the places we inhabited. It made for an inadvertent education in democratic citizenship.

Going to a ball game is different now, he continued. "Like most sports teams, the Minnesota Twins now play in a corporate-named stadium, called Target Field, replete with luxury skyboxes offering 'gourmet dining and bar service' and 'exclusive, dedicated concierge service,' where VIPs can watch the game in air-conditioned comfort far removed from the common folk in the stands below. So much for the soggy hot dogs and shared democratic experience. In the age of the skybox, it is no longer the case that everyone gets wet when it rains."

Sandel sees something similar unfolding throughout our society. "Today, people of affluence and people of modest means live increasingly separate lives. We live and work and shop and play in different places. We send our kids to different schools. I call this the 'skyboxification of American life.' This marks a departure from the Minnesota of our youth. It is corrosive of citizenship and democratic equality. At the time, we scarcely noticed this democratic civic landscape. It formed the background conditions of everyday life. It is more evident in retrospect, now that it has become a distant memory."

Sandel's thoughts were echoed by another St. Louis Park alum, Norman Ornstein—the political scientist, resident scholar at the American Enterprise Institute (a Washington, D.C., think tank) and one of the most oft-quoted analysts on American politics and Congress. His books include *It's Even Worse Than It Looks: How the American Constitutional System Collided with the New Politics of Extremism*, *The Permanent Campaign and Its Future*, and *Intensive Care: How Congress Shapes*

*Health Policy*, all co-written with Thomas E. Mann. Norm is five years older than me, and was actually born in Grand Rapids, Minnesota, where his dad moved from Canada to open a men's clothing store. His mom, though, was from North Minneapolis, and the family moved to the city when Norm was four and stayed until he was nine, when he attended the St. Louis Park Hebrew school and middle school. Then the family moved to Canada for a few years, and then moved back to St. Louis Park. He graduated from high school at age fourteen and entered the University of Minnesota as a sophomore at age fifteen.

When I asked what impact growing up in St. Louis Park had on him, Ornstein began our conversation by pulling out of his wallet his ticket from the seventh game of the 1965 World Series, October 14, in which the Minnesota Twins, the American League champions, lost to the Los Angeles Dodgers at Metropolitan Stadium. Broke my heart. Norm's too. But that loss wasn't all that stayed with us all these years. What he saw growing up, said Ornstein, was a politics driven by a passion for social justice, a passion for fair play and civility, and a "pragmatic looking-for-political-solutions" expectation by the public and "a deep respect for institutions." It is little wonder, therefore, he added, that his own career as a political scientist has been shaped around "working to protect, enhance, and improve institutions of government and educating the public about how to participate. I don't think I would have had that passion if it were not for how much of it was inculcated into me as a child growing up in Minnesota."

## *That Thing in the Water*

I am not naïve about my childhood—or about Minnesota or St. Louis Park. There was a lot that was also wrong about that time and place where I grew up. Racism was still rife. Sexism was still rampant—if many of my teachers were amazingly talented women, it was in part because the full world of work was not yet open to them. Gay rights were on virtually no one's agenda, and that left so many people hiding in closets. Those, alas, were norms prevalent throughout the country in those days and it's good that we have replaced them.

But if you were lucky enough that your life was not constrained by these prejudices, it was hard not to be impacted by all that was also right about Minnesota and St. Louis Park back then. And in my case it was hard not to carry with me for a lifetime a sense of optimism that human agency can fix anything—if people are able and ready to act collectively. And it was hard to leave there and not carry with me for a lifetime an appreciation of how much a healthy community can both anchor and propel people.

St. Louis Park was exactly what the political philosopher Edmund Burke was describing in his classic *Reflections on the Revolution in France* (1790), when he hailed the community, or what he called the "little platoon," as the key building block and generator of trust for a healthy society.

"To be attached to the subdivision, to love the little platoon we belong to in society, is the first principle (the germ as it were) of public affections," Burke wrote. "It is the first link in the series by which we proceed towards a love to our country, and to mankind. The interest of that portion of social arrangement is a trust in the hands of all those who compose it; and as none but bad men would justify it in abuse, none but traitors would barter it away for their own personal advantage."

St. Louis Park and Minnesota, when they were at their best, offered many of their citizens the opportunity to belong to a network of intertwined "little platoons," communities of trust, which formed the foundation for belonging, for civic idealism, for believing others who were different could and should belong, too. The world today gives us so many reasons and tools to hunker down and disconnect. St. Louis Park and Minnesota gave many of us who grew up there the opposite—it gave us reasons to believe that we could and should connect and collaborate, that pluralism was possible, and that two plus two often could add up to five.

In retrospect, though, I also realize what a relatively small distance we had to travel to bridge the economic and cultural gaps between us. That is not true today. In this age of tightening global interdependence and intimate contact between more diverse strangers, the bridges of understanding that we have to build are longer, the chasms they have to span much deeper. And that only makes the need for community building and healthy communities that can anchor diverse populations much greater.

Is that a bridge too far in too many places? I honestly don't think so—with the right leadership. But before I could even consider how we rise to this steeper global challenge, I needed this refresher course. I needed to go back and reconnect with that time and place in my life where politics worked, where community spirit was real, where public institutions were respected, where my friends were my friends, not "followers" on Twitter or icons on Facebook, and, yes, where when people really get mad at a reckless driver who almost kills them, they almost honk.

THIRTEEN

# *You Can Go Home Again (and You Should!)*

I seem to have a thing with parking garage attendants.

I was doing book research back in Minnesota in early 2016. I had rented a Hertz car and on January 9 I returned it to the airport that morning to catch my flight back to Washington, D.C. It was absolutely bone-chilling cold. I was wearing a heavy down parka. When I dropped off my car at Hertz, there was only one attendant on duty and he immediately flashed me a smile. His name was Qassim Mohamed, forty-two, and he had helped me at least once before. He was a news junkie and had engaged me about politics. I had not seen him for a while, though, and I couldn't remember if he was Arab or African. We chatted a bit as he went through the paperwork for my rental, and at the end I said to him, "Remind me where you're from?"

"Somalia," he said, "but we feel home here now."

What a nice thing to say, I thought. I hadn't asked him anything about how he felt about living in Minnesota. He just volunteered that he felt at "home." But there was one more thing he wanted me to know about his new home. His head was covered by the hood of his Hertz jacket—and as we spoke we could see each other's breath—so it was with a big grin that he added, "Different weather."

Different weather from Somalia for him—but now the same Minnesota for the both of us. What a remarkable place, I thought later. Four decades after I left, I can still return and feel at home; and a decade after he arrived, a Somali refugee can feel at home, too.

Our little exchange immediately reminded me of a conversation I had had with the former vice president Walter Mondale the previous August. I had taken him to lunch at a fish restaurant in the office tower of his law firm in downtown Minneapolis. A man of enormous decency and integrity, Mondale is one of my favorite people. We talked a lot about how enduring some of these Minnesota–St. Louis Park values had become. As we got up to leave, Mondale, his gait slowed by his then eighty-seven years but his mind as sharp as ever, remarked to me: "You know, it sustains itself—there is a continuity. Humphrey is gone, but the elements he started are alive in a new generation twice removed from him."

Coming back to Minnesota and St. Louis Park nearly forty years after I had left for college and a career, it was obvious to me that Mondale was right—and then some. With seventeen Fortune 500 companies choosing to headquarter there, and with the website Patch of Earth declaring that the Twin Cities had topped a combination of the seven major national ranking lists of "best cities" in which to live and raise a family, something must still be working, especially when you remember this place is a frozen tundra five months of the year.

But the question that tugged at me, and I asked over and over was: What was that "it" that was being sustained? I needed to know because I wanted to bottle "it" and share "it." Nothing, it seemed to me, would be more useful in this age of accelerations. Having returned home to reconstruct what had worked in the past to make my community an inclusive place that could anchor and propel many of its citizens, I wanted to understand what was still working today—and that is what this chapter is about.

I eventually concluded that the "it" starts with the fact that Minnesota, and even little St. Louis Park, has and had a critical mass of leaders who year in and year out came to politics and power in order to govern. They squabble and gridlock as much as any in the country (and even occasionally throw up the odd wrestler, like Governor Jesse Ventura, to take a hand at making a mess of the place), but at the end of the day, more often than not, they've forged compromises for the greater good of the community. Yes, that is what lawmakers are expected to do, but the venomous polarization that has swept across American politics in the last two decades has made that no longer the norm, or even the expectation, in Washington, D.C., anymore.

At the same time, there was and remains an unusually high degree of public-private collaboration in Minnesota, and St. Louis Park, where a critical mass of businesses view themselves as not just employers, but citizens, who have a corporate obligation to help fix local socioeconomic ills and whose executives are expected to actually volunteer in the community to do that. Again, what a stark contrast to Washington, D.C., where big business, post-2008, has disappeared from the national scene and debate—partly due to self-inflicted moral gunshot wounds by Wall Street bankers, partly because big business has been unfairly demonized post-2008, and partly because big American multinationals now have so many customers and employees overseas that their very sense of "American citizenship" has been diluted. As a result, they've largely given up trying to shape the national agenda on the big issues such as education, trade, and immigration the way they once did.

Additionally, the public in Minnesota and St. Louis Park have come to expect both politicians and business leaders to engage in these best practices; politicians are expected to compromise in the end, and corporations are expected to contribute to the community.

"CEOs here make clear that they want things to happen and for the two parties not to always be in blocking mode," said Lawrence Jacobs, of the Humphrey School of Public Affairs. "It's not Kumbaya inside the legislature, but the culture is that it's not acceptable to just be a blocker and ignore the reality."

All of these positive aspects of "it" over time built a lot of "social capital"—that is, trust—between and within the public and private sectors, and that trust has filtered back to reinforce these positive habits, enabling them to be sustained. Do I even have to mention what a contrast that is to Washington, D.C., where there is zero trust between the parties, or between them and the private sector, so the great engine of American growth—our public-private partnerships to promote research, infrastructure, immigration, education, and rules that incentivize risk-taking but prevent recklessness—has pretty much ground to a halt?

If one is to be honest, though, there was also another not-so-pretty "it" that made Minnesota work: the meme of "Minnesota nice" brushed under the rug systemic racism in housing and policing, particularly regarding African Americans. While the African American minority community in Minnesota was relatively small, it had a history of activism dating to at least the early 1960s. There were race riots in Minneapolis in 1967—and also a black power movement, among other mobilizations.

Nevertheless, stubborn de facto racial segregation in housing and employment, which continues to this day, kept enough blacks—and Native Americans—out of sight of enough whites for many whites to assume that things were in fact "Minnesota nice" for everyone. Recently, the shooting deaths of two unarmed black men by white police—one in North Minneapolis in November 2015 and one in suburban St. Paul in July 2016—helped to pierce that veil. So did a 2015 study by the American Civil Liberties Union that found that "black people in [Minneapolis] are 8.7 times more likely than white people to be arrested for low-level offenses, like trespassing, disorderly conduct, consuming in public, and lurking. Native Americans . . . are 8.6 times more likely to be arrested for low-level offenses than white people." Indeed, *The New York Times* reported that Philando Castile, the thirty-two-year-old school cafeteria worker who was shot and killed by a white policeman near St. Paul when he reached for his license after being pulled over, had previously been "pulled over by the police in the Minneapolis–St. Paul region at least 49 times, an average of about once every three months, often for minor infractions."

The good news is that Minnesota today has a much deeper public awareness of the aspects of "it" that have been working all these years—and that need to be preserved—as well as of the problems that cannot be overlooked any longer. African Americans and Native Americans are no longer willing to tolerate separate and unequal schools or mistreatment by police—and, to the state's credit, neither are many whites. When you add it all up, though, it means that the integration and community-building challenge faced by Minnesota today is now more difficult and more necessary.

It is more difficult because it involves integrating not only African Americans, Native Americans and Latinos in larger numbers, but also more traumatized populations such as Somalis and Laotian Hmong, who have fled to Minnesota from the World of Disorder. It also involves integrating African Americans who have "immigrated" to Minnesota from dangerous and disorderly neighborhoods in Chicago, Indianapolis, and Detroit.

To put it another way, I left Minnesota and St. Louis Park in 1973 to discover the world, and when I returned four decades later I found that the world had come to Minnesota and St. Louis Park. To be specific, St. Louis Park High School is now 58 percent white, 27 percent black, 9 percent Hispanic, 5 percent Asian, and 1 percent Native American, with the black population equally split between African Americans and Africans, mostly Somali Muslims, who immigrated to Minnesota in the last two decades and found St. Louis Park one of the most welcoming communities in which to settle—just as my Jewish parents did in the 1950s. Of the white students, the majority are Protestant and Catholic and about 10 percent are now Jewish. My high school, which had almost no Muslim students in my day, now has more Muslims than Jews. They serve halal meals in the cafeteria, and down every hallway you see young women who are covered.

The same demographic shift happened in the Twin Cities. Today, the Minneapolis Public Schools have 67 percent students of color, including Hispanic and Native Americans, and in St. Paul, that number is now 78 percent, with the single largest group being Hmong. For the entire Twin Cities metro area, the lower the grade, the more students of color, so the trend line will make it even more diverse. Roughly one hundred different languages are now spoken in the Minneapolis school system. The Twin Cities Metropolitan Council forecasts that two of every five adults in the Minneapolis–St. Paul area will be a person of color by 2040. In other words, this diverse population will be the pool from which Minnesota's Fortune 500 companies, start-ups, and small businesses will be drawing more and more of their workers.

At the same time, not every Somali has managed to make Minnesota "home" as much as my friend Qassim at Hertz. On November 19, 2015, CBS News reported that a new study by Congress found that "more than 250 Americans have attempted to join ISIS, and one in four of them is from Minnesota . . . The Cedar Riverside community in Minneapolis . . . has the largest Somali population in the country. Many came as refugees in the 1990s." The unemployment rate in Cedar Riverside is 21 percent, three times the state average. "And an alarming number of young Somali men from this neighborhood have left to join extremist groups. Since 2007, two dozen have joined Al-Shabab in Somalia."

If rising to the challenge of integration is more difficult than before, it is also, as noted, more important than ever before—because this is the same challenge now faced by communities all across America (and

Europe, for that matter). We are becoming a majority-minority nation, the expanding World of Disorder will only increase that trend—and it's all happening when the skills needed for all middle-class jobs are rising and lifelong learning will be required to keep them. In other words, Minnesota and St. Louis Park are not outliers anymore—they are microcosms of the central challenge for America today: Can we still keep making *e pluribus unum*—out of many, one—in the age of accelerations?

That's what I came back home to find out. And right now I'd say the jury is still out. I am not making any predictions; this is hard stuff, a lot harder than integrating Scandinavians and Jews in the 1960s. But here is the best news that I found coming back home: a lot of people there of different colors and faiths clearly want to get caught trying to make this work for another generation, and to get caught trying to make Minnesota truly "nice" for a lot wider circle of citizens than in my day.

Amory Lovins likes to say whenever people ask him if he is an optimist or a pessimist, "I am neither an optimist nor a pessimist, because they are each just different forms of fatalism that treat the future as fate and not choice—and absolve you from taking responsibility for creating the future that you want. I believe in applied hope."

I found a lot of people in Minnesota and St. Louis Park, from diverse backgrounds, still eager to apply hope—eager to innovate at the community level to fortify their eye in the age of accelerations—without knowing how the story will end.

Let's take a quick tour, starting at St. Louis Park City Hall.

## *What Are You Gonna Try Next?*

It is August 2015, and I am sitting in a conference room with the then St. Louis Park mayor Jeff Jacobs; the city manager, Tom Harmening; and the city's chief information officer, Clint Pires. Jacobs had been mayor since 1999 and on the city council since 1991. He is an unusual mix of Andy Griffith, Machiavelli, and Yogi Berra. That is, he has learned a ton about politics and human behavior through the window of a small-town city council, and he is capable of condensing his wisdom into memorable one-liners that Yogi and Machiavelli would have admired.

The local newspaper, the *Sun Sailor*, collected a few on December

9, 2015, to commemorate his retirement, including: On the city council, "Our job is to have seven people come together to disagree, then do it again the next week." When the power went out in a harsh Minnesota thunderstorm, Jacobs said: "Told my kids they need to watch TV by candlelight." And my own personal favorites: "I've always wanted to walk into a fire station and yell 'movie.'" And "Littering has two parents—the guy who dropped it and the guy who walked past it. People here will pick up a Mountain Dew can." Finally, "I am Republican by birth and a Democrat by choice—and now I have no time for either."

Since we've been talking a lot about what St. Louis Park got right, I began by asking what was the biggest thing they got wrong. All three of them smiled knowingly and proceeded to tell me the story: In 2006, after scores of public hearings, endless hours of research and debates, the city council voted to make St. Louis Park the first town in Minnesota to offer free public Wi-Fi. It is exactly the kind of thing St. Louis Park would do. After a close vote, the city council chose Maryland-based Arinc Inc. to build what would be the country's first citywide, solar-panel-powered, wireless Internet service. Soon afterward, these wireless radio towers went up all over St. Louis Park—with their signature solar panels at the top.

And then the first winter arrived.

Snow and ice piled up on the solar panels and didn't melt off according to plan. The whole system failed. Overnight it turned into a giant white elephant that had to be scrapped after eight months. The city eventually filed suit against Arinc for the amount of the project—$1.7 million—not chump change for my little town.

The day after they took all the solar panels and poles down, Jacobs recalled, "I stood up at a Chamber of Commerce meeting—and one of the poles was in my backyard—and what I said was this: 'Ladies and gentlemen, [installing that system] was decided on a four to two to one vote. And do you want to know who was the idiot who cast the deciding vote for that? That would be me.' We had a council member at the time who was an engineer, Loren Paprocki, and he had said during our meetings [deciding on the system], 'I just can't support this. I don't think it will work.' And he said something that I will never forget as long as I live. He said: 'We robustly debated it. I want you guys to know I am not going to support it. I just want you all to know that until this passes, I will be against it. But once it passes I will be 110 percent for it, because I don't want it to fail. [Afterward], he was the last guy in the world to say 'I told you so.'"

The job of the council, Jacobs added, "is to get together and debate and discuss—but to do it in a way that preserves the relationship so we can get together next week and do it again." And the key to that, he added, was always trusting the community with the truth—"telling the community that [the solar Wi-Fi] is going south" as soon as that was apparent.

But to my mind the most revealing aspect of this incident was related by Pires, who oversaw all the technical aspects and had an actual heart attack soon after it failed. Pires recalled for me the day they announced the system was being dismantled—and before his heart gave out: "After we made the announcement, I went over for lunch to a coffee shop next to City Hall. It was called the Harvest Moon. And a guy at the counter there recognized me. He said: 'Aren't you the Wi-Fi guy?'"

And then the man said something that blew Pires away. He said: "Too bad the company failed to get it to work. What is the city going to try next?"

Too bad that didn't work out. What is the city going to try next?

"I never forgot that," Pires told me. "The community can tell when you are trying to work for them and being responsive."

That is trust at work. Contrast that with what goes on in Washington today. Can anyone imagine any senator or representative saying to any president from another party regarding any issue today: "Too bad your idea didn't work. I know you meant well for the country. What should we try next?"

In 2011, U.S. taxpayers had to write off $535 million in federal guarantees, extended by the Obama administration, to a solar-panel venture start-up, Solyndra, whose technology also failed to deliver. This led to years of Republican recriminations, investigations, and allegations. We should not just shrug off the loss of $535 million, but venture investing isn't called "venture" for nothing; some projects are going to fail. The larger point is that in Washington, D.C.—no matter what the issue or the party—you are guilty today until proven innocent. In a healthy community, you are innocent until proven guilty, and even then people will cut you slack if they think you made a good-faith effort.

"Sometimes the wings come off the plane," said Mayor Jacobs, "but people will accept that—if they think you're trying to get to outer space. We were trying to do the right thing. The community was extraordinarily accepting of that. If you are terrified all the time of being excoriated in the press for some minor screwup, well, I have news for you, all progress happens in fits and starts . . . The space project would never have happened after the first rocket blew up if people did not accept that." If you want to change the way people view government, "you have to change the way government views people. If you view them as a necessary evil, they won't trust you—that is how they will view you."

But government also has to do the little things well, added Jacobs, "because they are not little—the stop signs, the curbs, the sidewalks, mowing the parks—[they are] what make people feel like they are living in a community . . . We have only one stock in trade—it is not building sidewalks or plowing the streets—it is trust, and if you lose that, you have nothing."

One of the reasons St. Louis Park has generated such a high level of trust is because it has taken civic engagement of the kind Michael Sandel talks about extremely seriously. It packs a lot of democracy into a small place. With only forty-seven thousand people, it has not only a city council, but thirty-five identified neighborhoods, and thirty of them have their own neighborhood associations, which the mayor and city manager use to build consensus and generate trust for all big decisions.

The city council in St. Louis Park is nonpartisan, although voters know each council member's tendencies when he or she runs. "When you run as a Republican or Democrat you automatically get typecast with a certain set of ideas," Jacobs told me. "[But] for the people subject to the decisions we make, what we do is less important than how we do it—the process we go through so people will trust it . . . There is a lot of transparency. If we don't communicate with the public before we make decisions, we hear about it," because, referring to the thirty neighborhood councils, "we have thirty little city councils inside our city."

The city provides two-to-three-thousand-dollar-a-year grants to each neighborhood to create its own neighborhood board and hold picnics and other events to create a spirit of inclusion, such as carving out a neighborhood garden or green space. You can't get the grant money, though, unless you've organized a neighborhood board with a president and treasurer.

"We've had other cities come in and study it and try to copy it," said Jake Spano, who took over for Jacobs as mayor in 2016. It's all about "getting to know your neighbors and what they see for the neighborhood . . . I grew up in Lawrence, Kansas, very liberal. I could not tell you about a single neighborhood there, other than the one I grew up in. But in St. Louis Park, not only do I know my neighborhood, but I know all the other ones . . . and not only do I know those neighborhoods, I know the leaders of those neighborhoods."

Once a year the city council has a neighborhood forum where all the leaders of every neighborhood get together and discuss things like how you pull off a successful neighborhood garage sale or block party or build a community garden, and everyone exchanges their best practices. "This did not happen overnight," explained Pires. "This was a twenty-year-plus evolution. It started with neighborhoods wanting to create spaces for community gardens, finding plots and ways to collectively maintain them." From that kernel, other forms of collaboration emerged, and eventually they ended up with a "tapestry of trust," said Pires, within the neighborhoods and between them and the city council.

One of the most important jobs in St. Louis Park city government today is the neighborhood coordinator—a full-time staffer—who interfaces with them all. Jim Brimeyer, who had long served on the city council and was city manager in the 1990s, told me that they valued those neighborhood councils so much that when the state cut back aid to local governments in the early 2000s, "we cut cops and firefighters and public works people—but we would not cut the neighborhood councils coordinator."

These councils are not only critical for improving governance in general; they become even more important as St. Louis Park's population becomes much more international and nonwhite—and not just primarily secular Christians and Jews. St. Louis Park, explains the city manager, Tom Harmening, still has

> a long way to go to make sure that everyone in our community who looks different is at the table. This building and our police station are ninety-five percent white. When we do our jobs we

> think about it from the perspective of a white middle-class person. We do not reflect the community we represent, but we are trying to represent the community . . . I don't know what it is like working a third shift and having your twelve-year-old have to take care of your six-year-old. We are well intended but very clumsy and we don't know what we don't know and I feel uncomfortable or not sure of how I should ask questions . . . But we're working on it. We now have one night each week in the summer where Somali women can come to the [recreation center] and swim in the pool without men. We do the same for orthodox Jewish women, so they can enjoy their community facilities in their own way.

Indeed, before I get up to leave City Hall, Harmening wants to make sure I understand: "St. Louis Park is not a suburb," he says. "It's a community."

When I shared some of these stories with Michael Sandel, he remarked that this was precisely what Alexis de Tocqueville so admired in America as a visitor from the Old World in the 1830s. "Tocqueville, one of the keenest observers of American democracy, noticed that participation in local government can cultivate the 'habits of the heart' that democratic citizenship requires," said Sandel. "The New England township, he wrote, enabled citizens 'to practice the art of government in the small sphere within their reach.' And that reach extends as the sphere expands. Civic habits and skills learned in local associations and neighborhood councils equip citizens to exercise self-government at the state and national level. Although Tocqueville did not make it to St. Louis Park, he would have recognized the civic virtues that led Minnesota-bred politicians to national political prominence."

## *St. Somalia Park*

While we were sitting around City Hall having this discussion in August 2015, the St. Louis Park High School student council was meeting in the room next door, so I asked whether the schools had maintained their standards and were still funded by the community at the level they needed to be and always used to be.

"In the last twenty-five years," said Jacobs, "we've had seven or eight property tax increases [to improve the public schools] and they all typically pass in the range of seventy to thirty"—70 percent for and 30 percent against—"even though only about thirteen to fifteen percent of the households here have kids in [the K–12] public schools anymore. There has always been a connection between the city and the schools. If your schools are not any good, it doesn't matter how well your streets are plowed. And if your roads are falling apart, if your housing stock is dilapidated, your government is dysfunctional so your quality of life suffers—your schools will follow suit."

The next day I went over to St. Louis Park High to meet with the superintendent of schools, Rob Metz. He's worked in St. Louis Park for nineteen years, as an elementary school principal, a high school principal, and a superintendent. I ask him, how has this place remained so progressive over three generations, spanning Swedes, Jews, Latinos, African Americans, and now Somalis? His view is that when St. Louis Park back in the 1950s and 1960s learned to absorb and accept the sudden wave of Jewish immigrants, with their emphasis on education, it changed the town forever. Now that the new wave consists of Africans from Somalia and Ethiopia, Latinos, and African Americans, that imbedded habit of inclusion just got applied to them.

"There have been different waves of openness and acceptance—racial and religious," said Metz, "but as each wave comes, that [impulse toward] acceptance has never left. In one generation it might be religious or racial or sexual orientation—but whatever the wave the school district and city say, 'You come in and be part of this.' And there has never been an ounce of 'stay out.' And in surrounding districts there is not such a welcoming mood. What held this place together were its values of openness . . . If you start building walls and keeping people away, that comes back to haunt you."

Because of this inclusive impulse, added Metz, "all our academic success is pretty close to where it was back in the 1960s—with a completely different group of kids." Indeed, *The Washington Post*'s assessment of America's Most Challenging High Schools from 2015 listed St. Louis Park as the sixth-ranked high school in Minnesota.

The diversity "is incredible now, but the energy behind education has not changed," added Brimeyer, the former city manager. There are now forty-some languages spoken in the St. Louis Park schools, and "yet they still perform above the average, which is not easy dealing with that kind of diversity."

He then adds a small point that reveals something much larger—how the culture got embedded back in the 1950s and then just kept being passed down from one leader to the next. The school district's boundary and the city's boundary are the same, explained Brimeyer, so they cooperate on everything and never do bond issues in the same year. "When I took over as city manager," he said, "the school superintendent called me and said, 'This is how we do business here—we cooperate on everything on community education. If we do a bond issue for schools, you don't have the city do one for infrastructure the same year," and vice versa. "And when a new school superintendent came in, I sat him down and said, 'This is how we do business here . . .' And when I left he called my successor as city manager and said, 'This is how we do business here . . .' "

One thing you never, ever hear in St. Louis Park is someone running for city council proposing to cut the school band or art classes to avoid raising taxes for the schools, he added. "We just say [to voters] this is our brand and it is a winning brand and help us keep it going. We all see our part in it." It helps that Minneapolis has had a pretty consistently strong economy to provide the economic substructure for this.

That attitude has carried over into a lot of leeway for the schools' academic leaders. "Not only is it expected that we take risks and innovate, but if it fails we will regroup and go again—finger pointing is not part of the culture," added Kari Schwietering, the assistant principal at St. Louis Park High School. "The community has your back. We had one of the first Spanish immersion programs in the state. The community expects you to be the first—not wait to see what everyone else is doing. That would not be St. Louis Park. We may get it wrong, but we are expected by the community to be first."

Like the city, Metz and the high school principal, Scott Meyers, believe in hyperrepresentation. The high school has a student council that is predominantly white, but it also has a black male leadership group, a female leadership group, a Latino one, and an African–Middle Eastern one. "These groups meet every other week and talk about their responsibility to the school," said Meyers. "They elect captains and if they have a grievance they come and see me." In the wake of the police shooting in Ferguson, Missouri, students staged a walkout and created a group called Students Organizing Against Racism, or SOAR. "If kids have a voice, with mentoring from teachers, it can make a huge difference," said Meyers. "They cannot be coming to a school and feeling like they are visiting someone else's school."

Metz remarked that when he was the high school principal he got to "talk to a lot of seniors when they go out the door and almost always their biggest regret is that they didn't mix with more kids. They have a sense that they got to go to this school where you have all these different racial and religious groups and they might not have another [experience like that]—and they kind of get that when they are leaving. They say, 'I wish I would have branched out more.'" Every year the graduating seniors at Park High leave a message for the incoming freshmen. "A common theme," said Meyers, is "'Get out and talk to your classmates, because I waited too long.'"

One afternoon Metz and Meyers gathered St. Louis Park High School's student leadership together for a dialogue with me. For someone such as myself, who had only one African American in his graduating class of 1971, the rainbow of faces and colored head scarves was almost blinding. Benetton ads have nothing on this group. What was more remarkable, though, was the honesty with which they talked in front of one another about their school, their differences, and what they knew was a pretty unusual place. I typed their words as fast as I could. Below is a kind of "phrase cloud" from the conversation.

African American girl student: "I'm queer," she began, explaining that in a science class her teacher invited her one day to talk about her sexuality. "I was impressed at how kids showed me respect . . . It made me be proud to be at Park." Somali girl student: "I am Somali. There are still cliques here. I don't notice a lot of tension, but there is definitely a separation if you look at the lunchroom. There are a lot of tables full of Somali students and others full of Caucasians and some groups don't interact, but even if you are not constantly interacting, I feel comfortable talking to anyone here." White girl student: "My least diverse classes are the upper-level ones. We have a large achievement gap and have a ways to go, but if we are talking about socially, there are divisions, but it does not have a large race correlation. It is more who you have classes with. We have all, like, grown up together. I have known her (pointing to an African girl) since second grade. She came from Ethiopia. We grew up with each other and we are not going to change our views just because the outside world is telling us to do so. I feel there is work in progress [here] and that we are doing it and we are coming along." White girl: "Being in the diverse school that has so many clubs and groups and talking about social justice issues makes you really aware of 'white privilege.' I nannied for a twelve-year-old girl, and her friend said to me when she heard that I went to St. Louis Park, 'Oh, pretty sketchy over there.' I said, 'No, it's not Minnetonka'"—another predominantly white town nearby—"and I was so grateful I grew up in the district I did." Latino girl: "I grew up in Nevada and came to Minnesota. I grew up with a lot of Hispanic people around and I came to St. Louis Park and it was a whole different atmosphere, and at first I was scared and tried really hard to fit in. When I was a freshman there were few Hispanic people, but after a few weeks being here you could just feel that everyone knew each other. It was really different in a good way. It was really diverse."

That is the sound of pluralism being built—the hard way, one encounter at a time. In an America becoming a minority-majority country, it's the only way we're going to live, and thrive, together. But every day is still a learning process for all parties. Les Bork, the principal of the St. Louis Park Middle School, where I was a student in the 1960s when it was virtually all white, remarked: "In 1985 we had five black students and now forty percent are students of color, and it was a rough transition. I had families of color who came in who were very accusatory that their kids were not achieving because we were racist. Now it is easier; now there is no dominant culture. The dominant culture is inclusion."

Again, it is all about chasing, grasping, losing, and rebuilding that elusive thing called trust. "Almost all my complaints come from e-mail and I never respond by e-mail," added Bork. "I always call and then we meet face-to-face and I give them my cell phone. [Parents] are shocked at that, [because] they so want to talk to a person," but it so rarely happens. When he actually calls them back, said Bork, "they are almost always taken aback. I am extending trust to them before they extend it to me."

## *Caribou Coffee*

I am sitting at Caribou Coffee in St. Louis Park, asking a question that I never dreamt I would ask. I am asking Sagal Abdirahman, eighteen, a Somali girl who covers and who graduated from Park High in 2015, if she has ever been to a bar or bat mitzvah.

"I got invited to one bat mitzvah party," she answered without hesitation. "Honestly, I thought it was fun—and I liked the dancing."

Welcome to St. Louis Park—the 2016 version. Sagal and her older sister, Zamzam—a twenty-one-year-old Park High grad now studying biology at the University of Minnesota, while Sagal is in her first year at Augsburg College—went almost entirely through the St. Louis Park school system after their mother moved there and found a job as a driver for an insurance company over a decade ago. Both girls won college scholarships from the St. Louis Park Rotary Club and from the Page Education Foundation, named for Alan Page, the former Minnesota Vikings football player who went on to serve on the state supreme court.

I asked Sagal what her biggest impression was, growing up in St. Louis Park schools. "It makes all these opportunities very clear. If you want to do something you can do it—you just have to ask."

Both girls attended the Park High School prom. "My best friend's dad is a pastor," said Sagal, who, like her sister, prays in a mosque in South Minneapolis.

> They are very welcoming. I have known her since second grade and she helped me learn English. I have been to his church in Edina. I would like my kids to grow up in St. Louis Park. It is welcoming and it is not uncomfortable to grow up here. I feel like it is safe, you can still have fun here. The schools are good. It is just a good community as a whole. I feel like Edina is a little too white. I would not be comfortable there. I would feel I would be looked at differently and it would be awkward. I would feel I have to explain myself in a way I don't in St. Louis Park.

Added Zamzam: "I really like St. Louis Park. My mom was [once] thinking of moving to Minneapolis. I said, 'That is not going to happen at all.' I really like where we live in this quiet little neighborhood. It feels very inclusive. We know everyone. Minneapolis feels too much city-like to me."

Do they have a hard time finding halal food? I ask.

There are some stores that carry it, said Zamzam, "or we just grab kosher if we are in a hurry."

Did they face much discrimination? I ask.

"When we were younger maybe a little bit," said Sagal.

> There were not that many Somalis here then. But people were people. For the most part it was welcoming. There was a little bit of division between the general colored people and then the white people and the Jewish people and the Somali people. We were African only and not African American . . . It was complicated and there are those people you get along with. But obviously in English or history classes topics come up that you have to discuss, and they can get uncomfortable and sometimes people have their opinions. We all went to school in a civil manner but every once in a while there would be butting of heads.

I met the two sisters through Karen Atkinson, who runs Children First, a community-wide effort to raise healthy kids, founded by a couple of St. Louis Park businessmen. On every visit back to St. Louis Park, I discovered a new social organization started by someone in the community to help someone less fortunate in the community. That is the definition of a community.

Children First was launched in 1992, when the then school superintendent, Carl Holmstrom, spoke to the Rotary Club of St. Louis Park and shared the challenges facing young people and their families in the community. Two elderly entrepreneurs and Rotarians—Wayne Packard, who owned Culligan Water Conditioning and was in his eighties, and Gil Braun, who owned Braun's women's clothing stores, where my mom always shopped, and was in his seventies—put up the initial funds to create a partnership between the business, city, faith, health, and education communities to support St. Louis Park youth. They teamed up with the Search Institute and began using their "40 Developmental Assets for Adolescents" scorecard, an itemization of relationships, experiences, skills, and expectations that help young people thrive. The scorecard includes things such as: "Family life provides high levels of love and support . . . Young person receives support from three or more nonparent adults . . . Young person experiences caring neighbors . . . School provides a caring, encouraging environment . . . Parent(s) are actively involved in helping the child succeed in school . . . Young people are given useful roles in the community . . . Young person serves in the community one hour or more per week."

Those who have lots of assets do better in school, volunteer in the community, and live a healthier lifestyle. They also are less likely to be involved in risky behaviors. Those with fewer assets tend to fall behind or into trouble. The initiative is dedicated to raising scores for all youth.

"The name Children First is a bit deceiving because it's really about changing the behavior of adults," explained Atkinson. "The initiative unleashes the community's capacity to support our young people, asking individuals and organizations to use the forty assets as a guide. More than two hundred fifty volunteers have been trained, including neighbors, pastors, bank tellers, and firefighters. They each determine their own unique and intentional way to connect with kids." These range from a free clinic for kids that was established by a partnership of the school district and Park Nicollet Health Services to an elderly couple inviting neighborhood children to use the basketball hoop in their driveway!

Not surprisingly, there is a lot more poverty in St. Louis Park today than in the past—it is especially pronounced among recent African immigrants—and some kids cannot afford school supplies. But sure enough, some social organization popped up to try to help. Every year now, explained Superintendent Metz, before school starts, a group of elderly St. Louis Park residents get together and create bags of supplies—450 bags of school supplies in 2015—that they distribute at a local church, St. George's Episcopal, for needy kids. The program was organized by a retired teacher and her husband, who was a retired principal. It is part of a local nonprofit called STEP—St. Louis Park Emergency Program, formed in 1975 to help local residents in need of food or clothing or advocacy.

It's those little things that create trust between the newcomer and the longtime resident—the kind of trust you can draw on in a crisis when you need it most. In 2013, a field trip by students at Peter Hobart

Elementary School in St. Louis Park ended in tragedy. The students were visiting a fossil site in St. Paul, on a Mississippi River bluff, when the earth gave way, burying two of the St. Louis Park schoolchildren alive in a mudslide. The steep slope had been saturated by rain earlier in the week. One of the kids killed was of Somali descent and the other's family was from Guinea. On May 22, 2014, the school held a memorial for the two boys. The local TV station KARE picked up the story: "Two elementary school students killed in a field trip landslide were remembered Thursday on the one-year anniversary of the tragedy . . . Students and staff of the school formed a ring around the school building, dressed in the school district colors of orange and black. Inside the ring, the families of the two boys, Mohamed Fofana, ten, and Haysem Sani, nine, released a few white balloons after district superintendent Rob Metz asked for a moment of silence. The families of the boys received settlements from the City of Saint Paul and the School District because of the incident at a popular site for fossil hunting under a slope. Some of that money is being used to build a school [in Guinea] and an orphanage in East Africa."

## *Innovation Comes in Small Packages*

Time and again I saw proof just in little St. Louis Park of Gidi Grinstein's dictum that social innovation is happening all over the country today at the local level. Nothing new has to be invented—all that exists just needs to be scaled, or as my colleague David Brooks observed in his June 21, 2016, *New York Times* column: "The social fabric is tearing across this country, but everywhere it seems healers are rising up to repair their small piece of it. They are going into hollow places and creating community, building intimate relationships that change lives one by one."

People in the St. Louis Park community feel so strongly about their public schools that they created a foundation to provide teachers with supplemental support for special projects. My English teacher Mim Kagol retired from Park High in 2002 and doesn't even live in St. Louis Park any longer, but in a nearby suburb. Yet she still volunteers to work with the St. Louis Park Public Schools Foundation. "I ask myself why," Kagol said to me. "Every year we raise forty or fifty thousand dollars for the public schools in St. Louis Park. These people are so tied to their schools and the community. Some people my age (I am seventy), retired teachers, are generous donors to the Schools Foundation because the pension system treated them so well."

No small community, even St. Louis Park, is going to integrate Somali war refugees, Latinos from Nevada, or African Americans from the inner city overnight; the cultural and religious gaps are way too wide. There are still many people living parallel lives there. But I saw enough applied hope at work, enough social entrepreneurship seeking to fill the gap between the single family and the federal government, to want to live long enough to come back in twenty years and see how this story ends. Until then, I will give the last word to Jeff Liss—a professional photographer who graduated from St. Louis Park High in 1968 and still lives in the town:

> When I grew up here there was a huge middle class and we all sat at the same tables in the lunchroom and the socioeconomic differences did not seem to matter. My two daughters now go to the high school. They say that even with all the diversity, the school still runs really smoothly, just like it did when we were there. Other communities around here are not as accepting, but these values never left our community. They were unconsciously passed on. I never sat down with my daughters and said, "Be this way or that way." There is not an undercurrent, but an overcurrent, of acceptance that everyone has the right to pursue their goals and dreams. I am not sure if it is unique to St. Louis Park but it is definitely prevalent. I was at my daughter's soccer game in Hopkins the other day, and to listen to them talk back and forth with their teammates—and there are a lot of Somali kids—I thought: "We have a good thing going here and it has not changed that much."

## *The Itasca Project*

St. Louis Park, as I said, does not exist in a vacuum. Many people there work in Minneapolis, so what happens in the Twin Cities economy matters a lot; an expanding economic pie is not sufficient to produce a more inclusive society, but it sure helps. So I can't end this chapter without a few words about the most innovative, and today maybe the most important, community/economy building project minted in the Twin Cities. It's called the Itasca Project—a loose coalition of local business leaders, Fortune 500 executives, educators, local officials, and philanthropists who came together in 2003, during a bad patch in Minnesota politics (following the governorship of the former wrestler Jesse Ventura, from 1999 to 2003), to get the community back on track.

The collaborative spirit in the state had "fallen off," explained Mary Brainerd, president of HealthPartners, who chaired Itasca from 2003 to 2008. Minnesota was starting to imitate Washington, D.C., in the toxicity of its politics, which was a departure from the native political culture. "The two parties could not solve problems you needed to solve. Everyone was just focused on the short term—two years ahead and the next election—and people said, 'We cannot thrive in this environment,'" recalled Brainerd. "We needed evidence-based decision making."

Itasca's first goal was to drive the growth of the local economy—more recently, it's also sought to reduce the area's racial divides. Essentially, Itasca set out to do what American business elites at their best used to do locally and nationally—hold politicians' feet to the fire and make them compromise on the biggest issues, such as infrastructure, education, transportation, and investment—and then, later, hold themselves responsible for opening up their workforces to more minorities. With more African Americans, Laotian Hmong, and Somalis having moved to Minnesota in the previous two decades, the state's racial disparities, which used to be easily ignored, could no longer be—not morally and not economically. Itasca is not a political party, but if it were it would be a Mother Nature party—nonpartisan, agile, heterodox, hybrid, adaptive, and focused on owning best practices.

The group named itself after the lake and state park in northern Minnesota—Itasca—where Minnesota's progressive elites used to vacation together in the summer back in the old days: the Pillsburys, the Daytons, the Cargills, and the McKnights, to name the key families. They were an unusual group of civic-minded patricians, who also set their own example by mandating generous corporate donations to improve community life. I was only vaguely aware of Itasca until, out of the blue, the group was profiled by Nelson Schwartz in *The New York Times* on December 29, 2015, under the headline "They're in the Room When It Happens." Here is how the story began:

> A nondescript conference room on the 38th floor of Minneapolis's tallest skyscraper bears little resemblance to the brick clubhouse a few blocks away where members of the local elite have gathered for more than a century.
>
> But swap out the Oriental rugs and dark wood for a granite table and Aeron-style chairs, and it serves much the same function as the Minneapolis Club once did.
>
> Every Friday morning, 14 men and women who oversee some of the biggest companies, philanthropies and other institutions in Minneapolis, St. Paul and the surrounding area gather here over breakfast to quietly shape the region's economic agenda.
>
> They form the so-called Working Team of the Itasca Project, a private civic initiative by 60 or so local leaders to further growth and development in the Twin Cities. Even more challenging, they also take on thorny issues that executives elsewhere tend to avoid, like economic disparities and racial discrimination.
>
> Think of it as The Establishment 2.0: more diverse than the nearly all-white and male establishment of old, to be sure, but every bit as powerful, and just as invisible when need be . . .
>
> Itasca's impact is very real, however. And its consensus-oriented approach offers an alternative path at a time when politics nationally—and in many state capitols—seems hopelessly divided along partisan lines . . .
>
> The guest list for the weekly breakfasts downtown includes the mayors of Minneapolis and St. Paul, as well as local legislators, school superintendents and university officials.
>
> So when a proposal to raise the gasoline tax in 2008 to help rebuild roads and transit systems was vetoed by the Republican governor at the time, Tim Pawlenty, phone calls from Itasca's business leaders helped persuade enough Republican legislators to cross the aisle and override the veto.

> More recently, pressure from Itasca helped secure increased funding for the state's college and university system. Itasca also spearheaded the creation of a new regional agency to attract companies looking to move or expand, as well as an effort to encourage procurement chiefs at local giants like Target and Xcel Energy to buy more goods and services locally.

The story noted that "Itasca's work is among the reasons that the Twin Cities region has emerged as an economic powerhouse. At 2.9 percent, the metropolitan area's unemployment rate is well below the national level of 5 percent. At the same time, Minnesota has excelled in creating the kinds of higher-paid, knowledge- and skills-based jobs that provide entree into the middle class today." The story also pointed out that most big cities and towns have chambers of commerce and economic development offices, but

> what makes Itasca unique, participants say, is a commitment to hard data and McKinsey-style analysis, as well as a willingness to depart from the script that drives many private sector lobbies.
>
> "We're not just asking for lower taxes and less regulation," said David Mortenson, the current chairman of the Itasca Project. "If we're taking on education or income disparity as a group of business leaders, we want to be able to break some eggs."
>
> That's different from what happens in most other cities, said Mr. Mortenson, who earlier this year took over M. A. Mortenson, a nationwide construction firm founded by his grandfather.
>
> In Seattle, where Mr. Mortenson lived for nine years before moving back to Minneapolis in 2012, "most of the big tech companies viewed the city as a convenient location to house some of its workers," he said. "They didn't engage unless it affected their business . . .
>
> "Tech leaders are very philanthropic," he added, "but they disconnect it from their business."

The story concluded by quoting James R. Campbell, a local banker who served in top positions at Norwest banks and Wells Fargo before retiring in 2002, as asking and answering: Could Itasca be repeated elsewhere? "My answer is maybe," Campbell told the *Times*. But "there's a unique willingness to trust each other here."

Wanting to understand how this group worked, I tracked down one of the founders, Tim Welsh, a senior partner in the McKinsey & Co. team in Minneapolis. He recalled for me the first meeting of the Itasca group on September 12, 2003:

> We had twenty-five or thirty senior people in town. Governor Pawlenty came and we spent well over an hour doing introductions—everyone was so passionate about this community and that there was the ethos that we wanted to preserve. We all knew it, but could not put our fingers on it, but it was a sense that we are all in this together and shared a commitment to the common good . . . To get started on our efforts, we launched the first task force to focus on how to get the University of Minnesota to be more connected to the business community.

In recent years Itasca has put much of its focus on local inequality. There is a lot of work to do. A 2012 task force on socioeconomic disparities found that in Minnesota, African Americans with a BA had a 9 percent unemployment rate, while whites with a BA had a 3 percent unemployment rate. Minneapolis ranked just above Detroit—not a good place—in the "color gap"—the gap between the percentage of whites and blacks of working age (sixteen to sixty-four) who were employed. A 2015 study by the Center on Reinventing Public Education found the four-year graduation rate for blacks and Hispanics from Minneapolis high schools was among the worst in the country. Studies are projecting a hundred-thousand-person labor shortfall in Minnesota by 2018, and most of the jobs will require some postsecondary education, so the business community can't ignore these disparities any longer.

One of the ways Itasca members sought to fix this problem was to get behind Sondra Samuels, who heads the Northside Achievement Zone (NAZ)—a collaborative of forty-three organizations and schools aimed at closing the achievement gap. NAZ was founded in 2008 in Minneapolis, and is modeled on Geoffrey Canada's Harlem Children's

Zone. It uses a holistic web of family coaches and tutors, combined with academic and wraparound support, for 1,100 families, to keep 2,300 children in an education pipeline from early childhood to college. North Minneapolis has been designated a racially concentrated area of poverty where more than 50 percent of the residents are people of color and 40 percent live below the poverty line, and where schools have long been low-performing. A 2016 headline in the *Star Tribune* called it the "Battle Zone." You cannot build a healthy community, Samuels argued, when African American students in Minneapolis have a 52 percent four-year high school graduation rate.

"From the beginning, we recognized the importance of a two-generation approach," Samuels explained in an essay published in the *Star Tribune* on June 21, 2016.

> We work with both parents and their children to make lasting progress. Supporting the entire family to succeed is critically important, because when parents provide stable homes, their children are able to focus on learning.
>
> We also recognized that schools can't do it alone, so we surround students with a team that provides everything from extra academic opportunities, parent education, and early childhood services to behavioral health counseling, housing and career support. In partner schools where the supports are most layered for NAZ students, they are doing significantly better than their peers in reading.

Samuels was not born or raised in Minnesota—she moved to Minneapolis in 1989 to work for Ford Motor Co. in the sales division and lived in St. Louis Park for a few years before creating NAZ. Maybe because she was born and raised in New Jersey, she is not afraid to bluntly denounce the quiet racism that existed in Minnesota for too many years while, in the next breath, praising her Minnesota supporters in groups such as Itasca for being sincerely focused now on fixing the problem.

"I grew up in New Jersey and became totally consumed with racial justice in my late teens," Samuels told me over coffee one morning in downtown Minneapolis. "My father and mother came from the Jim

Crow South and were the descendants of slaves and sharecroppers, and they came north for the same reason immigrants come to America—to find a better life and opportunity that the South could not provide." Her father joined the longshoreman's union and went from low income to middle class and then later moved the family from the equivalent of North Minneapolis to St. Louis Park, or from Newark to Scotch Plains, thanks to fair housing legislation passed in 1968. When she would give vent to her passion to struggle against racial injustice growing up, Samuels recalled, "my father used to say to me, 'Sandy, when you find that country that's better than this one, you tell me and we'll go live there together . . .' I always got stumped by that."

Speaking of Minneapolis, she said: "We have some great disparities in this community—'Minnesota nice' tried to cover up a lot of racism." But, "while I can tell you a story of real disparities and how there was structural racism in Minneapolis—historical and present-day—that has gotten us to where we are, I can also tell you that today we have a business community like no other business community." Today, "people are stepping up and saying, 'This cannot happen on our watch . . .' It is game on." Working with Itasca members and other business leaders, said Samuels, "we are trying to be there for each other. That is what we lost in our country or never really had. We all share a vision that we are not going out like this and we are not going to let our kids go out like this."

NAZ has benefited from both private and public support. It received a five-year, $28 million Promise Neighborhood Implementation grant from the Obama administration; and Target and General Mills each committed $3 million a year for three subsequent years, to make sure NAZ continues to have all the resources it needs to have a chance to be successful.

While Samuels is buoyed by the financial support and the partnerships with groups such as Itasca, she knows that the Northside of Minneapolis cannot be transformed without attention to systemic racism that still needs fixing. She also knows that that fixing will not be transformative unless the area's largely African American families take their future into their own hands as well. The good news, lost in the headlines, is that there are many signs that is happening with the families in the NAZ, she argued:

> What makes me most hopeful is the ownership that African Americans in this community are taking—[their realization] that nobody is coming to save us. Partners are critical but we have to save ourselves—we have to change our community ourselves. I see families creating achievement plans and they are working their plans and they are showing up differently at their child's school and they are enrolling in parenting education classes like crazy. I have fathers saying to me, "I didn't know I was supposed to read to my child." I am seeing a real commitment to change on a personal level, and people asking, "How can I use who I am in this change to help my neighbor on my block?" Everyone has to do their part, but I see families in North Minneapolis saying, "It is up to us" . . . With the right support we can create a culture where people believe they are expected to succeed.

For its part, Itasca has understood that its members needed to go well beyond check writing and make a personal commitment to own the change as well. To do that, Itasca set up a yearlong leadership seminar that focused on increasing workforce diversity. Local CEOs were asked to explore their own prejudices (the group is predominantly white, but includes a few members of other races) and move their organizations to contribute to regional efforts to reduce employment disparities. The project was cochaired by MayKao Y. Hang and Brad Hewitt, CEO of Thrivent Financial. If you talk to Hang, forty-three, for just ten minutes, you understand how far Minnesota has come in the diversity realm since I grew up there—and how far it still has to go.

A Hmong refugee from Laos, she came to the United States in 1976 with her family. She was a first-grader at the time and moved to St. Paul in 1978. She went through the St. Paul public school system, which is now 31 percent Asian American (predominantly Southeast Asian), earned a BA from Brown, a master's degree in social policy and distributive justice from the Humphrey School of Public Affairs, and a doctorate in public administration from Hamline University. Today she is chair of the board of directors of the Minneapolis Federal Reserve Bank in her spare time. Most of her day is spent as president of the Amherst H. Wilder Foundation, a nonprofit organization dedicated to improving lives in greater St. Paul and beyond.

"I was asked three years ago to partner with Itasca to reduce socioeconomic disparities," Hang told me one afternoon in her St. Paul office. The data Itasca had assembled was not Minnesota nice: Minnesota had a tight labor market, and yet students of color with bachelor's degrees "were three times as unlikely to be hired as a white person," said Hang. "We had bias around hiring and that should not happen in a tight market like this. There were barriers to employment."

So Itasca and Hang set up a CEO forum that would help corporate leaders take a deep and honest look at themselves and their hiring practices. So many CEOs signed up that they were oversubscribed. "They would say to me: 'I care about diversity but I don't know how to do it.'" This is the second year and each CEO was challenged, said Hang, to ask him- or herself: "Am I self-aware about diversity?" "What is my personal transformation going to be?" "And what business plan am I going to put in place to change the practices of my organization?" Added Hang: "We help them hold up a mirror."

When the CEOs in the group start to share their life stories, said Hang, she also shares hers:

> I look like other CEOs, but I come from a very different culture and I go home after work to a clan-based community that has experienced displacement, trauma, and war. And as a Hmong woman I don't have a lot of power [in that setting]. I feel the social powerlessness and loss of status when I go home. So when someone who has never experienced that sees the world through my eyes, that allows us to create trust, to see what is similar and very different: "You are like me in this way and different in that way." It is a lot harder for me to judge someone if I have a relationship with them—and that is part of the trust building.

Mary Brainerd said Itasca's diversity training program had a profound impact on her health care company's hiring and its readiness to ask some basic questions: "Are black women getting mammograms as often as white women and are African American men getting colon screenings as often as white males? Now we measure that across the state."

Brad Hewitt, CEO of Thrivent Financial, a Fortune 500 insurance firm based in Minneapolis, and currently vice chair of Itasca, went through the training and concluded: "It changed me profoundly. We uncovered the unconscious biases we all had and now we're trying to get a hundred other CEOs into the same kind of process."

Thrivent grew out of the Aid Association for Lutherans and Lutheran Brotherhood insurance co-ops, founded in 1899 to serve German and Norwegian immigrants after a mill explosion killed many breadwinners and left their families destitute. "We were happily going along doing really well in our co-op serving Lutherans, meaning Swedish, Norwegian, German, and Finish immigrants," said Hewitt. These communities were a long way from Somalia. The Itasca diversity initiative "forces you to recognize that you are blind to unconscious bias and privilege and if you want to actually be more welcoming to others you have to work on those things systematically. Our [company] culture was very strong. Every year we had a big Christmas party with ludefisk [a dried whitefish and Scandinavian/Minnesota favorite]. It was just natural. We have about three thousand employees and had about one percent or less of color. Well, we just doubled that in eighteen months."

If you've been raised on ludefisk and it was a staple at your Christmas party, introducing halal food or other ethnic delights is an adjustment. "It's like learning another language," said Hewitt. "You don't always get it right the first time, but you don't let that stop you . . . The number-one thing when you're learning languages is that you have to be ready to be laughed at—we are learning a new language," he said. "I have gotten comfortable being laughed at when trying to promote diversity."

But embracing diversity in Minnesota today, as in parts of the rest of the country, is not simply a matter of overcoming hidden prejudices toward African Americans. It also involves integrating very different cultures such as Somalis and Hmong. I didn't encounter anyone in my research who wanted Somalis or Hmong to give up their cultural identities, any more than Norwegians or Jews had to, to become "Minnesotans." But there is a strong aversion in Minnesota, which I share, to the kind of globalist multiculturalism that took hold in Europe, where everyone is just left to go their own way and one day you wake up and discover the melting pot is busted and there is no real community. The Minnesota way is that everyone should maintain their customs, but there are certain *bedrock values*—regarding how you treat women, the rule of law, other faiths, public institutions, and community spaces—that are nonnegotiable.

Jonathan Haidt, a social psychologist at the NYU Stern School of Business, made the case for why in an essay in *The American Interest* on July 10, 2016, entitled "When and Why Nationalism Beats Globalism." "Having a shared sense of identity, norms, and history generally promotes trust . . . Societies with high trust, or high social capital, produce many beneficial outcomes for their citizens: lower crime rates, lower transaction costs for businesses, higher levels of prosperity, and a propensity toward generosity, among others . . . The trick . . . is figuring out how to balance reasonable concerns about the integrity of one's own community with the obligation to welcome strangers, particularly strangers in dire need."

Minnesota right now is wrestling with that trick—as are other communities in America. Michael Gorman, who heads Split Rock Partners, an investment fund, and is a founding member of the Itasca Working Team, eloquently shared with me how he sees the challenges and tensions around this issue in Minnesota today. (It is not as if Minnesota hasn't faced this challenge before: as Hewitt joked, until the 1960s the Lutheran Germans would not sell to the Lutheran Norwegians!)

"Most of us who grew up here identify with the tribe of Minnesota," said Gorman.

> There is something special about our civic culture that has developed over time. Minnesotans and Minnesota companies are distinctive in their level of engagement and commitment to the community, and their willingness to devote financial and human capital toward the public good. There is a sense that we can't let up. Minnesota still retains the elements of community and connectivity that have served the region well since pioneer days. With the arrival of recent immigrants from backgrounds very different than their predecessors from Northern Europe, however, the cultural pH is changing. Figuring out how to include Minnesota's new voices and perspectives while retaining the best attributes of the majority culture that has worked for a long time is a challenge.

From one side, he argued, the commons in Minnesota needs to be expanded and become more inclusive—the definition of being Minnesotan needs to broaden so that every person, regardless of their background, sees Minnesota as fertile topsoil in which to grow and prosper. But it cannot just be a one-way conversation.

"There also has to be assimilation from the new arrivals," said Gorman. "Our message has to be: 'We are glad that you are here and can't wait to see the contributions you will make to our community. That asks something of us. But it also has to ask something of you. What are you doing to embrace the existing culture, to be a part of this place you have chosen as your new home?'"

As a son of immigrants, Gorman is sensitive to this balancing act. "Immigrants in every era have taken comfort in the traditions and cultural touchstones from their homeland, particularly in the private domain. But regardless of our cultural heritage, all of us have to participate in American society. To do that successfully requires speaking English, gaining an education, and making a contribution. Most people, particularly those who immigrate in search of a better life, just want to live in a peaceful place and raise their children to be productive citizens. We should help them do that in every way possible."

Institutions have failed some of these new immigrant communities miserably in their own countries, Gorman added. Many have grown up in stressed and dysfunctional societies, or lived in refugee camps—"so trust is understandably in short supply. They have just been trying to survive. And the reason that is relevant is that we trust that institutions work here, that justice will be fair—we can count on our government to be largely noncorrupt; those are defining attributes of Minnesota. But to a new arrival, these may not be natural assumptions. We should be clear about how things work here, and back it up by making sure interactions with the community build their capacity to trust. There are many moments of truth, and everyone has to play their part."

New immigrants and native-born Minnesotans all need to act like we're on the same team, Gorman concluded: "Many parts of Europe have paid a heavy price for failing to integrate immigrants into mainstream culture. We should be very intentional about not making the same mistakes. It is all about building trust that our future together is preferable to one characterized by separation and isolation."

This is a very important conversation. It is often avoided by all parties, but it cannot be any longer, now that states such as Minnesota are receiving a large influx of immigrants from traumatized countries in the World of Disorder. That's why what happens here and in similar communities matters a lot now—and innovative social organizations, such as Itasca, are going to be critical to making this work.

But color me an optimist. As anyone from Minneapolis or St. Louis Park will attest, one of the favorite outdoor pastimes of the area is walking around the lakes that dot the Twin Cities, almost all of which are lined with beautiful walking trails and bike paths. (With Minneapolis's 22 city lakes and more than 170 parks, no resident lives more than six blocks from a park, according to the mayor's office.) As I've said, those lakes are one of the Twin Cities' great Mixmasters—you see every income, every race, every class walking around them. One day in the spring of 2016 I was walking Cedar Lake with my wife and friends and we bumped into three local African refugee community leaders—two from Somalia, one from Ethiopia—one of whom I had met at a University of Minnesota seminar. They were walking around the public paths on this warm May afternoon, just like we were, just like my mom and I had done hundreds of times over the years. Every once in a while a group of Somali women would also pass, walking the lakes wearing traditional Somali robes and head covers, but you could see their very modern Nike walking sneakers peeking out from below, almost like they were winking at you.

If I have to bet, I am going to bet on those lakes. I am going to bet on the basic decency that is still at the core of this community. I am going to bet on that decency expanding to embrace the people it has left out and behind, and on that embrace being reciprocated. Not because anything is "inevitable" but because I met too many people applying hope.

## *It Takes a Dining Room Table*

But it only works if you start with "a dining room table," said Tim Welsh, the Itasca cofounder and McKinsey partner.

"What we've discovered in Itasca is that a dining room table really matters," Welsh explained. "When issues were really difficult to deal with—literally—we would get all the key players around someone's dining room table." In 2006, when Itasca persuaded the legislature to overturn the then governor Pawlenty's veto of the transportation bill, it was after discussions around the Itasca member Charlie Zelle's dining room table, including key Republican legislators willing to vote against their own sitting governor. At the time, Zelle was the president and CEO of Jefferson Lines, a local bus company.

"Itasca does that regularly," said Welsh. "I just hosted two dinners of next-generation leaders, in my dining room, about what our generation wants to see the state become. You get them all together around a dining room table and they leave realizing, 'There are other leaders in the community who want the same things that I do at just a basic human level—that the community be safe and that everyone have better opportunities.' It is a check your ego at the door and check your politics at the door group."

It would be easy to simply classify (and possibly dismiss) the Itasca Project as just another well-intentioned group of civic-minded individuals. It is anything but. In fact, I would suggest that Itasca could be a model of how dialogue and community building happen in the age of accelerations—between businesses, governments, and key civic players. It is living by Mother Nature's killer apps—nimble, hybrid, heterodox, diverse, fact-based—unbound by partisan ideologies or other entrenched interests.

Indeed, Itasca is a thoroughly twenty-first-century network. It has no by-laws, board of directors, executive director, CEO, or office space—no formal structure in any sense. It's got a laughably bad website. Indeed, the group notes that it needs to exist only if there is work to be done—that is why it is called a "Project." It's made up almost entirely of volunteers. The volunteers are very senior leaders from nearly every part of the community—business, government, and nonprofit. The only full-time staff are two project managers seconded to Itasca by McKinsey. And because there is very little staff, these volunteer leaders actually do the work. It manages itself through a Working Team that meets nearly every Friday at 7:30 a.m. for ninety minutes. "Yes—quite senior volunteers meet nearly every Friday morning," noted Welsh. "All of them describe this as one of the most interesting meetings on their calendar, a meeting they 'genuinely look forward to.'"

Yet, despite its unusual structure, Itasca has made a tangible contribution to the economic and civic vitality of Minneapolis–St. Paul since 2003, argued Welsh. Besides its success in advancing the state's transportation infrastructure and working on minority inclusion and CEO diversity training with people such as Sondra Samuels and MayKao Hang, it has:

- Launched Real Time Talent, one of the most innovative workforce development initiatives in the country. It links the curriculum and training for more than four hundred thousand postsecondary students with the skill requirements of employers in the state (RealTimeTalentMN.org).
- Created the Business Bridge, which facilitates connections between the procurement functions of large corporations and smaller potential suppliers located in the region. As a result of this effort, participating businesses added more than $1 billion to their spending with local businesses in two years—a year ahead of their goal.
- Helped to build the case for investing more aggressively in higher education. By strengthening relationships between business and higher education leaders, and using a fact-based set of findings to justify investing more than an incremental amount, a coalition organized by Itasca helped increase spending in the state by more than $250 million annually.

In 2017, Itasca won a grant from the Bush Foundation to expand its dining room table, or rather tables. It will now be able to organize five hundred trust-building dinners a year to bring together community stakeholders from all political perspectives, as well as reach out and include leaders from more rural areas of Minnesota—the areas that voted overwhelmingly for Trump. Now that deep partisanship has begun to infect and paralyze state government in Minnesota as it has so many other places, "we have to think more about what we can do ourselves to promote inclusion and diversity in Greater Minnesota," explained Welsh.

But it is an even broader inclusion, he added—not just racial or religious, but also rural-urban. If the state capitol in St. Paul "can't bring us together *we* have to become the drivers of improvement in education or transportation. And because we have no bylaws, no charter, no

CEO, no lobbyist, no formal board, no year-end reviews, we can be very adaptive. We can get things going. We can bring in people really fast, because there is one thing that companies know how to do and that is scale things." Itasca is as good an example as one can find of a complex adaptive coalition practicing progressive localism.

Of course there are and will continue to be disagreements and different perspectives as Itasca's partners hash out the direction they want to go. That's not the point. That's a sign of health. The point, said Welsh, is that you don't get up from the table until you've worked those differences out so you can move ahead—and no grandstanding allowed.

"Trust doesn't just materialize," Welsh concluded. "It takes work. It requires a whole bunch of people to keep at it—to keep showing up, and that doesn't just happen magically."

# PART IV
# ANCHORING

FOURTEEN

# *From Minnesota to the World and Back*

The timing of this book was an accident—but it was an accident waiting to happen.

Many of the ideas that make up this book were roiling around in my head for a while, but it took a chance encounter with a parking attendant to inspire me to pull them all together—to do what Dov Seidman calls "pausing in stride": to stop and reflect and try to imagine some better paths that might help more people take advantage of this age of accelerations.

What has surprised me the most is how many unexpected things I've learned along my journey from Minnesota to the world and back to Minnesota—personally, philosophically, and politically.

As I alluded to earlier, I knew that what was pulling me back home to Minnesota and St. Louis Park was not simply an academic interest in the extraordinary politics of these places. What was pulling me back was a reaction to four decades of covering the Middle East and then Washington, D.C., and seeing how much these two arenas had come to mirror each other—and how little they resembled the places that had shaped me in my formative years.

My time in the Middle East led me to realize that, with a few rare exceptions, the dominant political ideology there—whether you were talking about Sunnis or Shiites or Kurds, Israelis, Arabs, Persians, Turks, or Palestinians—was "I am weak, how can I compromise? I am strong, why should I compromise?" The notion of there being "a common good" and "a middle ground" that we all compromise for and upon—not to mention a higher community calling we work to sustain—was simply not in the lexicon. So when I came back to Washington in 1988, after thirteen years abroad, it was with a certain eagerness to rediscover America. But over my nearly thirty years now of reporting from Washington, what I found instead was that with every passing year American politics more and more resembled the Middle East that I had left. Democrats and Republicans were treating each other just as Sunnis and Shiites, Arabs and Persians, Israelis and Palestinians did—self-segregating, assuming the worst of each other, and, lately, shockingly, never wanting one of their kids to marry one of "them."

This is awful and has become totally debilitating at *exactly the wrong time*. We have so much work to do. We need accelerated innovation in so many realms, and it can only happen with sustained collaboration and trust.

So, as I said, I went back to my roots in Minnesota to see if this place—where, in my memory at least, people practiced a politics based on that "common good" and where trust was more the rule than the exception—still existed. The place had certainly become more complicated than it was, but, all in all, I was not disappointed, for all the reasons that I have explained.

My most important political lesson, though, was how much the kinds of efforts being made to build a more inclusive St. Louis Park and Minnesota matter—not only for those who live there, but for every community in America today.

Just review some of the trends we're witnessing: There are now roughly fifty million students in K–12 public schools in America, and in 2015—for the first time ever—the majority were minority students: primarily African Americans, Hispanics, and Asians. At the same time, students on free and reduced-price lunch programs hit an all-time high in 2016. A report by the Georgetown University Center on Education and the Workforce predicted that by 2020, 65 percent of all jobs in the economy will require some postsecondary education and training beyond high school. Meanwhile, a research study by the University of Oxford's Martin School concluded in 2013 that 47 percent of American jobs are at high risk of being taken by computers within the next two decades.

What those numbers tell you is that, in this age of accelerations, everyone is going to have to raise their game in the classroom and for their whole lifetime. What those numbers tell you is that we truly cannot afford to leave any child behind anymore. What those numbers tell you is that pluralism matters more than ever going forward, because on current trends America will become a majority-nonwhite country over the next quarter century—yet without having worked out our racial issues. The foreshocks are being felt right now, and with more migrants fleeing zones of disorder, this problem will only become more acute across the globe. Therefore, societies that can truly make "out of many, one" will have so much more political stability, not to mention innovative prowess.

What those numbers also tell you is that leadership matters more than ever—at the political and personal levels—but a particular kind of leadership. At the national and local levels, we need a leadership that can promote inclusion and adaptation—a leadership that starts every day asking, "What world am I living in? And how do I engage in the relentless pursuit of the best practices with a level of energy and smarts commensurate with the magnitude of the challenges and the opportunities in this age of accelerations?" It is also a leadership that trusts the people with the truth about this moment: that just working hard and playing by the rules won't suffice anymore to produce a decent life.

That's why leadership also matters more now on the personal level. Back in the 1960s, in places like Minnesota, we had so much wind at our backs that "you needed a plan to fail." No more. Now you need a plan to succeed, a plan for lifelong learning and skills growth. That means more personal leadership, more of everyone taking ownership of their own future and embracing the "start-up of you."

It is not too late for any of us, let alone America, to manifest that kind of leadership. But as the environmentalist Dana Meadows used to say about mitigating climate change: "We have exactly enough time, starting now." And not a moment less, because the margin for mistakes or delays is shrinking on every front for every nation and every person. I repeat: when the world is fast, if you get off course—as a leader, a teacher, a student, an investor, an employee—you can find yourself with a very long road back. Small errors in navigation can have really serious consequences when the Market, Mother Nature, and Moore's law are all accelerating at this speed.

Finally, philosophically speaking, I have been struck by how many of the best solutions for helping people build resilience and propulsion in this age of accelerations are things you cannot download but have to upload the old-fashioned way—one human to another human at a time.

Looking back on all my interviews for this book, how many times in how many different contexts did I hear about the vital importance of having a caring adult or mentor in every young person's life? How many times did I hear about the value of having a coach—whether you are applying for a job for the first time at Walmart or running Walmart? How many times did I hear people stressing the importance of self-motivation and practice and taking ownership of your own career or education as the real differentiators for success? How interesting was it to learn that the highest-paying jobs in the future will be *stempathy* jobs—jobs that combine strong science and technology skills with the ability to empathize with another human being?

How ironic was it to learn that something as simple as a chicken coop or the basic planting of trees and gardens could be the most important thing we do to stabilize parts of the World of Disorder? Who ever would have thought it would become a national security and personal security imperative for all of us to scale the Golden Rule further and wider than ever? And who can deny that when individuals get so super-empowered and interdependent at the same time, it becomes more vital than ever to be able to look into the face of your neighbor or the stranger or the refugee or the migrant and see in that person a brother or sister? Who can ignore the fact that the key to Tunisia's success in the Arab Spring was that it had a little bit more "civil society" than any other Arab country—not cell phones or Facebook friends? How many times and in how many different contexts did people mention to me the word "trust" between two human beings as the true enabler of all good things? And whoever thought that the key to building a healthy community would be a dining room table?

That's why I wasn't surprised that when I asked Surgeon General Murthy what was the biggest disease in America today, without hesitation he answered: "It's not cancer. It's not heart disease. *It's isolation.* It is the pronounced isolation that so many people are experiencing that is the great pathology of our lives today." How ironic. We are the most technologically connected generation in human history—and yet more people feel more isolated than ever. This only reinforces Murthy's earlier point—that the connections that matter most, and are in most short supply today, are the human-to-human ones.

"We are powerfully connected electronically but increasingly disconnected interpersonally," explained Dr. Edward Hallowell, founder of the Hallowell Centers for Cognitive and Emotional Health and a leading psychiatric expert in the field of Attention Deficit Hyperactivity Disorder (ADHD). "People everywhere, not just patients, are starved for human connection, which I call 'the other Vitamin C,' and they don't even know they are starved for it or that it's missing. The beauty of this Vitamin C is that it's free and infinite in supply; the problem is people walk right past it."

Don't get me wrong: technology has so much to offer to make us more productive, healthier, more learned, and more secure. I am awed by the intelligent assistance I discovered in researching this book and the potential it has to lift so many people out of poverty and discover talent and make it possible for us to actually fix everything. I am hardly a technophobe. But we will get the best of these technologies only if we don't let them distract us from making these deep human connections, addressing these deep human longings, and inspiring these deep human energies. And whether we do that depends on all that stuff you can't download—the high five from a coach, the praise from a mentor, the hug from a friend, the hand up from a neighbor, the handshake from a rival, the totally unsolicited gesture of kindness from the stranger, the smell of a garden and not the cold stare of a wall.

I realize that in the dizzying moment we're experiencing right now, both blue-collar and white-collar workers in the developed and developing world feel like they are just one small step ahead of a machine or robot making their job obsolete. I understand that in such a transition it is much easier for humans to visualize what they will lose than all the benefits they will gain, or already have gained.

But it is impossible for me to believe that with so many more people now empowered to invent, compete, create, and collaborate, with so many more cheap and powerful tools enabling us to optimize social and commercial and governmental interactions, that we won't develop the capability to solve the big social and health problems in the world and that, in the process, we won't also find ways for humans to become even more resilient, productive, and prosperous as they are reinforced by intelligent machines.

I like the way John Hagel, the expert on flows, once put it in his blog: "The long arc of human history shows an amazing pattern of exponential performance improvement on a wide range of dimensions—reduction of violence, health, and prosperity. Over time, more and more of the global population has been able to participate in this performance improvement. Sure, we've faced deep setbacks along the way, but in every case we've somehow found a way to overcome those setbacks and return to the longer-term trend of performance improvement."

Along the way, Hagel added, "we've been dramatically supported . . . by technology innovation that has helped us to do more with less, culminating with recent decades of exponential digital technology performance improvement that shows no signs of slowing down . . . So, there's a real basis for hope. It's not just a figment of the imagination."

But that hope has to be applied hope; happy endings don't just happen because you wish they will. And what needs to be applied most now is imagination—the ability to reimagine our governing institutions, economic institutions, learning institutions, and religious values–generating institutions.

"Rather than simply attacking our institutions, or conversely, simply defending them, we need to come together in a quest to redefine them," concludes Hagel. "Redefine them to reflect the new context of a rapidly changing world, so that our institutions can continue to support us."

Transitions, though, are a real bitch.

The most dangerous time to be on the streets of New York City was when cars were first being introduced but horses and buggies had not yet been fully phased out. We're in that kind of transition now—but I am convinced that if we can just achieve the minimum level of political collaboration to develop the necessary social technologies to work through it, keep our economies open, and keep lifting learning for everyone, a better life will become more available than ever to more people than ever—and the second quarter of the twenty-first century could be an amazing time to be alive. The transition will not be easy. But human beings have made transitions like this before and I believe they can again. "Can" doesn't mean "will," but it also sure doesn't mean "can't."

## *A Tree Grows in Minnesota*

So let me truly end where I literally began.

On a research trip back home in the summer of 2015, I drove by our old house in St. Louis Park, at 6831 West Twenty-Third Street, where my parents first moved from North Minneapolis in 1956. I hadn't done it for years, but decided on the spur of the moment to swing by. At one level the neighborhood's tightly packed rambler homes looked remarkably the same as when I left mine for college and for work in the 1970s. Ours was still painted a light blue. But something also struck me as different, and I couldn't put my finger on it at first. My old neighborhood was totally familiar, but slightly unfamiliar. It took me a while to figure it out—and then it finally dawned on me: it was the trees.

They were all small and scrawny when I was small and scrawny. Ours was a spanking-new neighborhood when I grew up there. And now, a half century later, all the trees had grown tall and thick, with long branches, and they were full of leaves—so much so that the neighborhood was considerably more shaded. The light had changed slightly and it caught my eye, because it contrasted with the much brighter mental image I had been carrying around for so long, like an old picture stuck in the back of my wallet.

Those trees and I had both grown up and out from the same topsoil, and the most important personal, political, and philosophical lesson I took from the journey that is this book is that the more the world demands that we branch out, the more we each need to be anchored in a topsoil of trust that is the foundation of all healthy communities. We must be enriched by that topsoil, and we must enrich it in turn.

That prescription is easier to write than to fill, but it is the order of our day—the real über-task of our generation. It is so much easier to venture far—not just in distance but also in terms of your willingness to experiment, take risks, and reach out to the other—when you know that you're still tethered to a place called home, and to a real community. Minnesota and St. Louis Park together were that place for me. They were my anchor and my sail. I hope this book will inspire you to pause in stride and find yours.

And don't worry if it makes you late . . .

AFTERWORD

# *Still an Optimist*

While I was writing the original hardcover of this book over the three years ending in the summer of 2016, I was also watching the looming U.S. presidential election out of the corner of my eye and sometimes head-on. Who couldn't? But I never attempted to cover it in this book. I was truly focused on how the three accelerations had evolved and the long-term challenges and opportunities they were generating for people, businesses, and communities. Not surprisingly, though, when I began my book tour—just two weeks after the election of Donald Trump—the question often arose whether I thought there was a connection between Britain's decision to leave the European Union, Trump's victory, and the trends that I had written about.

My answer was an unequivocal "yes."

For a certain portion of the population in both America and the European Union, the world had indeed—as I worried in chapter 7—gotten "just too damned fast." The age of accelerations brought on so much automation, immigration, and competition and so many flows of new ideas and mores that a swath of the population began to feel unmoored, especially less-educated, working-class whites living in rural areas.

The challenge started at the office, where all of a sudden average was over. People discovered that just working hard and playing by the rules was not enough to ensure an average middle-class lifestyle. You had to become a lifelong learner to get a job and hold that job—but too many people were not equipped to make that change. They were prepared to show up, work hard, and do an honest day's labor—but they were quite content to be told what to do and quite sure that if they did what they were told to do well then job security, a decent wage, and the American Dream or British Dream would still be theirs.

At the same time, thanks to the age of accelerations, communities were rapidly going from monocultures to polycultures, due to a surge in migrants and refugees from Latin America into North America and from Africa, Eastern Europe, and the Arab-Muslim world into Western Europe. All of a sudden, the lady at the checkout counter was wearing a head covering that was not a baseball cap. And the language she was speaking with her coworkers was not English. The fact is, the European Union expanded too far, too fast, enabling large numbers of Eastern Europeans to flock to London faster than parts of British society could absorb them. The fact is, China's exports to the United States, and illegal immigration from Mexico and Latin America, came too far, too fast for some communities in America to adapt, culturally or economically. These societal changes happened at the same time the market and Moore's law were accelerating. Without proper surge protectors, some people really get burned.

The British-born writer Andrew Sullivan put it in family terms in a March 31, 2017, essay he wrote for *New York* magazine: his brother and father both voted in favor of Brexit. "I asked my brother to explain," Sullivan wrote.

> "It was really quite simple, he said. He believed that immigration into Britain was happening at too fast a pace. He saw overcrowded schools and hospitals, a groaning transportation system, an acute housing shortage, and a country that had been transformed so fast many of its inhabitants began to not recognize it at all. He wanted immigration to come down to more manageable levels—and, in the last election, he therefore voted Conservative, because that's what the Tories promised. They pledged to bring immigration down to the tens of thousands a year. But after a year in office, the immigration statistics showed no drop at all: Over 600,000 migrants were still entering the U.K. per year, with close to 300,000 from the EU, and there was no end in sight. That number was completely unprecedented

> before 2014, but had stayed at that level for three years in a row. For a comparison, EU immigration into Britain was a mere 44,000 in 1992 and 66,000 in 2003."

And, while all of this was happening, the acceleration of flows was also changing social norms faster than some could adapt. In the United States, this process brought new genders into restrooms, new kinds of couples into marriage, new rules of political correctness onto college campuses and new phrases into the lexicon—like "white privilege," "black lives matter," and "safe spaces." Practically every week came a news story about students confronting administrators or professors on college campuses for using a turn of phrase deemed demeaning or upsetting to some member of their community. Through social networks, every one of these stories, like the one about the Yale administration warning against wearing Halloween costumes that might insult one culture or another, seeped into the news, cheering some and leaving others seething that political correctness had become stifling.

Whatever one might personally feel about all these changes, one would have to be blind, deaf, and dumb not to see that they came at too rapid a clip for many white middle- and lower-middle-class men and women in exurbia and rural areas. They looked around, and it seemed like every group had been handed some kind of step stool by society to get a leg up in this age of accelerations—except them. And Trump and pro-Brexit politicians spoke to and for them—promising to slow down or reverse some of the changes that were making people so dizzy at school, so unmoored at work and so not-at-home in their communities.

In short, Trump's victory and Brexit were not just economic backlashes, they were also cultural backlashes by white working-class voters who felt they were being forgotten, left behind, and looked down upon all at the same time. Trump and Brexit were their "I am somebody" fist in the face of a system that had threatened their status and their livelihoods. Justin Gest, assistant professor of public policy at George Mason University and author of *The New Minority: White Working Class Politics in an Age of Immigration and Inequality* put the point succinctly in an April 20, 2017, *Washington Post* podcast with Jonathan Capehart.

"So much of Donald Trump's politics is symbolic," Gest explained. "They're symbolic in the sense that this is what people want to hear and if it doesn't get done, it's almost beside the point because he's elevating the prerogatives of his constituents to the national stage after having been relegated to the fringes of American politics for decades . . . The sense of having a voice suddenly, after feeling voiceless for so long, is powerful. It's not in their cultural interests to vote against him, no matter how little he has delivered to actually help them in any kind of material way."

Looking back, you can see how and why this Trump/Brexit moment coalesced in the early 2000s. For reasons described earlier in this book, the 1950s to the early 1970s were the golden years for the American middle class and in particular for low- and middle-skilled workers. In those postwar years, globalization had not set in yet, trade was more restricted, and unions were strong. America was the dominant industrial/manufacturing power coming out of World War II, which had devastated Europe and Asia, and the IT revolution hadn't yet reached full force, so there were many Americans, particularly men, who could enjoy the "high-wage, middle-skilled job." Machines then mostly amplified human power, rather than substitute for it, at least not at scale. This was the era when, as Rick Nolan, the congressman from Minneapolis, said: if you were an average male worker in Minnesota back then, "you had to have a plan to fail."

That world started to change in the early 1970s, with a combination of factors that steadily gained momentum right up to the early 2000s. The 1973 Arab oil embargo slowed economic growth and productivity. Meanwhile, globalization, trade expansion, and the IT revolution started to take hold in developed economies and began to eat away at middle-skilled work. In the late 1970s, China, led by Deng Xiaoping, unleashed its farmers and small manufacturers to grow and make things and export for their own profit, like good capitalists; Deng also created four special economic zones along the coast of China to attract foreign manufacturers. And, at the same time, Ronald Reagan and Margaret Thatcher began a process of deregulating and opening their economies, a trend that spread across the West. Bill Clinton continued to embrace trade and globalization by nurturing both NAFTA and the World Trade Organization in the 1990s, and it was on his watch that the dot-com boom boomed. Unions became weaker as competition from offshore labor steadily rose, and lower-skilled workers had to deal with a steady flow of unregulated immigration from Mexico and Latin America.

It was during this period from the 1970s to early 2000s, that "average" started to be over, and the high-wage, middle-skilled job began to be torn asunder—into either a high-wage, high-skilled job or a low-wage, low-skilled job. Globalization and technology simultaneously raised the skill level required for any middle-class job and, at the same time, lowered the barriers to entry for foreign goods and for foreign workers to compete with their American counterparts. Alas, not every worker was able or willing to raise his or her skill level to adjust to these new conditions.

Nevertheless—during this period from the mid-1970s to early 2000s—many households were able to hang onto the middle-class American Dream because of a number of cushions. We created a lot of housing-construction jobs through financial engineering of mortgages. We also extended credit through Visa cards and other forms of plastic, so middle-skilled workers could borrow their way into the middle class. We also lowered the cost of many basic goods through Walmart and Target and Costco. We also enabled a lot of middle-skilled people with less-than-stellar credit to secure mortgages, again through financial engineering—and as housing prices rose, they were able to borrow against those homes or ride their values up with inflation to stay in the middle class. Finally, and maybe most importantly, we also invited their wives to join the workforce and created many two-earner households. With all of these tricks and buffers and "subsidies," we kept the American Dream alive for a lot of people.

And then came the early 2000s—particularly the years 2001, 2007, and 2008—which left a lot of people feeling insecure both physically and financially. September 11, 2001, marked the first time the American mainland had been struck by a foreign enemy—and by a group of terrorists no less, who converted our own passenger jets into suicide weapons. But that was not all that happened in 2001.

China joined the World Trade Organization that year and fairly quickly administered a shock to a number of American communities, who saw their factories suddenly move to China or get wiped out by Chinese competitors. As economists David Autor, David Dorn, and Gordon Hanson explained in their important January 2016 study, *The China Shock: Learning from Labor-Market Adjustment to Large Changes in Trade*, the sudden surge of imports from China between 1999 and 2011 cost the United States 2.4 million jobs and the wages and employment opportunities for those who lost those jobs remained depressed, in large part because local and national governments failed to create policies to cushion them from the worst effects of this one-time trade shock and prepare them for better jobs.

Workers in these industries and affected regions didn't "go on to better jobs, or even similar jobs in different industries," Stony Brook University economist Noah Smith observed in a January 26, 2016, essay on Bloomberg.com. Instead, they shuffled "from low-paid job to low-paid job, never recovering the prosperity they had before Chinese competition hit. Many of them end up on welfare. This is very different from earlier decades, when workers who lost their jobs to import competition usually went into higher-productivity industries, to the benefit of almost everyone."

So true. One reason the normal adjustment to a sudden surge in trade did not happen was because of something else that was explained earlier in this book: the unleashing of the supernova in and around 2007, and the attendant surge in technology, software, machine learning, and automation. These developments wiped out many middle-skilled white-collar and blue-collar jobs right when the China trade surge was doing the same. So, middle-skilled workers who tried to jump to a new middle-skilled job to avoid getting hit by the China train got hit by the 2007 train coming the other way.

This third phase was exacerbated, we now know, by another train that slammed into middle-skilled workers in the early 2000s. It was a slow-moving train but a very powerful one. That was the steady shrinking of the percentage of our national income that goes to wages and wage-earners as opposed to machines and the owners of capital. Think of America as a two-person company—with just one worker and one owner. In 1900, explains James Manyika, who heads McKinsey Global Institute, the share of national income that went to wages for that worker in our two-person company was about 90 percent, and the owner and his machines got 10 percent.

That's because so much of the value-added work was being done by that worker's hands and relatively little by the machines of the owner. But by 1980, 76 percent of national income went to wages, and in the 1990s it fell to around 65 percent—and today it's down to the low 60s and maybe even the high 50s. The owner and his machines are doing much better and the average middle-skilled wage-worker much worse.

"For each unit of economic output, the mix of labor and capital has been steadily changing," explained Manyika. "So while labor is still the lion's share of the national income, its share has been declining over the last few decades. There is nothing malicious about it; it was not any conspiracy. Capital increasingly taking the form of machines and software is just contributing more than in the past." (No doubt the declining power of unions—thanks to the rising power of machines and software and the downward pressure exerted by overseas labor markets as globalization intensified—was both a cause and effect of this trend as well.)

So, the owner of the machine is doing better today and those who can only do wage labor are fighting over a shrinking pie, unless they have ever-rising skills. That is why—as noted earlier in the book—GDP growth, productivity growth, employment growth, and wage growth all used to move up together in the American economy in the last century—but today they are no longer as tightly linked. GDP growth rises and productivity rises because of the introduction of many more machines, robots, software, and artificial intelligence algorithms, but the number of workers needed for those jobs sinks and the wages of the workers who are left—unless they are high skilled—have flattened. And it happened pretty fast.

A June 2017 McKinsey Global Institute study, *Making it in America*, quoted U.S. national income data showing that between 1993 and 2004 real incomes rose for virtually all segments of the U.S. economy and only 2 percent of U.S. households saw their incomes flatten or decline.

But thanks to the 2008 recession—which crushed the value of people's homes and in many cases wiped out their mortgages and their ability to borrow against them—plus the accelerations in Moore's law and globalization—between 2005 to 2014 a whopping 81 percent of U.S. households saw real incomes flatten or decline. That is a staggering number! "For the first time, a majority of Americans were no longer progressing," said Manyika.

The McKinsey study concluded that there is now a yawning pay gap between workers with postsecondary education and those without it. Highly skilled workers who can make the most of today's new technologies and leverage global flows are thriving, which is why a small number of high-growth metropolitan areas have bounced back strongly since 2008. But in almost two-thirds of U.S. counties, the study noted, real median household incomes remain below their pre-2000 peak—while the costs of maintaining a middle-class life have continued to climb. "Couple that with widening inequality and you have a potent brew," said Manyika. The same trends have played out the same way in Britain, France, and Italy.

So it was no surprise when, in February 2017, Autor, Dorn, and Hanson produced a sobering follow-up study to their work on trade shocks. It was entitled *When Work Disappears: Manufacturing Decline and the Falling Marriage-Market Value of Men*. It found that "the structure of marriage and child-rearing in U.S. households has undergone two marked shifts in the last three decades: a steep decline in the prevalence of marriage among young adults, and a sharp rise in the fraction of children born to unmarried mothers or living in single-headed households. A potential contributor to both phenomena is the declining labor-market opportunities faced by males, which make them less valuable as marital partners."

In particular, the authors cited "trade shocks to manufacturing industries" which "have differentially negative impacts on the labor market prospects of men and degrade their marriage-market value along multiple dimensions: diminishing their relative earnings—particularly at the lower segment of the distribution—reducing their physical availability in trade-impacted labor markets, and increasing their participation in risky and damaging behaviors . . . The falling marriage-market value of young men appears to be a quantitatively important contributor to the rising rate of out-of-wedlock childbearing and single-headed childrearing in the United States." It's hard to quantify, but I also believe that a decline in the Protestant work ethic in some communities—of all races and ethnicities—has weakened society's ability to adapt to the age of accelerations. That decline is due in no small part to the rise of reality television, video games, social networks, and fake news, as well as the erosion of respect for science and learning in general and the readiness of so many more people to have children out of wedlock and not do the hard work of sustaining a family. J. D. Vance's *Hillbilly Elegy* demonstrated how this trend has cut across society. The ability to form and sustain stable families was a key ingredient of the great American middle class in the '50s, '60s, and '70s—and the increasing inability to form and sustain stable families is at the heart of the current education-to-work crisis in America today. At a time when you need to have a plan to succeed, and that plan needs to be updated every year, some people have learned bad habits or made some really bad choices about what's needed to thrive. All the things that matter in life and work require real time and energy to build, but in too many places in too many ways the signals being sent by popular culture in America today are that there's a quick and easy path to anything and everything. Why waste your time studying or analyzing facts when you can make up fake news and get paid for it or become an overnight success on YouTube with an edgy video or you can go viral with a smart-alecky tweet on Twitter?

In sum, 2001, 2007, and 2008 seem to have left more middle-class Americans feeling physically insecure, culturally insecure, economically insecure, and socially unmoored. Some, who were living in healthy communities, managed to overcome these trends; many others clearly slipped into a sense of helplessness, a loss of agency, and retreat into drugs and video games. Generally speaking, the federal government in America and the national government in Great Britain failed to put in place enough of the surge protectors, safety nets, and trampolines to soften the blows.

Trump and Brexit were born in that maelstrom. If you look at the recent dramatic shifts in American politics—from George W. Bush to Barack Obama to Donald Trump, the common denominator "is not that voters were just looking for change," Don Baer, who served as President Clinton's director of communications, remarked to me one day. "They were looking for help. They were looking for someone to help them." First they dared to try Obama as a radical change agent, and when that didn't do it, many of them were ready to try Trump or Brexit.

## *It's Community, Stupid*

So how can I still be an optimist?

Because while all these studies and stories about family and communal breakdown accurately describe today's social woes and political turmoil, they do not tell the whole story. A year after Brexit and Trump's election we started to see the backlash against the backlash, starting in Europe. Dutch voters, French voters, German voters, and some British voters—many of them young people who were thriving and too busy to vote for Brexit or Trump—made their views known with a vengeance. They did not want to disconnect in a connected world, which is what Brexit and Trump both offered. They didn't want to go backward to monocultures after growing up in increasingly diverse urban areas. They were born into the age of accelerations. They understood that the convergence of accelerating technological change with accelerating globalization increases the urgency and necessity of building resilience and propulsion for middle-skilled workers in new ways—and they want leaders ready to rise to the challenge. So time and again they voted for the "open" party in their respective countries over the "closed" party. And more and more they are looking to build that resilience and propulsion not from the top down but from the bottom up by trying to create networks of healthy communities, towns, and cities. And that is pretty much the new divide in global politics today—the open parties versus the closed parties, the wall people versus the eye people, the communities that are creating complex adaptive coalitions and rising from the bottom up and those where the bottom is falling out.

You would never know that this was the case, though, by the dystopian address President Trump delivered at his inauguration. He painted a picture of America as a nation gripped by vast "carnage"—a landscape of "rusted-out factories scattered like tombstones" that cried out for a strongman to put "America first" and stop the world from stealing our jobs. It was a shocking speech in many ways and reportedly prompted former President George W. Bush to say to those around him on the inaugural dais, "That was some really weird shit."

I am with Bush on that. Trump is right that we in America have an epidemic of failing communities. But what he misses is that we also have a bounty of thriving ones—not because of a strongman in Washington but because of strong leaders at the local level. Indeed, this notion that America is a nation divided between two coasts (where everyone is rising, modernizing, pluralizing, and globalizing) and a vast flyover interior (where jobs have disappeared, drug addiction is rife, and everyone is hoping Trump can bring back the 1950s) is highly inaccurate. From my travels throughout America, I would argue that the big divide is not between the coasts and the interior. It's between strong communities and weak communities. You can find weak ones along the coast and thriving ones in Appalachia, and vice versa. It's community, stupid—

not geography. St. Louis Park and Minneapolis just happened to be the ones I knew, so they were the foundation of this book. But I also knew from my travels in recent years that there were many other thriving localities. So, to make that point, in May 2017, I took a four-day car trip through the heart of flyover America—I drove from Austin, Indiana, down through Louisville, Kentucky, wound through Appalachia and ended up at the Oak Ridge National Laboratory in Tennessee.

I used the trip to write a long column in *The New York Times* highlighting the checkerboard of thriving and failing communities that is America today. The communities that are making it have all developed their own complex adaptive coalitions, their own version of the "Itasca Project."

I started in one of those hit-bottom places: Austin, Indiana, a tiny town of four thousand off Interstate 65, which was described in a brilliant series in the Louisville *Courier-Journal* as "the epicenter of a medical disaster," where citizens of all ages are getting hooked on liquefied painkillers and shooting up with dirty needles. The federal Centers for Disease Control and Prevention confirmed that Austin "contains the largest drug-fueled H.I.V. outbreak to hit rural America in recent history." Its 5 percent infection rate "is comparable to some African nations." Austin, the newspaper noted, doesn't just sit at the intersection between Indianapolis and Louisville but at the "intersection of hopelessness and economic ruin."

I chose to go there to meet the town's only doctor, Will Cooke, whose heroic work I learned of from the *Courier-Journal* series. Cooke's clinic, Foundations Family Medicine, sits at 25 West Main Street—opposite MarkO's Pizza & Sub, a liquor store, and a drugstore. Down the street was a business combination I'd never seen before: Eagle's Nest Tanning and Storage. It's the Kissed by the Sun Tanning Salon and a warehouse—both of which seemed to be shuttered, with the space available for rent. For generations Austin's economy was anchored by the Morgan Foods canning plant, but, as the *Courier-Journal* noted, "then came a series of economic blows familiar to many manufacturing-based communities. The American Can plant next to Morgan Foods shut its doors in 1986 after more than 50 years in business. A local supermarket closed. Workers left along with the jobs and poverty crept up among those who stayed."

The age of accelerations can be very hard on one- or two-company towns (in Russia they call them "mono-towns"). In the worst case, the one or two companies move away or close—as happened to Austin—and in the best case they stay but become more automated, so that the workforce shrinks and requires higher skills, all at the same time. Attracting new factories or service industries is hard for these towns because they don't have a critical mass of high-skilled workers needed for twenty-first-century industries. The workers' only hope is to commute to larger urban areas or help develop their own adaptive coalitions, but they often lack the visionary local leadership to do so.

Austin, Cooke explained to me, got caught in the vortex of declining blue-collar jobs at its only two factories, leading to a loss of dignity for breadwinners, depression, and family breakdown. All of this coincided with doctors' and drug companies' pushing of painkillers, and with too many people in the community failing to realize that to be in the middle class now required lifelong learning. "Thirty percent of students were not even graduating from high school," said Cooke. "Then you take high unemployment, generational poverty, homelessness, childhood abuse and neglect, and cloak that within a closed-off culture inherited from Appalachia, and you begin to have the ingredients that contributed to the H.I.V. outbreak."

Austin's insularity proved deadly for both jobs and families. "The close-knit, insular nature of the community worked against it, with the C.D.C. later finding up to six people shared needles at one sitting, and two or three generations—young adults, parents and grandparents—sometimes shot up together," the *Courier-Journal* reported. Starting in 2017, though, Cooke told me, the town's prospects began to improve, precisely because the community came together, not to shoot up, but to start up and learn up and give a hand up.

"The local high school has introduced college-credit classes and trade programs so people are graduating with a head start," said Cooke. Faith-based and civic groups have mobilized, providing community dinners called "Food 4R Soul" and even installing community showers for people without running water.

Addiction is often a byproduct of social breakdown and the sense of isolation that comes with it. Cooke feels hopeful because he sees the tide slowly shifting as "social isolation gives way to community." He added: "Only a healthy individual can contribute to a healthy family, and only a healthy family can contribute to a healthy community—and all of that requires a foundation of trust. That kind of change can't come from the outside; it has to be homegrown." I shared with him a point that Dov Seidman made in this book—that "trust is the only legal performance-enhancing drug." Dr. Cooke liked that a lot, and only wished he could prescribe it as easily as others had prescribed opioids.

Just forty minutes down the highway from Austin, I interviewed Greg Fischer, the mayor of Louisville, a city bursting with energy and new buildings—full of people in the "open" party. Fischer told me: "That 'Intifada' you wrote about in the Middle East is happening in parts of rural and urban America—people saying, 'I feel disconnected and hopeless about participating in a rapidly changing global economy.' Drug-related violence and addiction is one result—including in a few neighborhoods of Louisville."

But Louisville also has another story to tell: "We have 30,000 job openings," said Fischer, and for the best of reasons: Louisville has "a vision for how a city can be a platform for human potential to flourish." It combines "strategies of the heart," like asking everyone to regularly give a day of service to the city; strategies of science, like "citizen scientists" bearing GPS-enabled inhalers that the city uses to track air pollution, mitigate it, and warn asthma suffers; and strategies for job creation that leverage Louisville's unique assets. For example, one job-creation strategy leveraged the fact that Louisville is the worldwide air hub for UPS, to encourage a slew of new businesses to make "end of runway" products for rapid delivery. Another leveraged the fact that Kentucky is the Napa Valley of bourbon, an artisanal product that is now booming, to promote "bourbon tourism." Another led to a partnership with Lexington, home of the University of Kentucky, to create an advanced manufacturing corridor and another involved leveraging Humana's headquarters in Louisville, unleashing a lifelong wellness and aging-care industry.

Fischer said he would push more and more resources down to the cities—from the federal and state levels—precisely because they are, or have the potential to be, effective trust platforms that can experiment and move much quicker. "Trust leads to relationships, relationships lead to possibilities, and possibilities lead to action," concluded Fischer. Show me a community that understands today's world and is working together to thrive within it, and I'll show you a community on the rise—coastal or interior, urban or rural.

And I found more such communities as I moved south on Interstate 75 through Tennessee to Oak Ridge, home of the Manhattan Project facilities where the enriched uranium for the "Little Boy" atomic bomb dropped on Hiroshima was produced. Today, the Oak Ridge National Laboratory, which sprawls across two counties, is still involved with nuclear weaponry, but its supercomputer, one of the world's most powerful, and its hundreds of scientists help drive a broad array of research in energy, materials science, 3-D manufacturing, robotics, physics, cybersecurity, and nuclear medicine—research it now actively shares with the surrounding Appalachian communities, to spawn new industries and jobs.

Sitting on the spot where the K-25 Manhattan Project facility once stood, I interviewed Ron Woody, county executive for Roane County, where Oak Ridge is partly located, and Steve Jones, an industrial recruiter hired by the city of Oak Ridge to seek out companies interested in investing in the region or leveraging spinouts from Oak Ridge's labs. That kind of active community-entrepreneurship is a new thing for Roane County, where generations of people have known only a government job.

"Back in the 1980s you had the T.V.A. [Tennessee Valley Authority], and it had over fifty thousand employees. Now it has ten thousand employees," explained Woody, so "we were not diversified in our employment. We had to convince the public that we can't rely on the Oak Ridge lab and T.V.A. The Cold War is over. So our communities had to make a big transition from a lot of government programs to very few." In his region, many are starting to rise to that challenge, said Woody, "but progress is slow."

One of those success stories was luring a former three-time Tour de France winner, Greg LeMond, to open a 65,000-square-foot factory for his new company, LeMond Composites, which makes lightweight carbon fiber bikes, based on new materials pioneered at Oak Ridge. "The research Oak Ridge has done is going to change the way we make things," LeMond explained to me, as we sat in his new factory. "It is a really exciting future. My goal is that you will be able to go on my website and design your own bike out of carbon fiber."

Of course, just because workers are looking for employment and there are new jobs opening, that doesn't automatically mean local people can work in those jobs, explained Jones, the recruiter. Because of the opioid crisis, many people cannot pass a mandatory drug test—and years of working for the government have also left many unprepared for the pace of today's private sector.

"The two biggest issues we are dealing with are the soft skills and passing the drug test," explained Woody. "I thought the problem was that people needed more STEM skills." But that's not the case. It turns out that it's not that hard to train someone, even with just a high school or community college degree, to operate an advanced machine tool or basic computer. "Factory managers would say, 'I will train them and put them to work tomorrow in good jobs' requiring hard skills," said Woody. "The problem they have is finding people with the right soft skills." What are those soft skills? I asked. "Employers just want someone who will get up, dress up, show up, shut up, and never give up," Woody responded without hesitation. And there are fewer workers with those soft skills than you might think, he added.

When new companies come into the area today, noted Jones, who grew up on a farm, they ask specifically for young people who were either in a 4-H club or Future Farmers of America (now called FFA) because kids with a farming background are much more likely to get up, dress up, show up on time, and never give up in a new job.

Meanwhile, as discussed earlier, the Oak Ridge lab is partnering to embed top-level local technical talent as entrepreneurial research fellows so they can to absorb advanced manufacturing skills and then share them with local companies. Think about car dealers in the future who, instead of needing a huge lot with hundreds of cars in inventory, will just custom print the car you want—out of carbon fibers—by using giant 3-D printers. Auto manufacturers could pop up anywhere. Why not? "Our only inventory is carbon fiber pellets that cost two dollars a pound, and we can make any product out of them," said Lonnie Love, an Oak Ridge Corporate Fellow. "You won't need inventory anymore."

Over the last one hundred years, Love pointed out, we went from decentralized artisan-based manufacturing to centralized mass manufacturing on assembly lines. Today, with emerging technologies, we can go back to artisans, which will be great for local communities that spawn the leadership and workers needed to take advantage of these opportunities. We are going to see a world of micro-factories, and you can see them sprouting around Oak Ridge already.

"There's a new wave of kids coming up who love this stuff," said Love. "We can create mini-moonshots all over the place." The same lessons apply to the design of a good's parts as to its whole. Thom Mason, the director of the Oak Ridge lab, explained to me that high-performance computing "allows you to design and test out all the parts on the computer and only make those that you know will work. It is speeding up the iteration loop of physical manufacturing. You move all the trial and failure into the digital world—so you don't need to do all that costly tooling of prototypes—and then go straight to manufacturing."

I ended my little tour in Knoxville, Tennessee, where I met Mayor Madeline Rogero over dinner in the newly rejuvenated downtown square, a beehive of restaurants, public art exhibitions, theaters, shopping, and museums. "Until the mid-1980s, the old economic development model here was low wages and no unions. That model wasn't sustainable," said Rogero, the first woman mayor of Knoxville and a former organizer for Cesar Chavez's United Farm Workers union. "We wanted better schools, and you cannot build a great school system on the back of low-wage workers. So we started thinking about what are our unique assets and stopped selling ourselves as a low-wage town."

The whole region came together around that project and wove an adaptive coalition that could draw in investors based on the region's strengths. It's called "Innovation Valley," the mayor explained, and it markets the assets of the University of Tennessee in Knoxville, Oak Ridge National Laboratory, Pellissippi State and Roane State community colleges, as well as the workforce skills in the metro area and available infrastructure that can be utilized by technology companies. It has also stimulated a dialogue between employers and higher-learning institutions, to ensure they're meeting the labor force needs of the future.

But none of this is easy. There are real constraints that need to be overcome. The region has a shortage of both manufacturing and back-office workers. "We face the same workforce development issues that all metro areas in the U.S. are facing," explained Bryan Daniels, president of the Blount Partnership, one of Knoxville's regional development boards. "Our local law enforcement has described the prison populations as having approximately 65 percent opioid-related inmates." It is vital, therefore, for the community to develop programs to get this population back to being employable. As part of that, said Daniels, the Knoxville region is exploring new ways to get workers from outlying rural areas into the metro area labor force and help them acquire the "educational attainment they need to get their skill level up" for a modern economy. They are even studying "public-private partnerships to provide transportation for [rural] workers up to a two-hour commute radius," he said.

This is the real picture of America—cities rising and falling side by side. The task of politics today is to scale the institutional innovations that the thriving communities have developed to as many other places as possible as fast as possible. Top-down solutions are not irrelevant—it would be a huge help, for example, to have a national health-care system that enabled every worker to be mobile. But we cannot depend on these top-down initiatives to drive the scale of adaptation that will be required across the country. "Icebergs melt from the edge," says Gidi Grinstein, president of the Reut research and strategy group. "That is where the innovation has to start."

Rather than adopting top-down, hierarchical command and control structures, "successful movements are all organized around small, local action groups, typically ten to fifteen people, who work together to achieve impact in very different contexts," adds John Hagel. "These action groups are united by a loosely coupled network that enables them to seek help from others and to observe and learn from the diverse actions of each group what actions can achieve the greatest impact." That is exactly what the Itasca Project and its dinner tables aspired to do within Minneapolis and then across Minnesota, and the same kind of networks need to spread across the country.

Bottom line: This is not your grandparent's America, but it is also not Trump's America—that land of vast carnage and industrial wastelands.

It's actually Bill Clinton's America.

Clinton once famously observed: "There is nothing wrong with America that cannot be cured by what is right in America." That idea has never been more accurate—and necessary—than it is today. What is wrong with America is that too many communities, rural and urban, have broken down. We have to look that square in the eye. But what is right with America is the many communities and regions that are coming together to help their citizens acquire the skills and opportunities to own their own futures. We also need to look that square in the eye and learn to share and scale these success stories.

Only strong communities, not a strong man, will make America great again. And the fact that we still have many such communities, and that more will be rising, is why I am still an optimist.

## *West 23rd Street, Again*

Amplifying that message is how I hope to spend my remaining years as a columnist. I cannot think of anything more important. After this book came out in November 2016, the St. Louis Park Historical Society asked me to come back home in April 2017, to do a fundraiser in support of their effort to build a physical museum that would capture the essence of our community—so that it might inspire future generations and be shared by others. Of course, I said yes and we had a lovely gathering at the Jewish Community Center, where Susan Linee interviewed me about my journey from St. Louis Park to the world and back. As a thank-you gift, the historical society gave me the thing they knew I would value most: the green street sign from the street where I grew up: "West 23rd Street." Now I can always have it with me wherever I live. They knew that I'd been lugging it around in my soul all those years—from Beirut to Jerusalem, from Moscow to Beijing, from Senegal to Sydney—so they thought I might as well have the real thing.

And so let me end this book, again, by quoting, again, from *Jersey Boys*, the play I drew from to begin the section on St. Louis Park. This time, though, it's Frankie Valli talking, looking back on his career as the lead singer with the Four Seasons. Nothing better describes my own life/career journey.

"They ask ya," he says, "'What was the high point?' The Hall of Fame, sellin' all those records, pullin' Sherry outta the hat?' It was all great. But the first time the four of us made that sound, our sound . . . when everything dropped away and all there was was the music . . . that was the best. That's why I'm still out there singin', like that bunny on TV with the battery, I just keep goin' and goin' and goin' . . . chasin' the music, tryin' to get home."

# *Acknowledgments*

So many people were generous in sharing their time and insights to make my writing this book possible. I want to do my best to acknowledge each of them.

First and foremost, I need to—once again—thank the *New York Times* chairman and publisher, Arthur Sulzberger Jr., and Andy Rosenthal, the editorial page editor when I was working on this book, for allowing me to cut my column-writing duties in half in order to do all the research and interviews that were the foundation for this work. It would not have been possible otherwise. I joined *The New York Times* in 1981. It remains the greatest newspaper in the world, and my many and varied assignments there have given me a front-row seat to so much history and the opportunity to travel and learn in so many different settings. I am forever indebted to Arthur and his late father, Arthur Ochs "Punch" Sulzberger, for affording me that opportunity for nearly four decades.

I have been fortunate over my career to have developed a small group of friends who are the most amazing posse one could partner with to toss up ideas, bat them back and forth, sharpen them, and eventually bring them to a point where they can form the spine of a book. While I have dedicated this book to all of them, this particular book benefited in some specific ways that demand extra acknowledgment.

No one was more generous with his time, insights, and encouragement in helping me assemble this book than my friend and teacher Dov Seidman, the CEO of LRN and the author of the book *How.* Dov is a truly unique observer of the human condition, and I have learned so much from him about people and organizations and values, which is why he is quoted more than anyone else in this book. But his influence in shaping my thinking goes well beyond those quotes. Idea's that Dov first articulated through our endless walks and talks are suffused throughout this book. Lucky is the person who has Dov Seidman for a friend.

Once again, my teacher and friend Craig Mundie, the former senior executive at Microsoft and now an executive coach, stepped up to guide me through the latest generation of technology and make sure that I not only understood it well enough to explain it, but also, better yet, could explain it accurately! This is the fourth book of mine that Craig has helped with. To have Craig Mundie as your technology tutor is to have Babe Ruth as your batting coach.

Speaking of longtime tutors, this is the seventh book to which my friend Michael Sandel has contributed his insights, but this one was particularly fun, since he was present at the creation back in Minnesota when we were young boys in the same Hebrew school class. Michael's thoughts on the civic virtues that enrich and are enriched by a healthy community were particularly valuable.

Michael Mandelbaum, my coauthor on my last book and almost daily partner in chewing over the news and trying to understand it, has been sharing his ideas with me and sharpening my own for more than two decades. He listened to the reporting that went into this book, as he has for five previous ones, and always generously helped me think through the ideas.

Erik Brynjolfsson and Andrew McAfee, authors of *Race Against the Machine* and *The Second Machine Age*, had a big impact on my thinking as well, as I note in the book, and generously shared their insights with me.

And, of course, a heartfelt thanks goes to Ayele Bojia, the parking attendant at the underground public parking garage in Bethesda, Maryland, whose stopping me to ask about how to improve his blog set this whole book in motion! He is a good man, always struggling to make Ethiopia, the country of his birth, a better place for all.

Marina Gorbis was one of the very first people I talked to about the ideas in this book and the first to host me for a roundtable on the themes at her little jewel box, the Institute for the Future, in Palo Alto. She was always generous with her insights and her time.

Johan Rockström was kind enough to walk me through all the planetary boundaries while on a visit to his wonderful research center in Stockholm; he also proofed some of this text. There is no better teacher on the environment. My thanks as well to Hans Vestberg for hosting me at Ericsson on that same trip to Sweden.

John Doerr and his colleague Bill Joy were, as always, open to sharing their insights and improving mine on cross-country ski runs and hikes. Yaron Ezrahi, also on his seventh book with me, never failed to teach me something new and force me to think deeper about what I had already written. Alan Cohen never tires of tutoring me on the cutting edges of technology, and Moshe Halbertal does the same when it comes to the Middle East.

In addition, I had many rich conversations over the past two years, from which I benefited enormously, with Larry Diamond, Eric Beinhocker, Leon Wieseltier, Lin Wells, Robert Walker, K. R. Sridhar, Sadik Yildiz, P. V. Kannan, Kayvon Beykpour, Joel Hyatt, Jeff Bezos, Wael Ghonim, Nandan Nilekani, Gautam Mukunda, Rabbi Tzvi Marx, Rabbi Jonathan Maltzman, Russ Mittermeier, Glenn Prickett, Dennis Ross, Tom Lovejoy, Richard K. Miller, Jeffrey Garten, Moises Naim, Carla Dirlikov Canales, David Rothkopf, Jonathan Taplin, David Kennedy, Zach Sims, Jeff Weiner, Laura Blumenfeld, Kofi Annan, Peter Schwartz, Mark Madden, Phil Bucksbaum, Bill Galstos, Craig Charney, Adam Sweidan, Sadasivan Shankar, James Bessen, Robert Walker, Reid Hoffman, and James H. Baker, director of the Pentagon's Office of Net Assessment. I thank each of them for the time they shared making me smarter about everything from politics to ethics to the climate to geopolitics.

A hearty thanks to Ian Goldin at the University of Oxford for hosting me for three really stimulating days at the Martin School, and to Gahl Burt for doing the same at the American Academy in Berlin. A shout-out of thanks as well to Nader Mousavizadeh and his colleagues at Macro Advisory Partners in London for always being up for doing jazz on any subject.

I would not have understood the education-to-work channel without the generous and repeated tutoring of Byron Auguste, Karan Chopra,

Stefanie Sanford, and David Coleman—the absolute A-team of thinkers on the nexus between education and work. And a special thanks to Alexis Ringwald for sharing her insights on that subject from LearnUp and to Eleonora Sharef for doing the same with all she learned cofounding HireArt—and more.

From Minnesota, I am deeply indebted to Vice President Walter Mondale, the late Bill Frenzel, Senator Al Franken, Senator Amy Klobuchar, Sharon Isbin, Wendi Zelkin Rosenstein, and Norman Ornstein for taking time to share their insights. And a huge and special thanks goes to Larry Jacobs of the Humphrey School at the University of Minnesota, for not only hosting me but educating me about Minnesota politics today and reading portions of the book. I am also deeply indebted to Tim Welsh and his colleague at McKinsey & Co. in Minneapolis, Julia Silvis, for all the introductions they made and all their help in reading portions, sharing ideas, introducing me to the right people—like MayKao Hang, Brad Hewitt, Michael Gorman, David Mortenson, and Mary Brainerd—and enabling me to understand the Itasca Project. Sondra Samuels patiently tutored me on the good works of the Northside Achievement Zone and her important partnership with Itasca. Rob Metz, the superintendent of schools in St. Louis Park, and Scott Meyers, the high school principal, were enormously helpful in getting me together with both their students and their colleagues and sharing their own insights.

Another very special thanks to Jeanne Andersen, the engine behind the St. Louis Park Historical Society. Jeanne got me together with interesting members of the community, I drew on her historical writings, and she was kind enough to review the final draft. I am deeply grateful for all her help. The same goes for Children First's coordinator, Karen Atkinson, who introduced me to some great members of the Somali community in St. Louis Park and shared her insights, as did Paul and Susan Linee and other members of the Historical Society board.

My high school AP American history teacher, Marjorie Bingham, and my English teacher, Mim Kagol, are still teaching me four decades after I left their classrooms. I am so grateful for their helping me to understand St. Louis Park High then and now. How lucky am I to have had such extraordinary teachers and lifetime friends. The St. Louis Park mayors Jeff Jacobs and Jake Spano, the city managers Tom Harmening and Jim Brimeyer, and the technology leader Clint Pires were so much fun to talk to and to learn from.

A special tip of the hat, though, goes to my childhood friend Fred Astren, for carefully reading and contributing to parts of the manuscript with his insights, as well as other members of the Pennsylvania Avenue Poker Club for their input and lifelong friendship—Mark Greene, Howard Karp, Steve Tragar, and Jay Goldberg. We have all been friends from St. Louis Park for more than fifty years. Brad Lehrman, who used to bowl with me on Sunday mornings with our fathers, was also very generous in sharing his thoughts about the old neighborhood and making sure that I never bowled alone.

And, as always, my best friend, Ken Greer, and his wife, Jill, listened to and encouraged this project from the very beginning, often while walking around one of the lakes in Minneapolis. There is nothing more fun for me than sharing ideas with them.

On the corporate front, a huge thanks to Randall Stephenson, who heads AT&T, and his colleagues John Donovan, Ralph de la Vega, Bill Blase, and Krish Prabhu. Stephenson shared with me AT&T's human resources policies—in ways that were extremely helpful to my understanding of the world of work today—and their latest thinking on technology. No matter where he was in the world, John Donovan answered my follow-up questions on, as he would say, the first ringy-dingy.

IBM's Watson team—particularly David Yaun and John E. Kelly III—were incredibly generous in helping me navigate the wisdom of Watson on two visits to IBM.

At Google, I am particularly indebted to Astro Teller, who heads Google's X innovation center. The little graph Astro sketched out for me on the spur of the moment became a central theme of this book, and his rigor—and that of his colleagues Courtney Hohne and Gladys Jimenez—in making sure that I got the argument right was truly impressive. And a huge thanks to Sebastian Thrun for all that he taught me about education in multiple visits to Udacity.

My friend Andy Karsner not only brought me together with Astro but has been an all-purpose idea generator for different parts of this book and many columns. Being able to do idea jazz with Andy is one of the great joys.

At Intel, Gordon Moore, Brian Krzanich, Bill Holt, Mark Bohr, and Robert Manetta could not have been more helpful. Elliot Schrage of

Facebook and his colleagues Dan Marcus and Justin Osofsky were full of valuable insights. Tom Wujec and Carl Bass hosted me for an amazing day at Autodesk. From McKinsey & Co.'s Global Institute, James Manyika and his colleagues Susan Lund, Richard Dobbs, and Jonathan Woetzel, as well as Alok Kshirsagar from Bombay, provided wonderful research that enriched many aspects of this book. James, in particular, never failed to respond to an emergency call from me for insight into any number of subjects.

At Microsoft, Bill Gates, Satya Nadella, Brad Smith, and Joseph Sirosh all shared ideas with me along this journey that enriched my thinking. At Hewlett Packard Enterprise, Meg Whitman and Howard Clabo were very generous in taking me inside their thinking and innovations. At General Electric, a big thanks to William Ruh and Megan Parker both for the ideas they shared and for all the GE engineers to whom they introduced me. At Walmart, Doug McMillon, Neil Ashe, Dan Toporek, and their colleagues showed me in exacting detail every digital interaction that happened behind the scenes when I tried to buy a television from Walmart's mobile app. They also introduced me to the best ribs in Arkansas.

I am deeply indebted to Doug Cutting from Hadoop and Chris Wanstrath from GitHub for patiently walking me through the evolution of both of their companies and ensuring that I got every fact right. It took multiple visits and follow-ups with both for me to fully understand what they had each helped to create, and I am extremely grateful for their tutoring.

Qualcomm's cofounder Irwin Jacobs did the same on my two visits to his campus. He, his son Paul, and their whole team were enormously generous with their time. I owe a particular thanks to Joe Schuman and Nate Tibbits from Qualcomm for going the extra mile.

Gidi Grinstein literally spent hours with me sharing his impressive work on strengthening communities in Israel. Gidi is a special thinker and a wonderful friend, and his ideas on community have influenced me a lot. I would say the same for my multiple conversations with Hal Harvey, another dear friend and truly original thinker. I could not have done the chapter on Mother Nature and politics without the physicist Amory Lovins, a great teacher, who always combines his good humor with precision thinking.

•

And, as always, a special thanks to my golf partners Joel Finkelstein, Tom O'Neil, George Stevens, Jr., Jerry Tarde, and the late Alan Kotz. (Miss you, pal.) Last but not least, the amazing documentary team from *Years of Living Dangerously*—Joel Bach, David Gelber, Sydney Trattner, and John Pappas—who took me to places I never dreamt of going (and got me home).

This is my seventh book with Jonathan Galassi, the president and publisher of FSG, whose inspiration and support have been life-changing for me. There is nothing more that I can say. My literary agent, Esther Newberg, always in my corner, did her usual fine job handling all the details. Jonathan and Esther and I have been together now since 1988. I cannot imagine writing a book without them. My FSG editor on this project was Alex Star, who quietly but firmly brought his great touch and acute intelligence to making sure all the dots in this book both connected and shined brightly. His work improved the book with each iteration, as did the editing of Picador's Anna deVries and James Meader on this paperback. My tireless longtime assistant, Gwenn Gorman, was always there to support this project in any way I needed—from research to scheduling. I am lucky to have had her working with me for so long. My loving sisters, Jane and Shelley, were kind enough to fact-check all the stories of our youth in St. Louis Park.

But there is no one more deserving of thanks than my brilliant wife, Ann Friedman, who edited every page, made wonderful suggestions on organization and wording, and also made everything I did better. I worked on this book over three years and broke my shoulder in the middle of it all. So Ann put up with a lot, while trying to start her own museum, Planet Word. As Alexander Hamilton says in the musical *Hamilton* of his own spouse, she is "best of wives and best of women." And, of course, my daughters, Orly and Natalie, are always rooting for their dad, and always an inspiration.

With so many generous friends and family members from so many places for so many years, how could I not still be an optimist?

Thomas L. Friedman
Bethesda, Maryland (but really still from Minnesota)
August 2016

# Index

ALLEN LANE

*an imprint of*

PENGUIN BOOKS

## Also Published

Jonathan Losos, *Improbable Destinies: How Predictable is Evolution?*

Chris D. Thomas, *Inheritors of the Earth: How Nature Is Thriving in an Age of Extinction*

Chris Patten, *First Confession: A Sort of Memoir*

James Delbourgo, *Collecting the World: The Life and Curiosity of Hans Sloane*

Naomi Klein, *No Is Not Enough: Defeating the New Shock Politics*

Ulrich Raulff, *Farewell to the Horse: The Final Century of Our Relationship*

Slavoj Žižek, *The Courage of Hopelessness: Chronicles of a Year of Acting Dangerously*

Patricia Lockwood, *Priestdaddy: A Memoir*

Ian Johnson, *The Souls of China: The Return of Religion After Mao*

Stephen Alford, *London's Triumph: Merchant Adventurers and the Tudor City*

Hugo Mercier and Dan Sperber, *The Enigma of Reason: A New Theory of Human Understanding*

Stuart Hall, *Familiar Stranger: A Life Between Two Islands*

Allen Ginsberg, *The Best Minds of My Generation: A Literary History of the Beats*

Sayeeda Warsi, *The Enemy Within: A Tale of Muslim Britain*

Alexander Betts and Paul Collier, *Refuge: Transforming a Broken Refugee System*

Robert Bickers, *Out of China: How the Chinese Ended the Era of Western Domination*

Erica Benner, *Be Like the Fox: Machiavelli's Lifelong Quest for Freedom*

William D. Cohan, *Why Wall Street Matters*

David Horspool, *Oliver Cromwell: The Protector*

Daniel C. Dennett, *From Bacteria to Bach and Back: The Evolution of Minds*

Derek Thompson, *Hit Makers: How Things Become Popular*

Harriet Harman, *A Woman's Work*

Wendell Berry, *The World-Ending Fire: The Essential Wendell Berry*

Daniel Levin, *Nothing but a Circus: Misadventures among the Powerful*

Stephen Church, *Henry III: A Simple and God-Fearing King*

Pankaj Mishra, *Age of Anger: A History of the Present*

Graeme Wood, *The Way of the Strangers: Encounters with the Islamic State*

Michael Lewis, *The Undoing Project: A Friendship that Changed the World*

John Romer, *A History of Ancient Egypt, Volume 2: From the Great Pyramid to the Fall of the Middle Kingdom*

Andy King, *Edward I: A New King Arthur?*

Thomas L. Friedman, *Thank You for Being Late: An Optimist's Guide to Thriving in the Age of Accelerations*

John Edwards, *Mary I: The Daughter of Time*

Grayson Perry, *The Descent of Man*

Deyan Sudjic, *The Language of Cities*

Norman Ohler, *Blitzed: Drugs in Nazi Germany*

Carlo Rovelli, *Reality Is Not What It Seems: The Journey to Quantum Gravity*

Catherine Merridale, *Lenin on the Train*

Susan Greenfield, *A Day in the Life of the Brain: The Neuroscience of Consciousness from Dawn Till Dusk*

Christopher Given-Wilson, *Edward II: The Terrors of Kingship*

Emma Jane Kirby, *The Optician of Lampedusa*

Minoo Dinshaw, *Outlandish Knight: The Byzantine Life of Steven Runciman*

Candice Millard, *Hero of the Empire: The Making of Winston Churchill*

Christopher de Hamel, *Meetings with Remarkable Manuscripts*

Brian Cox and Jeff Forshaw, *Universal: A Guide to the Cosmos*

Ryan Avent, *The Wealth of Humans: Work and Its Absence in the Twenty-first Century*

Jodie Archer and Matthew L. Jockers, *The Bestseller Code*

Cathy O'Neil, *Weapons of Math Destruction: How Big Data Increases Inequality and Threatens Democracy*

Peter Wadhams, *A Farewell to Ice: A Report from the Arctic*

Richard J. Evans, *The Pursuit of Power: Europe, 1815-1914*

Anthony Gottlieb, *The Dream of Enlightenment: The Rise of Modern Philosophy*

Marc Morris, *William I: England's Conqueror*

Gareth Stedman Jones, *Karl Marx: Greatness and Illusion*

J.C.H. King, *Blood and Land: The Story of Native North America*

Robert Gerwarth, *The Vanquished: Why the First World War Failed to End, 1917-1923*

Joseph Stiglitz, *The Euro: And Its Threat to Europe*

John Bradshaw and Sarah Ellis, *The Trainable Cat: How to Make Life Happier for You and Your Cat*

A J Pollard, *Edward IV: The Summer King*

Erri de Luca, *The Day Before Happiness*

Diarmaid MacCulloch, *All Things Made New: Writings on the Reformation*

Daniel Beer, *The House of the Dead: Siberian Exile Under the Tsars*

Tom Holland, *Athelstan: The Making of England*

Christopher Goscha, *The Penguin History of Modern Vietnam*

Mark Singer, *Trump and Me*

Roger Scruton, *The Ring of Truth: The Wisdom of Wagner's Ring of the Nibelung*

Ruchir Sharma, *The Rise and Fall of Nations: Ten Rules of Change in the Post-Crisis World*

Jonathan Sumption, *Edward III: A Heroic Failure*

Daniel Todman, *Britain's War: Into Battle, 1937-1941*

Dacher Keltner, *The Power Paradox: How We Gain and Lose Influence*

Tom Gash, *Criminal: The Truth About Why People Do Bad Things*

Brendan Simms, *Britain's Europe: A Thousand Years of Conflict and Cooperation*

Slavoj Žižek, *Against the Double Blackmail: Refugees, Terror, and Other Troubles with the Neighbours*

Lynsey Hanley, *Respectable: The Experience of Class*

Piers Brendon, *Edward VIII: The Uncrowned King*

Matthew Desmond, *Evicted: Poverty and Profit in the American City*

T.M. Devine, *Independence or Union: Scotland's Past and Scotland's Present*

Seamus Murphy, *The Republic*

Jerry Brotton, *This Orient Isle: Elizabethan England and the Islamic World*

Srinath Raghavan, *India's War: The Making of Modern South Asia, 1939-1945*

Clare Jackson, *Charles II: The Star King*

Nandan Nilekani and Viral Shah, *Rebooting India: Realizing a Billion Aspirations*

Sunil Khilnani, *Incarnations: India in 50 Lives*

Helen Pearson, *The Life Project: The Extraordinary Story of Our Ordinary Lives*

Ben Ratliff, *Every Song Ever: Twenty Ways to Listen to Music Now*

Richard Davenport-Hines, *Edward VII: The Cosmopolitan King*

Peter H. Wilson, *The Holy Roman Empire: A Thousand Years of Europe's History*

Todd Rose, *The End of Average: How to Succeed in a World that Values Sameness*

Frank Trentmann, *Empire of Things: How We Became a World of Consumers, from the Fifteenth Century to the Twenty-First*

Laura Ashe, *Richard II: A Brittle Glory*

John Donvan and Caren Zucker, *In a Different Key: The Story of Autism*

Jack Shenker, *The Egyptians: A Radical Story*

Tim Judah, *In Wartime: Stories from Ukraine*

Serhii Plokhy, *The Gates of Europe: A History of Ukraine*

Robin Lane Fox, *Augustine: Conversions and Confessions*

Peter Hennessy and James Jinks, *The Silent Deep: The Royal Navy Submarine Service Since 1945*

Sean McMeekin, *The Ottoman Endgame: War, Revolution and the Making of the Modern Middle East, 1908–1923*

Charles Moore, *Margaret Thatcher: The Authorized Biography, Volume Two: Everything She Wants*

Dominic Sandbrook, *The Great British Dream Factory: The Strange History of Our National Imagination*

Larissa MacFarquhar, *Strangers Drowning: Voyages to the Brink of Moral Extremity*

Niall Ferguson, *Kissinger: 1923-1968: The Idealist*

Carlo Rovelli, *Seven Brief Lessons on Physics*